POLYMER SCIENCE AND TECHNOLOGY

ADVANCES IN POLYMER RESEARCH

POLYMER SCIENCE AND TECHNOLOGY

Additional books in this series can be found on Nova's website
under the Series tab.

Additional E-books in this series can be found on Nova's website
under the E-book tab.

POLYMER SCIENCE AND TECHNOLOGY

ADVANCES IN POLYMER RESEARCH

LOSIF SIDOROV
AND
FEDOR NIKITIN
EDITORS

Nova Science Publishers, Inc.
New York

For permission to use material from this book please contact us:
Telephone 631-231-7269; Fax 631-231-8175
Web Site: http://www.novapublishers.com

Library of Congress Cataloging-in-Publication Data

ISSN: 2165-0667

ISBN: 978-1-61942-223-0

Published by Nova Science Publishers, Inc. † New York

CONTENTS

PREFACE

In this book the authors present recent advances in polymer research. Topics discussed include polymer composite materials with photosensitive bis-crown-ether dye; organometallic polyamines as physical materials and biomaterials; a quick test for the assessment of weatherability of polyethylene and grafting of poly(lauryl methacrylate) from the surface of layered silicates using oligomeric azo-initiators.

Polymer composite materials (PCM) with immobilized photosensitive bis-crown-ether dye (BCED) show changes of the absorption and fluorescence spectra in the presence of glycine (as a typical amino acid) that may be of particular interest as an example of chemosensors for biologically active compounds.

A number of metal-containing polyamines have been synthesized employing the condensation process, generally one of the interfacial polycondensation processes. The majority of these have been synthesized from the reaction between mono and diamines and Group IVB metallocene dihalides and organotin dihalides. The major reasons for the synthesis of the Group IVB polyamines is to control light and as control release agents and more recently for electrical applications. Ruthenium-containing polyamines were synthesized as part of a solar energy conversion effort. The major reasons for the synthesis of the organotin polyamines is biological since many of these materials offer good ability to resist a wide variety of bacteria, cancer cell lines, and viruses.

This work presents an accelerated weathering procedure for the evaluation of the outdoor weathering performance of polyethylene (PE). Virgin PE Samples and PE compounded with fillers were exposed to simulated weathering conditions of radiation and moisture. A 500 W Halogen floodlight was used as the radiant energy source and was positioned 30 cm from the sample exposure rack. Samples were periodically exposed to incident radiation for 10 min. and then dipped in water at temperature of between 4° C and 10° C. Indices of degradation were developed based on some property changes observable during polymer degradation. These were monitored by visual examination of the samples and a rating system was used to assign quantitative measurements. Samples were tested for a total time period of 300 min. with visual examination and rating of deterioration done at 30 min. interval. The virgin PE degraded noticeably during the entire duration of the test and its degradation indices dropped sharply. The PE containing the fillers exhibited varying degrees of resistance to degradation throughout the test.

To improve the dispersion of filler in the polymer matrices to achieve exfoliated nanocomposites, a number of different methods have been used. Grafting of polymers from the surface of the filler is an attractive approach among these methods. To achieve polymer grafting, it is required to immobilize an initiator on the surface of the filler platelets. This initiator can then be used to graft polymer brushes from the surface of the filler. Initiators for this purpose can either be monocationic or bicationic, and offer their specific limitations and advantages. The use of bicationic initiators leads to generation of less amount of free polymer than the monocationic initiator but has also higher potential of quick termination of the radicals owing to their close proximity with each other. The use of living nitroxide mediated polymerization has also been demonstrated to delay the uncontrolled termination of the radicals which leads to extensive as well as controlled grafting of polymers from the filler surface.

The complexity of the brain and the membranous blood-brain barrier (BBB) has proved to be a significant limitation to the systemic delivery of pharmaceuticals to the brain rendering them sub-therapeutic and ineffective in the treatment of neurological diseases. Apart from this, lack of innovation in product development to counteract the problem is also a major contributing factor to a poor therapeutic outcome. Various innovative strategies show potential in treating some of the neurological disorders; however drug delivery remains the most popular. To attain therapeutic drug levels in the central nervous system (CNS), large, intolerable systemic doses are generally administered. The success of maintenance therapy in many neurological diseases depends on a number of variables, including the constant release of neurotherapeutics, a reduction in the dosing frequency, a greater antipsychotic drug bioavailability and ultimately improved patient compliance, many of which is not achievable by conventional oral or parenteral formulations.

Polymer phase (polymer matrix) behavior in nanocomposites in many respects defines the behavior of nanocomposite as a whole independently from the used nanofiller type (disperse particles, organoclay, nanotubes and so on). In connection with this it is necessary to account for polymer matrix structure changes at nanofiller introduction in initial matrix polymer. These changes can be realized with the aid of different processes, namely, crystallization, amorphous polymer phase structure change, interfacial regions formation. In its turn, such factors as polymer matrix chain flexibility, interfacial adhesion level, nanofiller particle shape and so on influence on characteristics and realization possibility of the mentioned processes. Hence, at polymer nanocomposites structure formation complex dynamics of polymer phase behavior in them is observed, that defines in the long run a nanocomposite properties. In this aspect particularly important is the role of interfacial regions, which are the same reinforcing element of structure in polymer nanocomposites as actually nanofiller. For semicrystalline polymer matrix nanofiller can play a nucleator role changing in reality the indicated matrices crystallinity degree. In the present review the quantitative relationships of the initial polymer characteristics and their modification at nanofiller introduction and their influence on nanocomposite final structure are considered.

The friction and wear properties of polymers are difficult to predict theoretically, due to the many parameters that influence the tribological behaviour. Knowledge about the tribological properties of self-lubricating properties of polymers is therefore mainly gathered from experimental testing, which is often closely related to the final application conditions. In contrast to traditional pin-on-disc test set-ups, we present the analysis of a contraformal cylinder-on-plate contact with initial line contact providing high contact pressures and promoting polymer transfer to the counter-face at running-in. Especially, different types of polyimide cylinders are tested because of their good thermal and mechanical properties and resistance against high initial contact pressures. The statistical variation in test results is discussed in relation to variations in the sample geometry.

There is a statistical variation of about 7 to 9% on dynamic coefficients of friction, about 10 to 12% on static coefficients of friction, about 8 to 9% on wear rates from weight loss and about 12 to 15% from wear rates from dimensional measurements. The on-line and off-line wear measurements can be combined to distinguish between real material loss, visco-elastic deformation, plastic deforma-tion and thermal expansion of the polymer samples. The transition bctwccn running-in and steady-state sliding conditions are further detailed, in relation to the visco-elastic deformation and variations in macroscopic contact situation into a more conformal contact during sliding.

Finally, an analytical expression for calculating the maximum polymer surface temperature T* has been developed for the present sliding configuration.

Newly developed hyperbranched polymers possess a compact, highly branched, three-dimensional structure, which has a high density of functional end groups and inherently low viscosity. The combination of these two properties, low viscosity and high reactivity, makes them attractive candidates for an overwhelming variety of applications. The experimental and theoretical investigation of the phase behavior of hyperbranched polymer systems is a crucial requirement for a successful introduction of new applications to highly competitive markets. In this context, thermodynamic models, which accurately account for the impact of polymer branching on the phase behavior of polymer systems, play a very important role.

The lattice cluster theory (LCT) is an extension of the well-known Flory-Huggins theory, especially in the calculation of the entropy of the lattice. Whereas the Flory-Huggins theory is limited to linear chains the LCT can be applied to arbitrary chain architecture. This situation permits the incorporation of the architecture in the thermodynamic functions useful for phase equilibrium calculations. The polymer architecture plays an important role in the physical properties of hyperbranched polymers. Additionally, the combination of the LCT with the density gradient theory allows the theoretical investigation of the interfacial properties between the demixed phases.

A blend of poly(ε-caprolactone) (PCL) and poly(styrene-co-acrylonitrile) (SAN) containing 27.5 wt% of acrylonitrile having the critical composition (80/20 PCL/SAN) was studied. This PCL/SAN blend having a lower critical solution temperature (LCST) phase boundary at 122°C offered an excellent opportunity to investigate, firstly the kinetics of phase separation above LCST (125-180°C), and secondly the kinetics of phase dissolution below LCST (50-115°C). The blend underwent a temperature-jump above LCST where spinodal decomposition (SD) proceeded, yielding a regularly phase-separated structure (SD structure). Then, it was quenched to the temperatures below LCST when the phase dissolution proceeded. Optical microscopy was used to observe the spinodal decomposition qualitatively while light scattering was used to characterize the phase separation and phase dissolution quantitatively. It was found that during phase dissolution the peak maximum moved towards a smaller angle (wavelength of concentration fluctuations increases) while the peak intensity decreased. This behavior was explained by a model. Also it was found that the fastest phase dissolution kinetics at 80°C, which was characterized by an apparent diffusion coefficient, was about 10 times slower than the kinetics of phase separation at 180°C. Crystallization after various levels of spinodal decomposition was observed by optical microscopy. Order parameter of the lamellae inside the spherulites was evaluated with the help of H_v light scattering. Transmission electron microscopy revealed interesting lamellar structure after spinodal decomposition.

In: Advances in Polymer Research
Editors: Losif Sidorov, Fedor Nikitin, pp. 1-5
ISBN: 978-1-61942-223-0

POLYMER COMPOSITE MATERIALS WITH PHOTOSENSITIVE BIS-CROWN-ETHER DYE

***Sergei Yu. Zaitsev*[1,2*]*, *Viktor V. Bondarenko*[1], *Marina S. Tsarkova*[1], *Dietmar Möbius*[2], *Natalia A. Lobova*[3], *Artem I. Vedernikov*[3] *and Sergei P. Gromov*[3]**

[1] Department of Organic and Biological Chemistry, FGOY VPO MSAVM&B, Acad. Skryabina Str. 23, Moscow 109472, Russia.
[2] Max-Planck-Institute for Biophysical Chemistry, Göttingen 37070, Germany.
[3] Photochemistry Centre, Russian Academy of Sciences, Novatorov Str. 7a, Moscow 119421, Russia.

ABSTRACT

Polymer composite materials (PCM) with immobilized photosensitive bis-crown-ether dye (BCED) show changes of the absorption and fluorescence spectra in the presence of glycine (as a typical amino acid) that may be of particular interest as an example of chemosensors for biologically active compounds.

Keywords: polymer composite materials, absorption and fluorescence spectroscopy, photosensitive bis-crown-ether derivative, optical chemosensor.

INTRODUCTION

The design of polymer composite materials is one of the promising directions of polymer and supramolecular chemistry, material sciences and engineering, having both fundamental and applied importance.[1-5] For example, ultrathin films of photochromic crown-ethers can serve as unique models for the study of molecular recognition phenomena, well-known for biological membranes, as well as promising chemosensoring materials.[6-8] Such materials[6,7] with reliable structural-mechanical and physico-chemical, as well as other

* Email: szaitsev@mail.ru

desirable properties, may be realized by synthesis of particular amphiphilic photochromic crown-ethers.[8-10]

Here, we report the preparation of polymer composite materials (PCMs) with immobilized photosensitive bis-crown-ether dye (BCED) and the investigation of their spectral changes by interaction with glycine (Gly) as a typical amino acid.

The synthesis of BCED, according to IUPAC 1,1'-[1,4-phenylenedi (methylene)] bis{4-[(E)-2-(2,3,5,6,8,9,11,12,14,15-decahydro-1,4,7,10,13,16-benzohexaoxacyclooctadecin-18-yl)vinyl]pyridinium}diperchlorate (structure see Figure 1) was published recently.[9,10] For the preparations of PCMs 9 commercially available polymers (purchased from "Aldrich-Sigma" or "Fluka") were chosen: cellulose acetate (CA), cellulose acetatepropionate (CAP), cellulose acetatebutyrate (CAB), cellulose acetatephthalate (CAPh), cellulose acetatehydrophthalate (CAHPh), polyvinylbutyral (PVB), polystyrene (PSt), polyvinylchloride (PVC) and polyvinylchloride carboxylated (PVC-carb.). A modified spreading technique[7] was used for the preparation of polymer films on glass plates. Only 3 polymers (CAPh, CAHPh and PVB) were selected for the preparation of PCMs based on their structural-mechanical and physico-chemical properties, such as reliable adhesion to the substrate and spreading ability for obtaining thin films (10-40 μm), optical transparency between 400 and 700 nm, storage stability, optimal wettability etc.

A typical amino acid, glycine (purchased from "Aldrich-Sigma"), was used for studying the chemosensoring properties of BCED. Absorption and fluorescence specta of PCMs with BCED were measured by using commercial devices: "Hitachi 330" (Japan) and "Shimadzu RF 5001" (Japan).

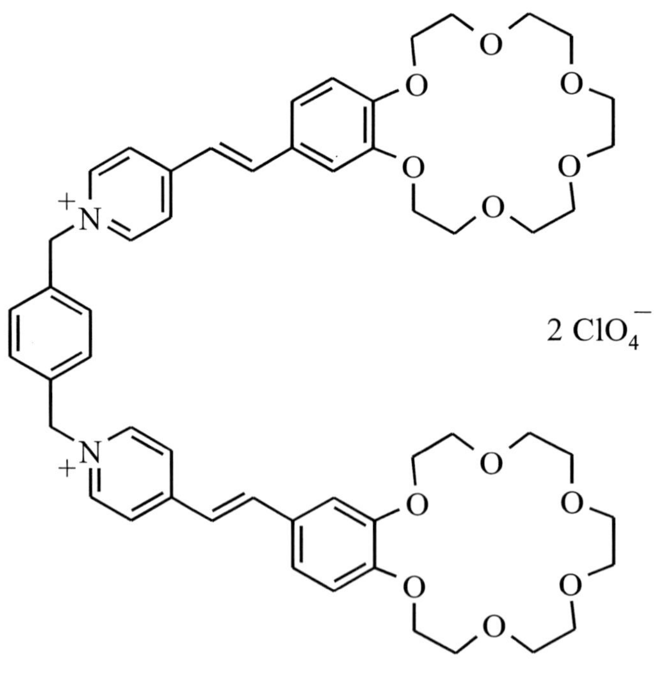

Figure 1. Structure of the photosensitive bis-crown-ether derivative (BCED).

RESULTS AND DISCUSSION

The preparation of composite materials, consisting of a polymer matrix (as thin polymer film) and a photosensitive bis-crown-ether derivative, was an essential part of our research because of the importance of PCMs for future technical application.[7] For comparing the spectral characteristics the PCMs, the position of the maxima (λ_{max}) of absorption and fluorescence in the corresponding spectra were used as key parameters (table 1).

Table 1. Absorption maxima (λ_{max}) and relative intensities (I_{max}) of PCMs based on CAHPh, CAPh, and PVB with BCED before and after (*) immersion into aqueous 1 mM glycine solution; shift $\Delta\lambda_{max}$ of the maxima and ratio I_{max} */ I_{max} of relative intensities

Parameters PCMs	λ_{max} nm	I_{max} a.u.	λ_{max}* nm	I_{max} * a.u.	$\Delta\lambda_{max}$ nm	I_{max} */ I_{max} %
CAHPh	401	0.359	408	0.360	7	100
CAPh	393	0.508	399	0.477	6	94
PVB	394	0.269	397	0.209	3	78

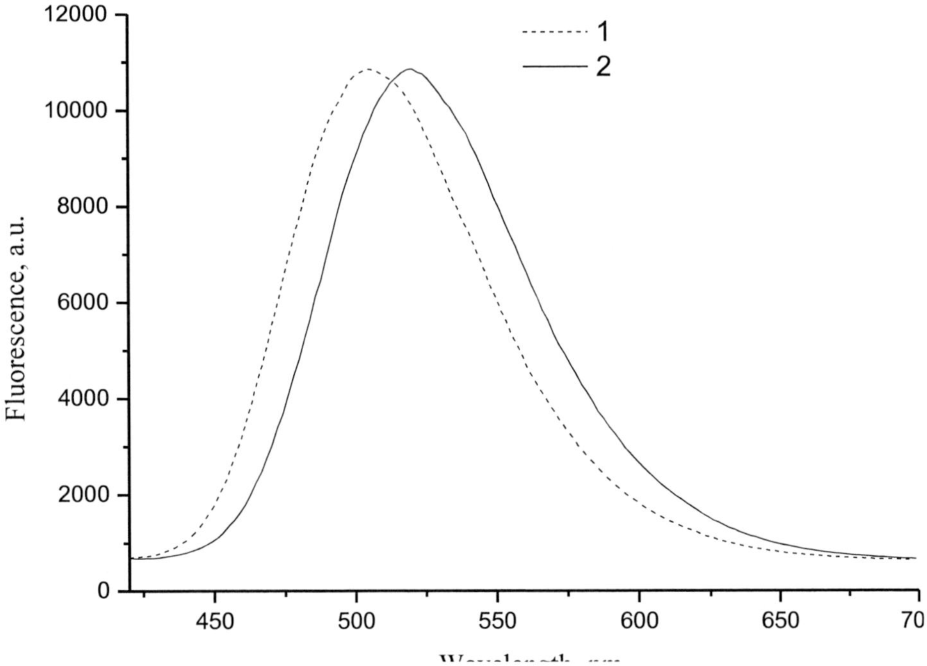

Figure 2. Fluorescence spectra for PCMs based on CAHPh and BCED before (dotted curve 1) and after (solid curve 2) immersion of the sample in 1 mM glycine solution.

The absorption maximum of PCMs based on CAHPh and BCED was at λ_{max} = 401 nm with intensity of 0.359 in dry film and after immersion in pure water. After immersion of the glass plates into aqueous 1 mM glycine solution, the absorption maximum had shifted to λ_{max}=408 nm with intensity of 0.360. The relative intensity of this band was unchanged, however, the observed shift of $\Delta\lambda_{max}$ = 7 nm to longer wavelengths is evidence for the interaction of BCED with Gly (table 1).

In the case of PCMs based on CAPh or PVB and BCED the absorption maxima were shifted on 6 and 3 nm, whereas the relative intensities of absorption were decreasing by 6% and 22%, respectively Thus, the observed shift of λ_{max} to the longer wavelengths is evidence for the interaction of BCED with Gly.

As can be seen from the Figure 2, the fluorescence maximum for PCMs based on CAHPh and BCED was about λ_{max}= 505 nm with intensity of 10832 (dotted curve). The immersion of the sample into aqueous 1 mM glycine solution shifted the fluorescence maximum to λ_{max} = 520 nm with relative intensity of 10824 (solid curve). This observed shift of $\Delta\lambda_{max}$= 15 nm of the fluorescence maximum to longer wavelengths is clear evidence for the interaction of BCED with Gly. The intensity of this fluorescence band was unchanged (Table 2).

Table 2. The fluorescence maxima (λ_{max}) and relative intensities (I_{max}) of PCMs based on CAHPh, CAPh, and PVB with BCED before and after (*) immersion of the samples into aqueous 1 mM glycine solution; shift $\Delta\lambda_{max}$ of the maxima and ratio I_{max} */ I_{max} of relative fluorescence intensities

Parameters PCMs	λ_{max} nm	I_{max} a.u.	λ_{max}* Nm	I_{max} * a.u.	$\Delta\lambda_{max}$ Nm	I_{max} */ I_{max} %
CAHPh	505	10832	520	10824	15	100
CAPh	514	6410	521	6980	7	109
PVB	494	6224	496	12300	2	198

In the case of PCMs based on CAPh or PVB and BCED the fluorescence maxima were shifted on 7 and 2 nm, and the intensities of fluorescence increased by 9% and 98%, respectively (table 2). Thus, the shift of λ_{max} to the longer wavelengths observed with these PCMs is additional evidence for the interaction of BCED with Gly.

Thus, the largest change (shift of $\Delta\lambda_{max}$= 15 nm to the longer wavelengths) by immersion of the samples into aqueous 1 mM glycine solution was found for the fluorescence spectra of PCMs based on CAHPh and BCED as compared to absorption spectra (shift of $\Delta\lambda_{max}$ = 7 nm). This demonstrates the advantages of fluorescence over absorption detection of the interaction of BCED with Gly in future chemosensoring optical devices.

In a previous study[11], the compound BCED has been incorporated in monolayers at the air-water interface by spreading a mixed solution of BCED and octadecanoic acid (ODA) at the stoichiometric molar ratio BCED:ODA = 1:2. Due to the electrostatic interactions between the carboxylate groups of dissociated ODA molecules and the pyridinium groups of BCED, this molecule has been transferred with the ODA matrix by the Langmuir-Blodgett technique to glass plates. Absorption and fluorescence spectra of such transferred monolayers clearly showed that BCED is present in the monolayer. The band maxima were at λ_{max} = 398 nm (absorption) and λ_{max} = 532 nm (fluorescence). BCED in the two-component monolayer interacted with diamino propane in the aqueous subphase as concluded from surface pressure-area and surface potential-area isotherms. In monolayers transferred in the presence of diamino propane in the aqueous subphase, the absorption and fluorescence bands were shifted to longer waves with maxima at λ_{max} = 406 nm (absorption) and λ_{max} = 540 nm (fluorescence). Therefore, it is reasonable to assume interactions between the ammonium group of glycine and the macrocycles of BCED leading to the observed band shifts in the composite films after immersion of the PCM nto an aqueous 1 mM glycine solution. In addition, the electrostatic interactions between the carboxylate group of glycine and the positively charged pyridinium rings of BCED reinforce binding of glycine to BCED.

We anticipate that the interaction of BCED with Gly leading to such a spectroscopic changes is a promising base for the design of chemosensoring materials that are capable of optical detection of amino acids in solution.

This work was supported by the Russian Foundation for Basic Research (RFBR), the Russian Ministry of Science and Education, the Max Planck Institute of Biophysical Chemistry (Germany).

REFERENCES

[1] Fendler, J. H. Membrane *Mimetic Chemistry: Characterizations and Applications of Micelles, Microemulsions, Monolayers, Bilayers, Vesicles, Host-Guest Systems and Polyions*. Wiley: New York, US, 1982; pp 85-125.

[2] Lehn, J.-M. Supramolecular Chemistry. *Concepts and Perspectives, VCH: Weinheim,* New York, Basel, Cambridge, Tokyo, DE, 1995; pp 201-257.

[3] Steed, J. W. and Atwood, J. L. *Supramolecular Chemistry*, Wiley:Chichester, UK, 2000; pp 19-83.

[4] Balzani, V., Credi, A. and Venturi, M. Molecular Devices and Machines. *Concepts and Perspectives for the Nanoworld*. 2nd Ed.; VCH: Weinheim, DE, 2008; pp 15-95.

[5] Zaitsev, S. Yu. Supramolecular Nanodimensional Systems at the Interfaces. *Concepts and Perspectives for the Bionanotechnology*; URSS: Moscow, RU, 2009; pp 106-138.

[6] Lednev, I. K. and Petty, M. C. *Adv. Mat.* 1996, 8, 615–627.

[7] Zaitsev, S. Yu. *Supramolecular systems at the interfaces as models of biomembranes and nanomaterials; Nord-Computer:* Donetsk, UA, 2006; pp 142-175.

[8] Gromov, S. P. and Alfimov, M. V. *Russ. Chem. Bull.* 1997, 46, 611-632.

[9] Vedernikov, A. I., Ushakov, E. N., Lobova, N. A., Kiselev, A. A., Alfimov, M. V. and Gromov, S. P. *Russ. Chem. Bull.* 2005, 54, 666-671.

[10] Ushakov, E. N., Alfimov, M. V. and Gromov, S. P. *Russ. Chem. Rev.* 2008, 77, N.1, 39-59.

[11] Zaitsev, S. Yu., Zarudnaya, E. N., Möbius, D., Bondarenko, V. V., Maksimov, V. I., Zaitsev, I. S., Ushakov, E. N., Lobova, N. A., Vedernikov, A. I., Gromov, S. P. and Alfimov, M. V. *Mendeleev Commun.* 2008, 18, 270-272.

In: Advances in Polymer Research
Editors: Losif Sidorov, Fedor Nikitin, pp. 7-29
ISBN: 978-1-61942-223-0

ORGANOMETALLIC POLYAMINES AS PHYSICAL MATERIALS AND BIOMATERIALS

Charles E. Carraher, Jr.* and Amitabh Battin
Florida Atlantic University, Department of Chemistry and Biochemistry, Boca Raton, FL 33431 and Florida Center for Environmental Studies, Palm Beach Gardens, FL 33410.

SUMMARY

A number of metal-containing polyamines have been synthesized employing the condensation process, generally one of the interfacial polycondensation processes. The majority of these have been synthesized from the reaction between mono and diamines and Group IVB metallocene dihalides and organotin dihalides. The major reasons for the synthesis of the Group IVB polyamines is to control light and as control release agents and more recently for electrical applications. Ruthenium-containing polyamines were synthesized as part of a solar energy conversion effort. The major reasons for the synthesis of the organotin polyamines is biological since many of these materials offer good ability to resist a wide variety of bacteria, cancer cell lines, and viruses.

Keywords: Metallocenes, ruthenium-containing polymers, metallocene-containing polymers, solar energy conversion, control-release, light control, organotin, arsenic-containing polymers, organosilicon-containing polymers, tin-containing polymer, pancreatic cancer, herpes.

INTRODUCTION

Metal-containing polymers are synthesized not only to create new materials but also to take advantage of the new, often essential, properties introduced because of the presence of the metal-containing moiety. This is the case with the metal-containing polyamines described in this paper.

The polyamines described here were generally synthesized employing one of several interfacial polycondensation processes. The interfacial process was popularized by Morgan and Carraher in the 1960s and 1970s and is today employed in the industrial synthesis of polycarbonates and aromatic polyamides generally known as aramids [1-4]. This process

* Email: carraher@fau.edu

requires what is referred to as high energy reactants, namely reactants that allow reaction to occur at somewhat low activation energies, generally in the range of 10-20 Kcal/mole (40 to 80 KJ/mole). For polyamine synthesis, the amine-containing reactant is dissolved in one solvent, typically water, along with an added base such as sodium hydroxide. The metal-containing reactant is dissolved in a largely water immiscible liquid such as hexane. The two phases are brought together with rapid stirring and the polymer is typically formed within less than a minute.

The metal-containing polyamines are named with the metal-containing moiety treated as a methylene unit.

Here, the paper is divided into two major sections related to the particular metal that is present in the polyamine. The initial section deals with polyamines containing transition metals, generally Group IVB metallocenes. The second section describes polyamines containing main group metals, generally Group IVA and VA metals.

TRANSITION METAL METALLOCENES

General

Interest in metallocene compounds began in the 1950s when two independent groups prepared ferrocene by reaction of iron II chloride with cyclopentadienyl magnesium bromide and by reaction of reduced iron with cyclopentadiene in the presence of potassium oxide.[5,6] Ferrocene is a true sandwich compound with the iron resting between two flat-plained cyclopendadiene groups.[7-12]

By comparison, the Groups IV B and V B metallocenes have the cyclopentadiene groups facing the metal atom but present in a distorted tetrahedral arrangement as **1**.[7]

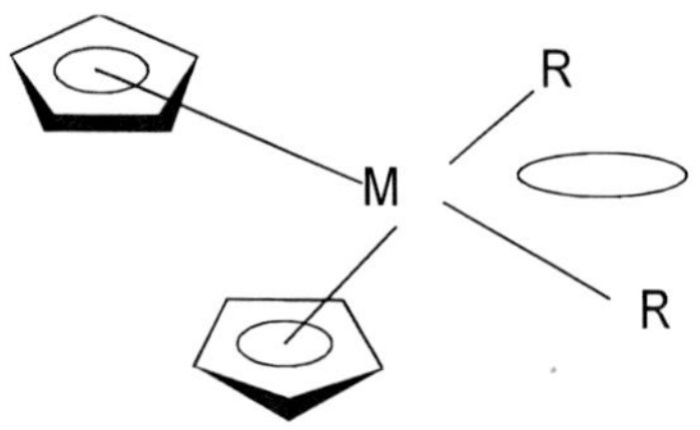

1.

Cotton and Wilkinson [7] describe Group IVB metallocene compounds as containing 9-coordinate bonding with the hybrid orbitals being derived from one-s, three-p and five-d orbitals. Each pi-Cp ring involves three hybride orbitals. The remaining three orbitals consist of two equivalent $spd_{x^2-y^2}$, d_{z^2} orbitals (overlapping the halides, oxygen, nitrogen, etc.) and one sp vacant orbital as shown above, **1**.

Over 2 million entries are given when entering the terms hafnocene and polymers in SciFinder Scholar.[13] Almost all entries for Group IVB and VB metallocenes connected with polymers are associated with their use as stereoregulating catalysts. These metallocenes are very important Ziegler-Natta catalysts and are also important catalysts in the current generation of so-called soluble catalysts.[14]

While the amount of interest in these metallocenes is great, the amount of work involving their incorporation in polymers is relatively small with much of it involving metallocenes being present as part of a supported catalysis system. This may be due to the impression that the mechanism of catalytic activity involves having at least one unoccupied orbital available about the metal atom. For instance, Grubbs and co-workers in 1977 suggested, based on work done with olefin hydrogenations, that the two chlorides in titanocene dichloride are absent in the active catalyst.[15,16]

Halides are more electronegative than the metallocene Cp_2M portion so they can undergo condensation reactions where the metallocene dihalide acts as a Lewis acid. When the Lewis base is mono-functional the products will be monomeric as shown below for a metallocene dichloride reactions with a primary amine.

$Cp_2MCl_2 + 2H_2NR \longrightarrow Cp_2M(RNH)_2 + 2HCl$ Equation 1.

But when the Lewis base, here a diamine, is difunctional a chain is formed.

$Cp_2MCl_2 + H_2NRNH_2 \longrightarrow \text{-(-(}Cp_2)MHN\text{-}R\text{-}NH\text{-)-} + 2HCl$ Equation 2.

Group IVB Metallocene Polyamines

A number of titanocene polyamines have been synthesized including those derived from aromatic and aliphatic diamines, **2**.[17-20] Polyamines were synthesized using both the regular interfacial synthetic process and the aqueous solution process where both the diamines and metallocene dichloride are simply dissolved in water and mixed with rapid stirring, about 18,500 rpm. The interfacial syntheses gave generally better product yields ranging from 20 to 90 % yields while the aqueous solution syntheses gave product yields ranging from 3 to 30%. Yield was studied as a function of stirring rate. For the interfacial systems, percentage yield continued to increase as stirring rate was increased from 9,100 to 21,400 rpm consistent with the increased stirring rate creating fresh surface for polymer to be formed during the reaction window of time. By comparison, for the aqueous solution system, percentage yield remained approximately constant throughout this stirring range consistent with reaction being dependent on only the reactants coming in contact with one another. The activation energy for addition of the Cp_2Ti^{+2} to an amine is about 10 kcal/mol (40 kJ/mol) consistent with this idea. The reaction was also studied as a function of pH. The percentage yield was approximately constant from a pH of 7.4 to 12. This is consistent with the active diamine special being the un-ionized RNH_2.

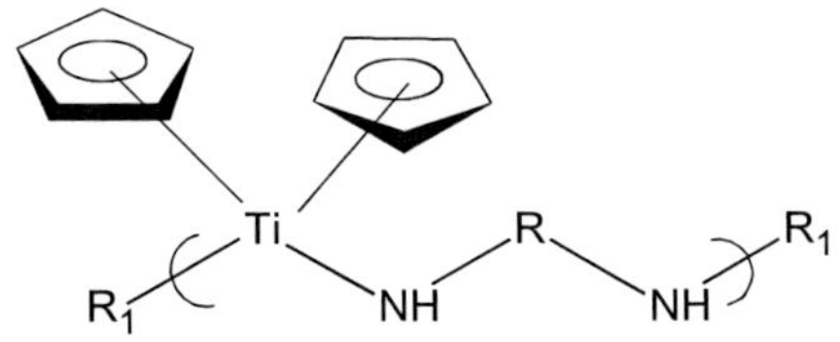

2.

No product was formed in the absence of added base or buffer consistent with the importance for removal of generated hydrogen chloride and/or importance of the base to assist in the removal of a proton from the amine as it reacts with the metallocene halide. Those derived when organic bases as triethylamine are employed in place of sodium hydroxide show at least some triethylamine end groups, **3**.

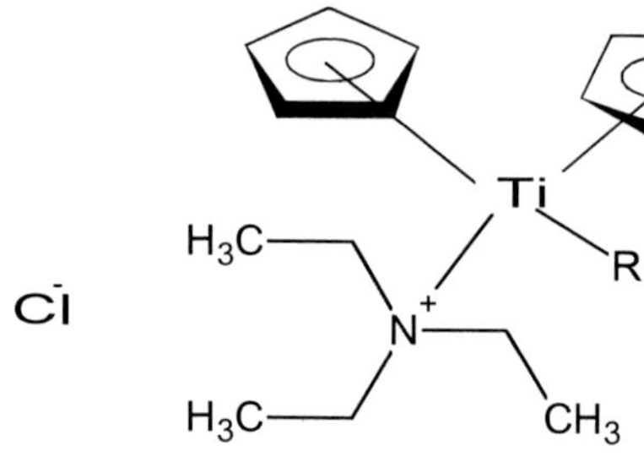

3.

Both the interfacial and aqueous solution systems showed an increase in yield as the amine to titanocene dichloride ratio increased indicating the importance of having sufficient amine present at the reaction site. Thus, in order to achieve high product yields, relatively high concentrations of diamine should be employed.

Product was also formed using the inverse interfacial system where the diamine is in the organic layer and the metallocene dichloride in the aqueous layer. The metallocene-containing aqueous layers should be made just prior to the reaction to eliminate unwanted reactions such as the formation of titanocene hydroxides and oxides.

Crosslinked products were made using a variety of tetra-amines.[20] Product yields were in the mid-range, about 20 to 40%. The products generally showed better than 60% weight retention to about 900 °C in nitrogen, but in air retention was less with some of the products having mostly titanium dioxide as the final solid. Research was halted on these materials when it was found that some of the tetra-amines were cancer causing agents.

As part of our effort to increase food production in third world countries, Group IVB polymers containing the diamine plant growth hormone kinetin in the backbone were synthesized.[21,22] Of interest is that the zirconocene product was a semiconductor with a (bulk) electrical conductivity of about 600 kilo-ohms, well within the range of being a semiconductor. The product yields were similar for all three Group IVB metallocene dichlorides, all within the range of 55 to 63%. The resulting polymers have a repeat unit as shown in **4** for the zirconocene product.

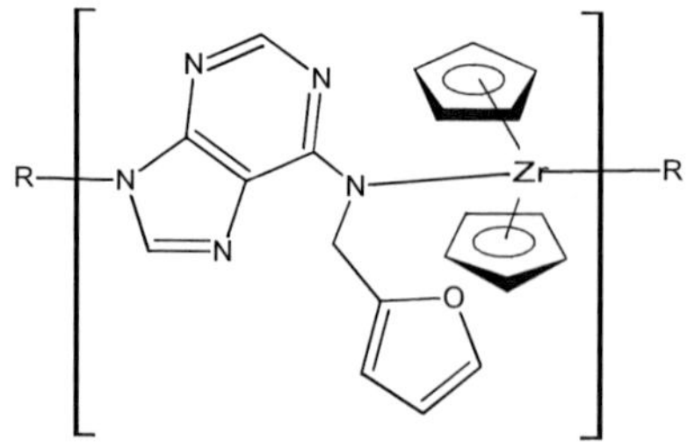

4.

As part of our effort to assist in the restoration of the Florida Everglades we investigated the ability of talc samples containing ppm of the kinetin-containing polymers to influence sawgrass seed germination. In the wild, sawgrass seed germination is reported to be from zero to 2%. Even in the laboratory the percentage germination of sawgrass seeds was low before our efforts, also near zero. Treatment of randomly selected sawgrass seeds for the control was zero, but for the polymers, in somc cases, over 50%. This would allow the Everglades to be re-seeded using a water boat or airplane employing simple casting of treated seeds rather than the present practice of hand planting single seedlings.

We also investigated the ability of these talc-kinetin polymer mixtures to influence the seed germination of damaged and old seeds.[23-25] The polymers generally showed increased germination rates for all seeds tested. Significant increases in germination were found for many of the treatments. The increases were modest for mustard with the largest germination rates found for the higher concentrations (1000 ppm) for the polymers with averages being 32 % (Ti), 34 % (Zr), and 40 % (Hf) in comparison to the control value of 23 %. For the damaged Jagger Wheat, the highest yields were found for kinetin, 10 ppm itself (25 %) and the titanocene-kinetin polymer, 1000 ppm (20 %; compared to a control value of 9 %). These increases represent a doubling in the percentage germination in comparison to the control. The largest differences were found for turnip seeds where all of the treatments gave at least a doubling in the germination rate with several giving about a six to eight fold increase. The highest increase was for the zirconocene-kinetin polymer, 1000 ppm (27 %).

There appeared to be little correlation between the concentration of the treatments and germination rate probably because the precise affect on germination is a complex mixture of factors including polymer degradation rate releasing the kinetin. It is important to note that slight variations in the structure of the cytokinin renders the PGH inactive.[24-29] Since the polymers containing kinetin exhibit PGH-like activity, the polymers probably release kinetin in an active form as they degrade. While degradation studies have not been performed on the present materials, similar condensation polymers have molecular weight half lives (time required for the average molecular weight to be halved) in HMPA and DMSO solution of days to months but molecular weight studies have not been done on simple exposure of polymers to water. Further, the situation is more complex since microorganism degradation may also occur.

The kinetin-bound polymers generally outperformed kinetin itself. Binding the kinetin with polymers will result in additional cost so that use of kinetin alone might be recommended on this account. Counter, kinetin-containing polymers can be stored in the open under room conditions for over two years without any noticeable change (by IR examination) while the PGHs themselves must be stored under refrigeration.

As noted before, it has been found that any variation in the structure of a plant growth hormone, PGH, will render it inactive.[24-29] Thus, while the structure of the active PGH was not determined it is assumed that the active agent is kinetin itself whenever PGH activity was found. For the polymers this amounts to degradation of the polymer chain releasing kinetin. It is believed that the metallocene moiety degrades to generally non-toxic metal oxide. Thus, the polymers are environmentally “friendly” degrading to environmentally acceptable compounds.

In summary, in almost all cases, polymer-containing PGH treated seed groups gave greater germination percentages in comparison to the control. Thus, such treatment may be advantageous to increasing the germination of older and damaged seeds increasing the number of viable plants from such seed feedstocks. Further, several grams of polymer are sufficient to treat millions of seeds.

In all cases seedling development appeared independent of the particular seed treatment. This is consistent with the favored use of "natural" PGHs incorporated in polymers with respect to their lack of production of negative side effects hopefully avoiding unintended negative consequences.

Recently it was found that a titanocene polyamine was found to increase its conductivity about 10^5 when doped with iodine. [30,31] This is the first example of a traditional condensation polymer whose conductivity was significantly increased through doping.

The polymeric material is derived from the interfacial condensation polymerization between titanocene dichloride and 2-nitro-p-phenylenediamine (Figure 1). This product was initially synthesized by us in 1975 employing both the interfacial and solution polymerization techniques.[19]

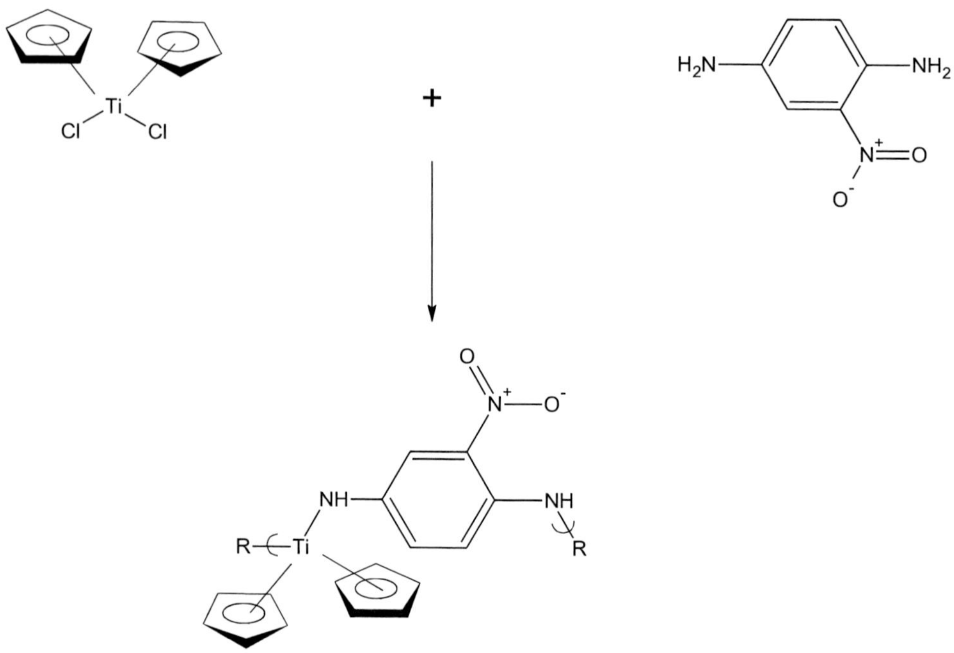

Figure 1. Synthesis of polymer from reaction of titanocene dichloride and 2-nitro-p-phenylenediamine where R is simple chain extension.

The foundation for these efforts was laid by MacDiarmid, Heeger, and Shirakawa (for which they received the Nobel Prize in 2000) and coworkers and much work continues today.[32-39]

We have just begun investigating the results of polymer doping. While there are a number of different doping procedures employed in the production of conducting polymers, the most widely employed is the exposure of the polymeric material, often in a compacted disk, to elemental iodine vapors.[32-39] During our initial efforts we found that one polymer did respond giving relativey large increases in its conductivity when doped with iodine vapors. As noted before, this polymer is obtained from reaction between titanocene dichloride and 2-nitro-p-phenylenediamine (Figure 1). The titanium atom in titanocene dichloride has a vacant orbital that can accept electrons allowing resonance to occur through it. The polymer can exhibit whole-chain resonance as shown in Figure 2 so its structure is consistent with the general structural criteria for conductivity. The product employed in the current study was synthesized in 45 %. It has a weight average chain length (degree of polymerization) of 310 units.

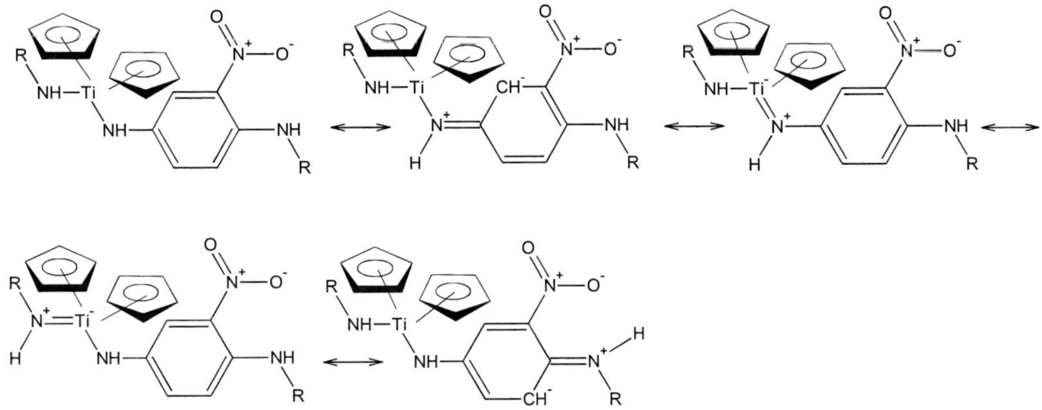

Figure 2. Resonance structures for 2-nitro-p-phenylenediamine unit where R = chain extension.

It was found that simply exposure of a solid pellet of the polymer to iodine vapor allowed the bulk conductivity to increase up to 10^5. It was also found that this process was cyclic. Thus, the pellet was exposed to iodine vapor resulting in an increased electrical conductivity. The pellet was then heated to remove the surface iodine and the conductivity decreased. The same pellet was again exposed to iodine vapor and again the electrical conductivity increased.

We also incorporated a number of dyes as part of the backbone of polymers. These polymers are generally referred to as polydyes. Our overall effort was to manage light, to either concentrate light or to disperse it. "Normal" and laser light is of concern to us. One goal is to couple known light harvesting units in a polymer chain with those that disperse light. Group IVB metallocenes are known to "attract" or harvest light, mainly in the visible and UV regions. They are sometimes referred to as UV "sinks" because of their ability to absorb UV radiation without begin degraded. This ability to absorb light and then give it off again has many potential uses including allowing energy absorbed in a coating to be re-emitted at a specific wavelength that would allow the coating to cure at greater depths and more rapidly. The absorption of light energy by a coating containing the polydye may also be part of an overall effort to harvest radiation to contribute to the heating and supplying of energy (including electrical energy) to buildings and dwellings.

Another attraction for including dyes into polymers is the (potential) permanent, non-leaching nature of such polymer-containing dyes allowing a material that contains this polydye to remain colored for longer times. Even within solids, such as plastics, migration and loss of colorants occurs, especially over a long period of time. The ability to retain color or other important additive property is especially important in applications such as in marine coatings where leaching and subsequent loss is aided by the ever present aqueous environment.

A number of dyes were examined including those derived from diamines such as nigrosine, **5**.[40-49]

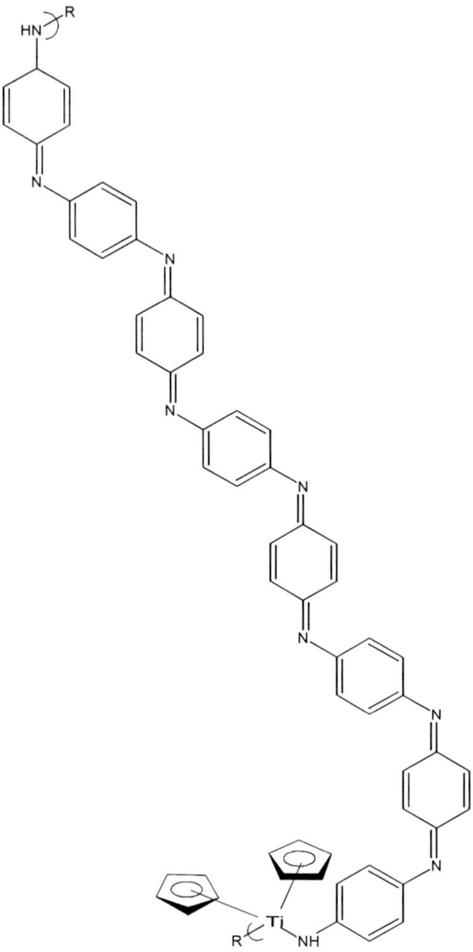

5.

The polydyes are generally soluble in dipolar aprotic liquids such as hexamethylphosphorus triamide (HMPA), N, N'-dimethylformamide (DMF), dimethyl sulfoxide (DMSO), and triethylphosphate (TEP). Polydye solutions were used to impregnate a number of items including plastics (polyethylene, polycarbonate, polypropylene, poly(vinyl chloride), polystyrene, nylon 6,6, SAN-styrene copolymer), cloth (cotton, denim, and 50-50% polyester fabric), paper, films, and coatings.

In one set of experiments, comparison was made with simple solutions containing the dyes themselves. After several days to months, the color of the dye solutions containing only the dyes themselves were greatly decreased. By comparison, the polydyes retained decent color for a year under fluorescent lighting. Exposure to outdoor light hastened deterioration of the color of both the dye and polydye solutions. Thus, incorporation into a polymer appears to lock in the dye structure and gives it some added stability to ordinary in-door light. In the dark polydye, solutions have remained fluorescent and colored now for about 25 years.[49]

Studies were undertaken using an Argon Ion laser operated at 514.5 nm (visible-green) as the high energy source. Optically transparent films were cast. The presence of polydye in the samples caused significantly less (almost ten-times shorter) "burn-through" times compared with those containing an equal amount of the dye itself.[48] In this study, the polydyes acted to concentrate the radiation dramatically decreasing the "burn-though" times. By comparison, similar samples except exposed to energy from a carbon dioxide laser operating in the infrared region (1.06 microns; 1060 nm).[40] Here burn-through times were significantly greater. Under these experimental conditions, the polydyes acted as dispersing agents. The films contained from 10 to 100 ppm polydye.

In another set of experiments using the carbon dioxide lazer, wood was coated with latex paint containing about 200 ppm polydye. The wood was exposed to laser light in the IR region. In comparison to non-impregnated latex coatings, the polydye-containing samples exhibited equal and greater burn-through times. Similar results were found for plastics doped with polydyes.[40,49]

Behavior was dramatically different depending on the range of employed radiation. Thus, in the IR region the polydyes act as dispersing agents protecting the material that is coated with it or that contains it. In the visible-UV region the polydye acts as a concentrating agent, concentrating the effect of the radiation.

Each tendency offers industrially important behavior. For instance, some chip fabrication and construction might be enhanced through addition of some polydye material that allows fabrication through use of lasers employing lower energies. The use of lower energies would result in lower temperatures that, in turn, would protect the chip from thermal induced deterioration. A similar advantage is achieved when cutting or etching occurs. Counter, in some applications greater stability is needed to resist potentially damaging laser radiation.

RUTHENIUM CONTAINING POLYAMINES

We have synthesized a number of ruthenium-containing polymers, mostly in a search for converting solar energy or other light energy into usable energy under the heading of solar energy conversion. It is of interest to note that our work in this area involves two different types of reactions, the coordination reactions involving reactions with diamines and condensation reactions involving thiols. While the present paper focuses on condensation polymers, we will briefly describe efforts involving the synthesis of ruthenium-containing polyamines employing the coordination process.

Keen, Salmon, and Meyer [50] reported the synthesis of monomeric amines from reaction of ruthenium(II) bis-2,2'-bipyridine dichloride, DBR, with diamines where the reaction is a coordination reaction rather than the condensation reaction that occurres with the thiols.

$$RNH_2 + Ru(bpy)_2Cl_2 \longrightarrow (RNH_2)_2Ru(bpy)_2Cl_2 \qquad \text{Equation 3.}$$

Reaction with diamines gave the analogous polymeric products. [51-55]

$$H_2N\text{-}R\text{-}NH_2 + Ru(bpy)_2Cl_2 \longrightarrow \text{-}(\text{-}H_2N\text{-}R\text{-}NH_2\text{-}Ru(bpy)_2\text{-})_n\text{-}, 2Cl \qquad \text{Equation 4.}$$

These efforts have two major aims; first, modification of the accepting and emitting energies of the ruthenium center, and second, to assist in attracting light that can be subsequently harvested by the ruthenium center, thus increasing the converted quantum yield of harvest light. The current effort is aimed at incorporating dyes into the ruthenium-containing polymers.[51,52]

In one series, products were formed from reaction of cis-dichloro-bis(2'2-bipyridine)ruthenium II, $Ru(bipy)_2Cl_2$, with the dye suspended in an 80 % by volume methanol aqueous solution. This mixture was refluxed on a steam bath for about six hours, after which the methanol was allowed to evaporate. The reaction mixture was cooled over night. The solution was washed with benzene removing unreacted diamine and washed with chloroform removing unreacted $Ru(bipy)_2Cl_2$. The remaining liquid was removed under vacuum giving the product as a solid, generally in good yield.

The product from $Ru(bipy)_2Cl_2$ and 6-butoxy-2,6-diamino-3,3'-azodipyridine, diazopy, **6**, is produced in good yield, 89%, and is soluble in DMSO, methanol, ethanol, and slightly soluble in chloroform. It is insoluble in benzene, carbon tetrachloride, acetone, and hexane. It has a molecular weight of 1.4 x 10^5 Daltons which corresponds to a degree of polymerization of about 180.

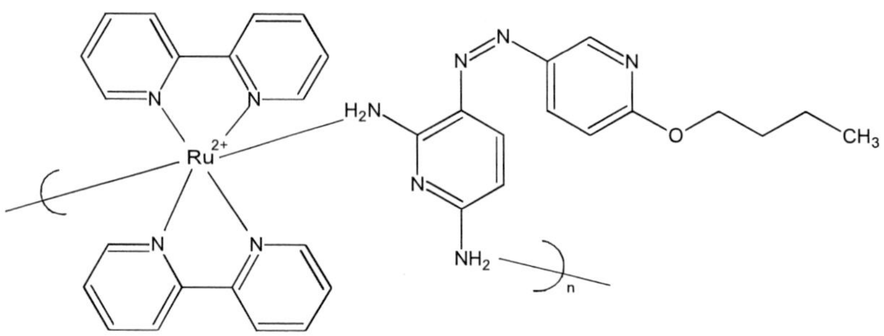

6.

Here we describe only the UV-VIS results since these most closely align with the use of such materials in light harvesting. Spectra were obtained in DMSO.

The ruthenium starting material has an octahedral geometry. The ruthenium monomer exhibits bands at 557, 379, and 300 nm with the 300 nm band being the lambda maximum band, which is in the middle of the ultraviolet region. These bands are associated with eg to t_2g d-electron transition. Substitution of the chloro ligands by amines should cause a change in the energy for d-electrons to undergo an eg to t_2g transition.

The diazopy dye itself has two bands at 291 nm and the lambda maximum band is at 454 nm. The polymer product shows a number bands with the bands appearing at 495, 437, 355, 328, 295, 266, and 252 nm. There is a broad band on the higher side of 500 nm that probably corresponds to the 557 nm band found in the ruthenium monomer. The lambda maximum band for the dye that appeared at 454 nm is now at 437 nm, still in the blue region of visible light. The lambda maximum band for the ruthenium monomer that appeared at 300 nm is now at 295 nm for the polymer.

For the product of $Ru(bipy)_2Cl_2$ and diazopy all the bands from the two reactants are essentially the same except for the lambda maximum band for the dye at 454 nm which is missing. However, the bands are broader for the polymer and the 454 nm band is now probably lost within this broadening caused by the $Ru(bipy)_2$ moiety. The product also contains a large peak at 295 nm. The 295 nm band is the lambda maximum for the product as is the 300 nm for the starting ruthenium material. There are also broad peaks at 355 nm and

495 nm with a less intense peak at 252 nm. Interestingly, the lambda maximum for the dye itself at 454 nm now corresponds to a much smaller peak at 437 nm.

A similar product was formed from the reaction of $Ru(bipy)_2Cl_2$ and N,N'-bis(3-aminophenyl)-3,4,9,10-perylenetetracarboxylic diimide, **7**. This product was formed in 85 % yield and had a DP of 33,000 by light scattering photometry.

The broadness of the spectral bands may be viewed as positive with respect to the polymer accepting a wide range of wavelengths of light. Emission spectra were not taken of the materials.

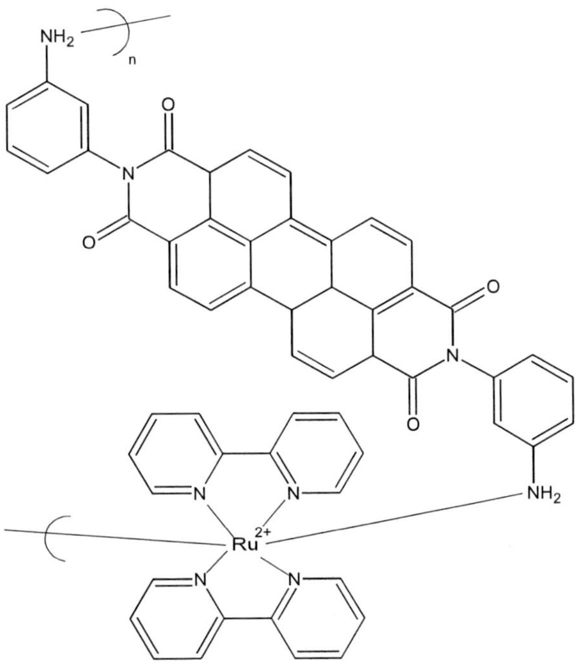

7.

Similar products will continue to be studied as the search for inexpensive energy continues.

MAIN GROUP POLYAMINES

Organosilicone Polyamines

Organosilicone polyamines were initially synthesized by us in 1969 (**8**). These materials were generally of low molecular weight and often were tar-like rather than solid. This is probably due to the high rate of hydrolysis of organosilane halides in comparison to organotin halides with hydrolysis effectively terminating chain growth.

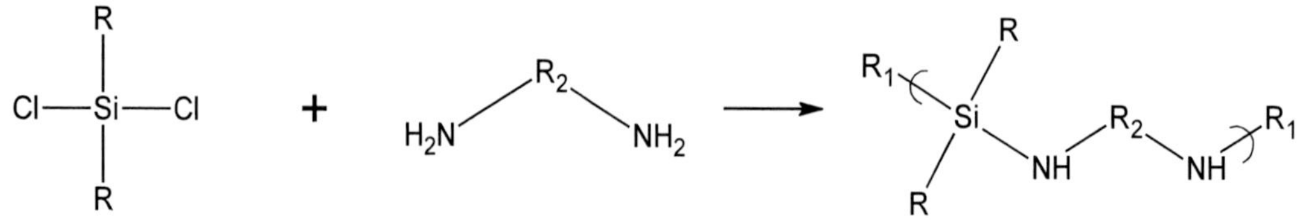

8.

Along with the traditional diamines, low molecular weight products were formed from reaction with hydrazine (**9**). [56]

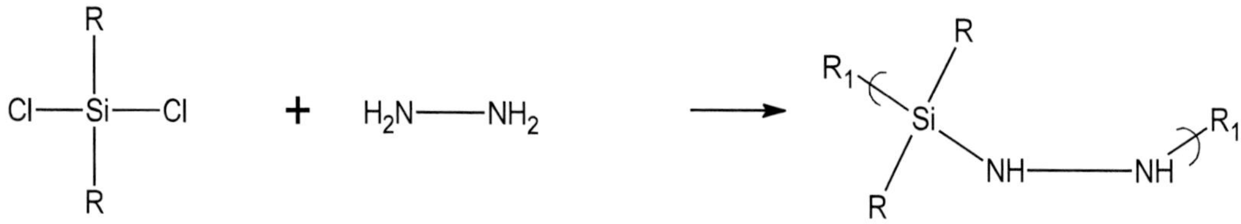

9.

A modified interfacial system was developed employing 2,5-hexanedione or acetonitrile in place of the usual aqueous phase in an attempt to reduce hydrolysis. Also, various bases were employed along with sodium hydroxide. The best results were obtained when employing triethylamine as the added base. [57]

Other hydrazines were also employed employing the non-aqueous interfacial system. [58] Yields to about 100% were obtained but similar to other organosilicone polyamines, yield generally decreased as the stirring time increases. This is consistent with a lack of good aqueous stability for the Si-N bond. Chain length generally increased with stirring time probably indicating that the first polymer chains to be degraded were the lower mass chains leaving the higher mass chains resulting in an apparent increase in chain length.

Additional studies were undertaken including developing other polymerization systems and investigating reaction variables. [59,60]

Organotin-Containing Polyamines

The first organotin compound was prepared by Sir Edward Frankland in 1849.[61] The topic of organotin polymers has been reviewed elsewhere.[62,63]

There are a larger number of organotin compounds in commercial use than for any other metal. [64] This has resulted in an increase in worldwide production of organotin compounds over the last few decades. This production exceeded fifty thousand tons in 1992 and accounts for about 7% of the tin metal used. [64]

Organotin polymers are important industrially serving as poly(vinyl chloride), PVC, heat stabilizers, in film for food packaging, and in PVC articles.[62,63] They are also used as antiseptic, antifouling, and anti-mildew agents in industry and agriculture and as additives in paint formulations. Films have excellent transparency.

PVC is unstable under exposure to light and heat resulting in discoloration and embrittlement. In the early 1940s it was found that this degradation could be prevented by addition of certain organotin derivatives. Presently, about 70% of the commercial organotin compounds are employed as PVC stabilizers in the form of mono and dialkyltin derivatives. Today, many of these stabilizers are based on organotin polymers made by Carraher and co-workers. [63]

Much of the more recent drive towards inclusion of organotin into polymers is the result of federal legislation that prohibits use of so-called unbound or monomeric organotin compounds in paints and coatings. Organotin monomeric compounds were widely used as antifouling and antimold agents but through migration they inhibited and killed nearby plant and animal life. (Chemically) bound organotin was permitted in the legislation. Thus, there is activity to create non-migrating chemically bound organotin materials, namely polymers that contain organotin moieties.

While much of our effort involves the synthesis of organotin polyamines from reaction with diamines, organotin polymers have been synthesized from reaction with preformed amine-containing polymers. Using polyethyleneimine, the products will be linear as the monohalo organotin is reacted with the polyethyleneimine, **10**, but they will be cross-lined when dihaloorganotin reactants with two functional groups are employed, **11**.

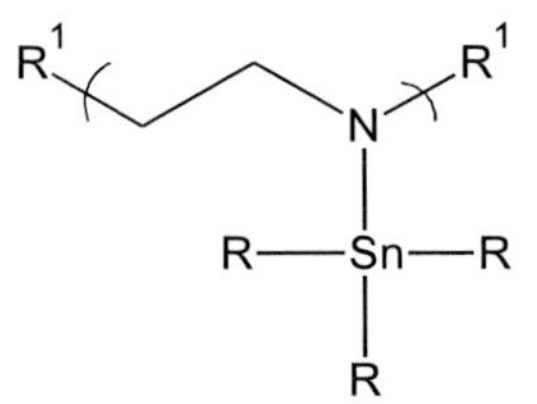

10

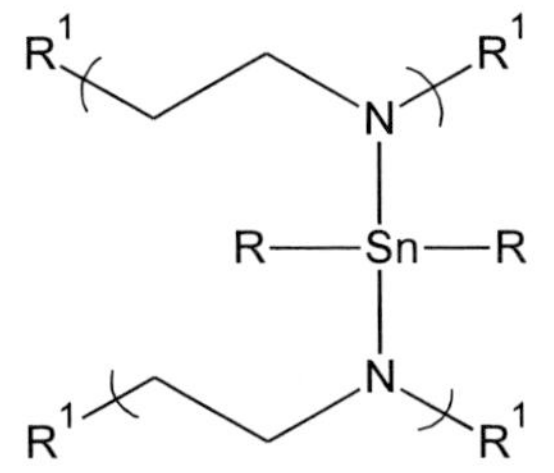

11

As in the case with the Group IVB metallocenes, we also synthesized a number of plant growth-containing polymers including polyamines derived from kinetin. The repeat unit for the product derived from dimethyltin dichloride is given below, **12**. The product is then an organotin polyamine. The use of these materials allowed the sawgrass seed germination rate to increase from the 0-2% range to over 20%. We have now systems that will give sawgrass germination rates of over 60% that makes wholesale seed distribution by air boat or airplane a practical way of seeding large areas in need of sawgrass restoration.[65-68] The organotin-containing products offer an advantage over the metallocene product in that they are resistant to degradation of many molds, yeasts, and bacteria because of the presence of the organotin moiety.

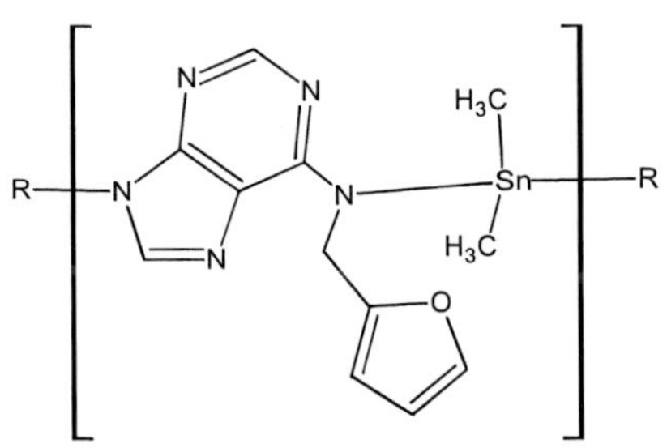

12.

These products also exhibit an ability to increase the germination of damaged and old food crop seeds.

Because of the increased worldwide production or organotin compounds for commercial use considerable amounts of organotins have entered various ecosystems. Inorganic tin compounds are often said to be non-toxic, but in fact their toxicity ranges from being non-toxic to moderately toxic. Organotin compounds offer varied toxicities from being mildly toxic to highly toxic with most of the compounds falling in the mildly to moderately toxic range. The toxicity and ecosystem relationships have been recently reviewed. [64]

While organotin-containing polymers offer a wide variety of biological activities we will focus on their use in the inhibition of bacterial and yeast at this juncture of the paper.

The general microorganism biological activities of organotin compounds has been studied since the 1950s. As noted above, generally inorganic tin compounds are non-toxic or only slightly toxic towards mammals, insects, bacteria, and fungi whereas organotin compounds show varying biological activities. For alkylorganostannanes, toxicity varies depending on the alkyl group. In general an increase in the alkyl chain length gives a decrease in toxicity.[69] The particular pattern for methyl, ethyl, propyl, and butyl varies with test organism. For tri-n-alkyltin acetates, the methyl organotin compounds are the most active for insects and mammals; for fungi and bacteria the propyl and butyl compounds are the most active. Further, activity decreases as the number of alkyl groups decreases as follows $R_4Sn > R_3Sn > R_2Sn > RSn > Sn$ all for tin IV compounds. For most cases our findings are consistent with these general trends and long alkyl chains such as n-octyl are generally biologically inactive with respect to bacterial inhibition.[70-94]

The organotin materials are suitable for treatment of infections (here mainly topical), contaminated sites, use as a preventative agent, and in the treatment of water sources. The organotin-containing drugs are rapidly (generally within 30 seconds) synthesized in good yield employing readily available reactants. Thus, ready availability on a gram to tons scale of target drugs is achievable. They can be used internally or topically as additives to creams, cleaning detergents and soaps, coatings (paints), plastics, paper, etc. [89] They can be generally be handled without need for gloves or other protective ware but it is always wise to be cautions because the individual toxicity of individual products varies with the person and product. Further, if the reactants have not been adequately removed, they can present adverse effects. These polymers typically have shelf-lives in excess of several years.

Following are microorganisms that have been successfully inhibited by condensation organotin-containing polymers including organotin polyamines.

E. coli,	B. subtilis,
B. catarrhalis,	S. epidermidis,
E. aerogenes,	N. mucosa,
K. pneumoniae,	A. calcoacetius,
A. flavus,	A. niger,
A. fumagatus,	Penicillin sp.,
Trichoderma reesei,	Chaetomium globosum,
P. aeruginosa,	S. aireis,
C. albicans,	T. mentagrophytes,
Staph MRSA,	S. cerevisiae,
B. subtilis,	Al. faecalis

Most of these products have been incorporated into paper, plastics, textiles, and the like with only some loss in biological activity.[89]

We will now move to the use of organotin polyamines as potent anticancer drugs. [74,74,96-100] Recently we synthesized a wide variety of organotin polymers derived from 4,6-diaminopyrimidines (Figure 3).[101] The molecular weights ranged from 3.5 x 10^4 to 3.7 x 10^6 and product yields from 47 to 88 %. The polyamines were synthesized for two main reasons. The first reason was to determine the biological activities of the products. The second was to determine the electrical properties of the polyamines. Because of the variety of products formed it should be possible to relate such factors as electronic nature of the pyrimidine, chain length and steric nature of the polymers to their electronic and biological activity. In general, the products exhibit good inhibition of cancer cell lines including those derived from human lung, bone, breast, prostrate, and colon cancers. Following is a brief discussion of these results.

Different measures are employed in the evaluation of cell line results. Here we use the two most widely employed- growth inhibition 50%, GI_{50}, values which is the lowest concentration where growth is inhibited by 50% and the Chemotherapeutic Index, CI_{50} which is a measure of the amount needed to inhibit 50% cell growth, GI_{50}, for the normal cell line, here WI-38 cell line, divided by the amount needed to inhibit 50% cell growth for one of the cancer cell lines. It is to be noted that different researchers generally emphasize one of these measures over the other with neither measure universally accepted. Thus, results from both of these measures are presented.

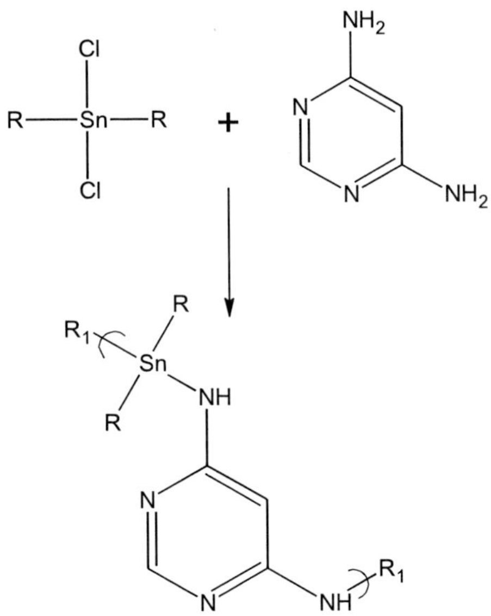

Figure 3. Reaction of organotin dichlorides with 4,6-diaminopyrimidine.

One reason for synthesizing this series of organotin polymers is to investigate if a clear trend can be established with respect to the electronic, or some other, nature of the pyrimidine. Two thiol-containing pyrimidines were employed because of the need for sulfur in the body and the general lack of sulfur-containing sources. This may encourage inclusion of the thiol pyrimidine units at important junctures in the cell growth cycle. Substitution on aromatic rings, in general, is consistent with groups such as hydroxyl, aliphatic, aromatic, and thiol releasing electrons to the ring resulting in a site that is electropositive. Conversely, the presence of groups such as nitro, nitroso, and halides withdraw electrons from the ring resulting in an electronegative environment for the pyrimidine ring. Many of the pyrimidines used in this study contain both electron withdrawing and electron donating substituents. The GI_{50} concentration values for the polymers are in the same general range of 2 to 0.04 micrograms/mL regardless of the identity of the pyrimidine. All of the polyamines offer reasonable inhibition of all of the cancer test cell lines and are in the same general range as cisplatin, the most widely employed chemo drug.

The second measure is the 50% chemotherapeutic index, CI_{50}. As noted before, the chemotherapeutic index is the concentration of the compound that inhibits the growth of the healthy cell by 50% divided by the concentration of the compound that inhibits the growth of the cancer cell by 50%. Larger values are desired since they indicate that a larger concentration is required to inhibit the healthy cells in comparison to the cancer cells or stated in another way, larger values indicate some preference for inhibiting the cancer cells in preference to the normal cells. In general, CI_{50} values larger than 2 are considered significant.

Both, the standard, dibutyltin dichloride, and all but two of the polymers, show at least one instance of a CI_{50} value of 2 and greater. Some show large, about 50% of the compounds, values much greater than 2. Also of interest is the behavior towards 2RA. 2RA is a fully transformed WI-38 and this couple is often taken as an indicator of the relative toxicities of drugs to a healthy cell compared with a cancer cell. Seven of the nine polymers exhibited CI_{50} values greater than one for the ratio of WI-38/2RA, and four of the nine polymers displayed CI_{50} greater than 2 which is consistent with the polymers being more toxic to the cancer cell line derived from the healthy cell line than the healthy cell line itself. Of the three polymers that generally exhibited the lowest GI_{50} values only one showed more than one instance where the CI_{50} values were equal to and greater than 2. Thus, there is little correlation between the relative toxicities measured by these two common measures. Further, there appears to be little correlation between the electronic nature of the pyrimidine and either GI or CI values. It is of interest that both of the pyrimidines that contain thiol groups exhibit low GI and high CI values.

The results also do not show a relationship between chain length or steric nature and ability to inhibit cancer cell growth.

Pancreatic cancer afflicts close to 32,000 individuals each year in the United States and 168,000 worldwide, and nearly all patients die from the ravages of their disease within 3 to 6 months after detection. It is the fourth leading cause of cancer death. Treatment of pancreatic cancer is rarely successful as this disease typically metastasizes prior to detection. Current therapies consist of surgery and, possibly, radiation and chemotherapy. Standard chemotherapy for patients with locally contained cancer includes gemcitabine. Gemcitabine has been demonstrated to improve the quality of life through better pain control, adequate performance status, decreased analgesic consumption, shrinkage of tumor, and prolonged survival. Radiation therapy is usually ineffective except as an adjunct to chemotherapy or as a palliative measure. There is no effective chemotherapy for metastasized pancreatic cancer. 5-Fluorouracil (5-FU) is the most widely used agent for single drug therapy in the treatment of pancreatic cancer. 5-FU is an antimetabolite that interferes with cellular processes essential for cell division and cell growth. As a result, 5-FU has the greatest effect on rapidly growing cells.[102-104]

Considering the lack of substantial impact on survival of any chemotherapy regimens, all patients with pancreatic cancer should be considered for enrollment in clinical trials. If this is not possible, gemcitabine appears to be emerging as the standard treatment due to its apparent favorable impact on disease-related symptoms, such as pain control and performance status.[105-108]

Three of the 4-6-diamiopyrimidine polymers exhibited CI_{50} values of 20 and greater towards two strains of pancreatic cancer, APSC-1 and PANC-1. [109] There polymers are derived from reaction of dibutyltin dichloride and 4,6-diamino-2-mercaptopyrimidine, 2-chloro-4,6-diaminopyrimidine, and 4,6-diamino-2-hydroxypyrimidine.

Some of the organotin polypyrimidines also exhibit excellent inhibition of both the HSV-1 virus and the Vaccinia virus. [110] The HSV-1 or herpes simplex virus is responsible for at least 45 million infections in the US or one of every five adolescents and adults. The Vaccinia virus is the viral strain responsible for small pox and is considered one of the viruses that might be employed in a viral terror attack. It was also found that these pyrimidine polyamines exhibit, as solids, decent inhibition of a variety of bacteria and yeasts. [111]

The 4-6-diaminopyrimidine polymers have structures similar to polypyrrole and exhibit whole or entire chain resonance. Even so, their inherent electrical conductivities were all within the non-conductor to semi-conductor range with no relationship between conductivity and electrical nature of the pyrimidine, chain length, or steric hindrance.[112]

Group V Polyamines

Doak and Freedman [113] synthesized the monomeric organoantimony amines, **13**, from reaction of organoantimoney dihalides with monoamines.

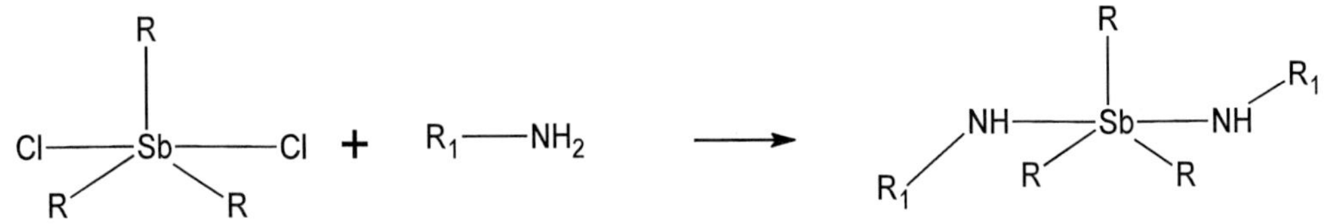

13.

The analogous polyamines,**14**, were synthesized employing the diamine.[114-117]

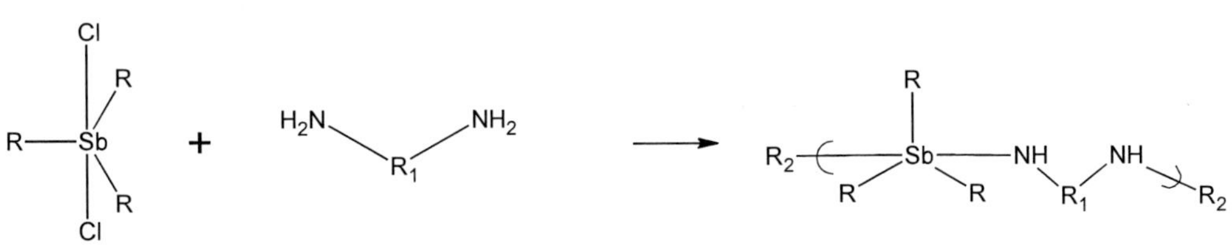

14.

Yields ranged from about 10 to over 90%. A wide range of diamines were employed including simple aliphatic diamines such as 1,6-diaminohexane, simple aromatic diamines such as p-phenylenediamine, to more complex diamines such as adenine, 2,6-diamino-8-

purinol, 4,4’-diaminodiphenylsulfon, Zineb, and 2,4-diamino-5(3,4-dimethoxybenzil) pyrimidine.

The molecular weights of the products were within the range of 10^3 to 10^6 Da. The product of triphenylantimony dichloride and 4,6-diamino-2-mercapto pyrimidine has a molecular weight of 5.7 x 10^5 corresponding to a DP of about 1,200. The product from triphenylantimony dichloride and 4,4-diaminodiphenylsulfon is only oligomeric with a molecular weight of 3 x 10^3 Da corresponding to a DP of 10.

The products were tested for their ability to inhibit a wide range of bacteria, fungi, and yeast. [114] The ability to inhibit ranged from essentially no inhibition against any test organism for the product from adenine and triphenylantimony dichloride to inhibition of all ten test organisms for the 4,6-diamino-2-methyl-5-nitrosopyrimidine and triphenylantimony dichloride product.

The products were also tested for their ability to inhibit a number of cancer cell lines including BHK-21, L929, and HeLa cells.[114] The products generally exhibited decent cell inhibition with GI_{50} values in the range of 5 micrograms/mL for the HeLa cells.

A number of other triphenylantimony dichloride-derived products were recently tested for their ability to inhibit Balb 3T3 cells as a measure of their ability as potential anticancer drugs. [116,117] This included those derived from cobalticinium-1,1’-dicarboxylic acid nitrate and thiopyrimidine. These materials showed GI_{50} values of about 10 micrograms/mL. By comparison, the most widely used anticancer drug, cisplatin, has a GI_{50} against the same strain of 3T3 cells of about 4 micrograms/mL. Thus, the antimony-containing polymers exhibit an ability similar to that of cisplatin against 3T3 cells.

The effect of use of phase transfer agents and crown ethers was studied for the synthesis of a number of antimony polyamines. [116] Comparison was made with the use of sodium hydroxide alone as the added base. Results were varied with some added catalysts resulting in an increase in product yield and chain length, others giving a lower yield and chain length, and still others showing no change in product yield and chain length. For instance, the use of tetraphenyl phosphonium iodide gave an increase from 35 to 45% in product yield and a threefold increase in molecular weight for the reaction of 2,6-diamino-8-purinol and triphenylantimony dichride. By comparison, the same phase transfer catalyst showed a decrease in yield from 38 to 28% and a four fold decrease in chain length for the reaction of 2,5-dichloro-p-phenylenediamine and triphenylantimony dichloride in comparison to the simple sodium hydroxide system.

A number of arsenic polypyrimidines have been synthesized. These polymers are good inhibitors of a number of bacteria.[118, 119]

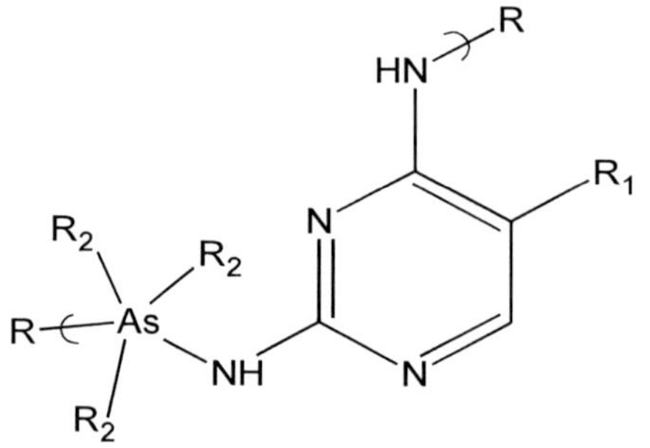

15.

FUTURE

While many metal-containing polyamines have been synthesized, many more are easily synthesized employing techniques already developed. It is important to remember that the polymers described in this report are easily and readily synthesized from commercially available reactants employing industrially utilized sequences. Further, synthesizes are generally rapid occurring in less than one minute. Thus, commercialization of any particular material is straightforward.

The concept of incorporating ruthenium-containing units within polymers for attracting solar energy is reasonable and additional work is merited towards this end. Unlike many competing efforts, the synthesis of these polymers is straight forward employing commercially available reactants. Further, the products offer reasonable solubility in a variety of solvents. Water solubility should be possible though the use of poly(ethylene glycol) solubilizing arms such as those employed to cause a number of otherwise insoluble conducting polymers to become water soluble. This approach is true for all of the products described in this paper.

As we continue to move towards nano parts, industrial use of polydyes to control ordinary and laser light also deserves further consideration since it requires the use of only minute, ppm, amounts of the polydye to bring about profound changes in the stability of the polydye-containing material. This is particularly true for the electronic industry with the incorporation or coating of parts can allow more precise fabrication of parts.

While much of the work dealing with cancer has focused on the use of organotin-containing polymers, the metallocene-containing polymers have also been found to offer decent inhibition of a variety of cancer cell lines. Both the organotin and metallocene-containing polymers generally have low toxicities in comparison to the monomeric drugs currently employed in chemotherapy today. Further, the biological testing results are consistent with the metallocene and organotin monomeric products acting at different junctures to inhibit cancer cell growth. In turn, the metallocene and organotin products appear to act at different junctures in comparison to other currently employed chemo drugs. Thus, mixtures containing either or both the metallocene and organotin drugs and currently employed drugs should have a positive effect through intersecting cancer cell growth at different junctures. The advantageous of employing polymeric anticancer drugs is well known.[for instance 120]

REFERENCES

[1] Morgan, P. W. Condensation Polymers by Interfacial and Solution Methods, *Interscience*, NY, 1965.

[2] Millich, F. and Carraher, C., *Interfacial Synthesis*. Vol I. Fundamentals, Dekker, NY., 1977.

[3] Millich, F. and Carraher, C., *Interfacial Synthesis*. Vol. II. Polymer Applications and Technology. Dekker, NY. 1977.

[4] Carraher, C. and Preston, J. *Interfacial Synthesis* Vol. III, Recent Advances, Dekker, NY, 1982.

[5] Kealy, T. and Pauson, P. *Nature*, 168, 1039, 1951).
[6] Miller, S. A., Tebboth, J. and Tremaine, *J. J. Chem. Soc.*, 632 (1952).
[7] Cotton, F. A. and Wilkinson, G. *Advanced Inorganic Chemistry*, Wiley, NY, 1988.
[8] Neuse, E. and Rosenberg, H. *Metallocene Polymers*, Dekker, NY, 1979.
[9] Wailes, P., Coutts, R. and Weigold, H. *Organometallic Chemistry of Titanium,, Zirconium, and Hafnium,* Academic Press, NY, 1974.
[10] Wilkinson, G., Stone, F. and Abel, E. *Comprehensive Organometallic Chemistry,* Pergamon, Oxford, UK, 1982.
[11] Neuse, E. W. "Metallocene Polymers", *in Encyclopedia of Polymer Science and Technology*, Vol 8 Wiley-Interscience, NY, 1968.
[12] Alaa, A. Abd-El-Aziz, Carraher, C., C. Pittman, J. Sheats, M. Zeldin, Macromolecules Containing Metal and Metal-Like Elements, Vol. 2. *Organoiron Polymers,* Wiley, Hoboken, 2004.
[13] Carraher, C., unpublished.
[14] Carraher, C., *Polymer Chemistry*, 6th Edition, Dekker, NY, 2003.
[15] Grubbs, R. H., Lau, C., Curkier, R. and Brubaker, C. *J. Amer. Chem. Soc.,* 99 (1977)
[16] Pittman, C. and Carraher, C., *J. Reynolds, Encyclopedia of Polymer Science and Engineering,* 2nd Ed., Wiley, 1987, Vol.10.
[17] Carraher, C. and Jorgensen, S. *Polym. Prepr.*, 16(1), 671 (1975).
[18] Carraher, C., *P. Lessek, Europ. Polym. J.*, 8, 1339 (1972).
[19] Carraher, C. and Jorgensen, S., *J. Polym. Sci., Polym. Chem.* Ed., 16, 2965 (1978).
[20] Carraher, C., Feiffer, R. and Fullenkamp, P. *J. Macromol. Sci.-Chem.,* A10, 1221 (1976).
[21] Carraher, C. and Chamely, D., *Polym. Mater. Sci. Eng*., 85, 358 (2001).
[22] Carraher, C., Chamely, D., Carraher, S. and Stewart, H. , *Polym. Mater. Sci. Eng*., 85, 381 (2001).
[23] Carraher, C., Chamely, D., Carraher, S., Stewart, H. , W. Learned, J. Helmy, K. Abby, *Polym. Mater. Sci. Eng*., 85, 375 (2001).
[24] J. Peterson, Carraher, C., Salamone, A. and Francis, A. M. *Polym. Mater. Sci. Eng*., 81, 149 (1999).
[25] Carraher, C., Chamely-Wilk, D., Carraher, S., Barot, G., Stewart, H. and Learned, W. *J. Polym. Mater*., 24, 149 (2007).
[26] Metraus, J. P. *Plant Hormones and Their Role in Plant Growth and Development,* P. Davies, Ed., pp 296-317, Kluwer, Boston, 1987
[27] Guerinot, M. and Salt, D. *Plant Physiology*, 125, 164 (2001).
[28] Somerville, C. and Bonetta, D. *Plant Physiology*, 125, 168 (2001).
[29] Trewavas, A. *Plant Psychology*, 125, 174 (2001).
[30] Battin, A., Carraher, C., *Polym. Mater Sci. Eng.*, 99, 368 and 371 (2008).
[31] Battin, A., Carraher, C., *J. Polym. Mater.,* 25, 23 (2008).
[32] Chiang, C. K., Park, Y. W., Heeger, A. J., Shirakawa, A. J., Louis, E. J. and MacDiarmid, A. G. *J. Chem. Phys*., 69, 5098 (1978).
[33] Shirakawa, H., Edwin, L. J., MacDiarmid, A. G., Chiang, C. K. and Heeger, A. *J. Chem. Soc. Chem. Comm*., 16, 578 (1977).
[34] Chiang, C. K., Druy, M. A., Gau, S. C., Heeger, A., Louis, E. J., MacDiarmid, A. G., Park, Y. W. and Shirakawa, H. *J. Am. Chem. Soc*., 100, 1013 (1978)

[35] Chiang, C. K., Fincher, C. R., Park, Y. W., Heeger, A. J., Shirakawa, H., Louis, E. J., Gau, S., MacDiarmid, A. G. and Alan, G. *Polym Comm.*, 31, 1098 (1977)
[36] Pekker, S., Morin, A. and Beniere, F. *Polym Comm.*, 31, 75 (1990)
[37] Epstein, A. J., Rommelmann, H., Abkowitz, M. and Gibson, H. W. *Mol. Crystals Liquid Crystals,* 77, 81 (1981).
[38] Mathai, C., Saravana, S., Anantharaman, M., Venkitachalam, S. and Jayalekshmi, S. *J. Physics D: Appl. Physics*, 35, 2206 (2002).
[39] Sajeev, U., Mathai, C., Saravana, S., Ashokan, R., Venkatachalam, S. and Anantharaman, M. *Bull. Mater. Sci.*, 29, 159 (2006).
[40] Carraher, C., Foster, V., Linville, R., Stevison, D. and Venkatachalam, R. *Adhesives, Sealants, and Coatings for Space and Harsh Environments,* (L-H. Lee, Ed.) Plenum, NY, 1988.
[41] Carraher, C., Schwarz, R., Schwarz, M. and Schroeder, J. *Org. Coat. Plast. Chem.,* 42, 23, (1980).
[42] Carraher, C., Schwarz, R., Schwarz, M. and Schroeder, J. *Org. Coat. Plast. Chem.*, 43, 798 (1981).
[43] Carraher, C., J. Kloss, *Polym. Mater. Sci. Eng.*, 64, 229 (1991).
[44] Carraher, C., Schwarz, R., Schroeder, J. and Schwarz, M. *J. Macromol. Sci.-Chem.,* A15(5), 773 (1981).
[45] Carraher, C., Schwarz, R., Schroeder, J. and Schwarz, M. *Interfacial Synthesis*, Vol. III, Recent Advances (Carraher, C., J. Preston, Eds.,), Dekker, NY, 1982.
[46] Carraher, C., Foster, V., Linville, R. and Stevison, D. *Polym. Mater. Sci. Eng.*, 56, 401(1987).
[47] Carraher, C., Li, A., Kloss, J. and Lombardo, A. *Polym. Mater. Sci. Eng*, 70, 38 (1993).
[48] Carraher, C., Kloss, J., Medina, F. and Taylor, A. *Polym. Mater. Sci. Eng*, 68, 253 (1993).
[49] Carraher, C., unpublished results.
[50] Keene, F., Salmon, D. and Meyer, T. J. *J. Amer. Chem. Soc.*, 98(7), 1884 (1976).
[51] Carraher, C. and Zhang, Q. *Metal-Containing Polymeric Materials*, Plenum, NY, 1996, 109.
[52] Carraher, C. and Taylor-Murphy, A. *Polym. Mater. Sci. Eng.*, 76, 409 (1997).
[53] Carraher, C. and Taylor-Murphy, A. *Polym. Mater. Sci. Eng.*, 86, 291 (2002).
[54] Carraher, C. and Zhang, Q. *Polym. Mater. Sci. Eng.*, 73, 398 (1995).
[55] Carraher, C. and Zhang, Q. *Polym. Mater. Sci. Eng.*, 71, 505 (1994).
[56] Carraher, C., *Macromolecules*, 2, 306 (1969).
[57] Carraher, C., *J. Polym. Sci. Pt.* A-1, 8, 3051 (1970).
[58] Carraher, C. and Wang, L. *Makromol. Chemie*, 160, 251 (1972).
[59] Carraher, C. and Greene, J. *Makromol. Chemie*, 130, 177 (1969).
[60] Carraher, C. and Greene, J, *Makromol. Chemie*, 131, 259 (1970).
[61] Franklin, E. *J. Chem. Soc.*, 2, 263 (1849);
[62] Carraher, C. Macromolecules Containing Metal and Metal-Like Elements Vol. 4. *Group IVB Polymers*, Wiley, Hoboken, NJ, 2005.
[63] Wei, R., Ya, L., Jinguo, W. and Qifeng, X. *Polymer Materials Encyclopedia* (J. Salamone, Ed.), CRC Press, Boca Raton, FL, p 4826;
[64] Hoch, M. *Applied Geochem.*, 16, 719 (2001).

[65] Carraher, C., Nagata, M., Stewart, H., Miao, S. Carraher, S., Gaonkar, A. Highland, C. and Li, F. *Polym. Mater. Sci. Eng.*, 79, 52 (1998).
[66] Carraher, C., Stewart, H., Carraher, S., Nagata, M. and Miao, S. *J. Polym. Mater.,* 18, 111 (2001).
[67] Carraher, C., Nagata, M., Stewart, H., Miao, S., Butler, C., Gaonkar, A., Barosy, C., Duffield, R. and Li, F. *Polym. Mater. Sci. Eng.*, 78, 36 (1998).
[68] Siegmann-Louda, D., Carraher, C., Chamely, D., Cardoso, A. and Snedden, D. *Polym. Mater. Sci. Eng.*, 86, 293 (2002)
[69] Omae, I. *Organotin Chemistry*, Elsevier, Amsterdam, 1989.
[70] Carraher, C., Butler, C., Naoshima, Y., Forste, V., Giron, D. and Mykytiuk, P. *Applied Bioactive Polymeric Materials*, Plenum, NY, 1990.
[71] Carraher, C., Butler, C., Sterling, D. and Saurino, V. *Polym. Mater. Sci. Eng.*, 66, 352 (1992).
[72] Naoshima, Y., Shudo, H., Uenishi, M. and Carraher, C., *J. Polum. Mats.*, 8, 51 (1991).
[73] Carraher, C., Saurino, V., Butler, C. and Sterling, D. *Polym. Mater. Sci. Eng.*, 72, 192 (1995).
[74] Carraher, C., Roner, M., Shahi, K., Ashuda, Y. and Barot, G. *JIOPM*, 18, 180 (2008).
[75] Carraher, C., Sabir, T., Roner, M., Shahi, K., Bleicher, R., Roehr, J. and Bassett, K. *JIOPM*, 16, 249 (2006).
[76] Siegmann-Louda, D., Carraher, C., Pflueger, F., Nagy, D. and Ross, J. *Functional Condensation Polymers*, Kluwer, NY, 2002.
[77] Pellerito, L., Maggio, F., Consiglio, M., Pellerito, A., Stocco, G. and Grimaudo, S. *App. Organomet. Chem*, 9, 227 (1995).
[78] Maggio, F., Pellerito, A., Pellerito, L., Grimaudo, S., Mansueto, C. and Vitturi, R. *Appl. Organomet. Chem*, 8, 71 (1994).
[79] Vitturi, R., Mansueto, C., Gianguzza, A., Maggio, F., Pellerito, A. and Pellerito, L. *Appl. Organome. Chem.*, 8, 509 (1994).
[80] Baratne Jankovics, H,. Nagy, L., Longo, F., Fiore, T. and Pellerito, L., *Magyar Kemiai Foly.* 107, 392 (2001).
[81] Vitturi, R., Zava, B., Colomba, M., Pellerito, A., Maggio, F. and Pellerito, L., *App. Organomet. Chem.,* 9, 561 (1995). (Contains listing of other related references.)
[82] Pellerito, L., Maggio, F., Fiore, T. and Pellerito, A., *App. Organomet. Chem.*, 10, 393 (1966).
[83] Pellerito, A., Fiore, T., Pellerito, C., Fontana, A., Di Stefano, R., Pellerito, L., Cambria, M. and Mansueto, C. *J. Inorganic Biochem*., 72, 115 (1998).
[84] Carraher, C. and Butler, C., US Patent 5, 043, 463 (Issued Aug. 27, 1991).
[85] Carraher, C. and Butler, C., US Patent 5,840,760 (Issued Nov. 24, 1998)
[86] Butler, C. and Carraher, C., *Polym. Mater. Sci. Eng*., 80,365 (1999).
[87] Carraher, C., Butler, C., Y. Naoshima, D. Sterling, V. Saurino, *Biotechnology and Bioactive Polymers*, Plenum, NY, 1994.
[88] Carraher, C. and Butler, C., *Biotechnological Polymers*, Technomic, Lancaster, PA, 1993.
[89] Carraher, C., Bytler, C., Naoashima, Y., Sterling, D. and Saurino, V. *Industrial Biological Polymers*, Technomic, Lancaster, PA, 1995, Chpt. 8.
[90] Carraher, C., Butler, C., Foster, V., Pandya, B. and Sterling, D. *Polym. Mater. Sci. Eng*., 68, 255 (1993).

[91] Carraher, C., Butler, C., Naoshima, Y., Foster, V., Giron, D. and Mykytiuk, P. *Applied Bioactive Polymeric Materials,* Plenum, NY, 1988.
[92] Carraher, C., Naoshima, Y., Butler, C., Foster, V., Gill, D., Williams, M., Giron, D. and Mykytiuk, P. *Polym. Mater. Sci. Eng*., 57, 186 (1987)
[93] Carraher, C., Butler, C. and Reckleben, L. *Cosmetic and Pharmaceutical Applications of Polymers*, Plenum, NY, 1991.
[94] Carraher, C., Rigdway, T., Sterling, D. and Butler, C., *Industrial Biotechnological Polymers*, Technomic, Lancaster, PA, 1995, Chpt. 13.
[95] Carraher, C., Butler, C., Reckleben, L., Taylor, A. and Saurino, V. *Polymer P,* 33(2), 539 (1992).
[96] Roner, M.,Carraher, C., Roehr, J., Bassett, K. *J. Polym. Mater.*, 23, 153 (2006).
[97] Barot, G., Roner, M., Naoshima, Y., Nagao, K., Shahi, K. and Carraher, C., *JIOPM*, 19, 12 (2009).
[98] Carraher, C.and Morie, K. *JIOPM*, 17, 127 (2007).
[99] Barot, G., Shahi, K., Roner, M. and Carraher, C. *JIOPM*, 17, 595 (2007).
[100] Barot, G., Shahi, K., Roner, M. and Carraher, C. *J. Polym. Mater.*, 23, 423 (2006).
[101] Carraher, C., Battin, A., Shahi, K. and Roner, M. *JIOPM*, 17, 631,2007.
[102] Wanebo, H. J. and Vexeridis, M. P. *Cancer*, 78, 580 (1996).
[103] Warshaw, A. L., Fernandez-DelCastillo, C., *N. Eng. J. Med*., 326, 580 (1992).
[104] Moertel, C. G. *Clin. Gastroenterol*, 5, 777 (1976).
[105] DeCaprio, J. A. and Myer, R. J. *Clin. Oncol.*, 9, 2128 (1991).
[106] Moore, M. and Andersen, J. *Proceedings ASCO*, 14, Abstract Number 473 (1995).
[107] Rothenberg, M. L., Burris, H. A. *Proceedings ASCO, 1*4, Abstract Number 470 (1995).
[108] Hui, Y. F., Reitz, J. *Am. J. Health-Syst. Pharm.*, 54, 162 (1997).
[109] Roner, M., Shahi, K., Barot, G., Battin, A. and Carraher, C., *JIOPM*, 19, 410, 2009.
[110] Shahi, K., Roner, M., Battin, A. and Carraher, C. *Polym. Mater. Sci. Eng*., 99, 365, 2008.
[111] Naoshima, Y., Nagao, K., Battin, A. and Carraher, C., *Polym. Mater. Sci. Eng*., 96, 405, 2007.
[112] Battin, A. and Carraher, C. *Poly. Mater. Sci. Eng*., in press.
[113] Doak, G., Freedman, L. D. *Org. Chem. Revs.,* 6 (1970) 574
[114] Carraher, C., Naas, M., Giron, D., Cerutis, D. R. *J. Macromol. Sci.-Chem.*, A19 (1983) 1101
[115] Carraher, C. and Naas, M. *Polym P.,* 23 (1982) 158
[116] Carraher, C. and Naas, M. *Crown Ethers and Phase Transfer Catalysis in Polymer Science*, Chpt. 7, Plenum, NY. 1984
[117] Carraher, C., *J. Polym. Mater.* 25, 35 (2008).
[118] Carraher, C., Langworthy, T. and Moon, W. *Polymer P.*, 17, 1 (1976).
[119] Carraher, C. and Moon, W. *Coat. Plast.*, 34, 468 (1974).
[120] Siegmann-Louda, D. and Carraher, C. Macromolecules Containing Metal and Metal-Like Elements. Vol. V. *Biomedical Applications*, Wiley, Hoboken, NJ, 2004.

In: Advances in Polymer Research
Editors: Losif Sidorov, Fedor Nikitin, pp. 31-38
ISBN: 978-1-61942-223-0

A QUICK TEST FOR THE ASSESSMENT OF WEATHERABILITY OF POLYETHYLENE

***Bolaji Aremo*[1]* *and Mosobalaje O. Adeoye*[2]**
[1] Centre for Energy Research and Development (CERD),
Obafemi Awolowo University, Ile-Ife Nigeria.
[2] Department of Material Science and Engineering,
Obafemi Awolowo University, Ile-Ife, Nigeria.

ABSTRACT

This work presents an accelerated weathering procedure for the evaluation of the outdoor weathering performance of polyethylene (PE). Virgin PE Samples and PE compounded with fillers were exposed to simulated weathering conditions of radiation and moisture. A 500 W Halogen floodlight was used as the radiant energy source and was positioned 30 cm from the sample exposure rack. Samples were periodically exposed to incident radiation for 10 min. and then dipped in water at temperature of between 4° C and 10° C. Indices of degradation were developed based on some property changes observable during polymer degradation. These were monitored by visual examination of the samples and a rating system was used to assign quantitative measurements. Samples were tested for a total time period of 300 min. with visual examination and rating of deterioration done at 30 min. interval. The virgin PE degraded noticeably during the entire duration of the test and its degradation indices dropped sharply. The PE containing the fillers exhibited varying degrees of resistance to degradation throughout the test.

Keywords: Polyethylene, weathering test, photodegradation, ultraviolet radiation.

1. INTRODUCTION

The degradation of polymeric materials in the outdoor environment is a very complex reaction process. Mechanisms for this degradation may include photo-oxidative degradation, thermal degradation, ozone-induced degradation, mechano-chemical degradation, catalytic degradation and biodegradation (Singh and Sharma, 2008). Oxidative photo-degradation of polymeric materials is a free radical reaction involving processes of initiation, propagation and termination (Ranby and Rabek 1975). When the polymer is exposed to ultraviolet (UV)

* Corresponding author: Email: bolaji_aremo@yahoo.com, Tel: (+234)8035024243

rays in an oxygen containing atmosphere free radicals are generated on the surface (Kumar and Gupta, 1998).

Hydrolysis can cause chain scission and consequent failure in several polymers (Salamone, 1999). This is a major problem with polymers at elevated temperature, though it can proceed slowly at ambient temperature with rain water as the source of water. Also, physical forces, such as heating/cooling, freezing/thawing, or wetting/drying, can cause mechanical damage such as the cracking of polymeric materials (Kamal and Huang, 1992).

Functional groups are added to some polymers like PE to protect them from photo-degradation and make them suitable for outdoor applications. A range of modified PEs has been developed with good resistance to photo-degradation. These include chloro-sulphonated polyethylene (CSM elastomer) and chlorinated polyethylene which have been successfully used in industrial roofing and reservoir covers (Hall 1996).

A number of outdoor and laboratory-based tests simulating long term outdoor weathering performance of PE have been developed in recent years. Some of these tests seek to infer long-term weather performance based on short term outdoor data. Basfar and Idriss (2006) reported a natural outdoor weathering test performed to investigate UV stability of PE thin films. This test however took 365 days to complete. The usually long duration of natural outdoor weathering tests and the need for quick, reliable tests of outdoor weathering performance has led to the development of a number of laboratory-based accelerated weathering tests. These tests provide an assessment of weather performances in a fraction of the time used for natural weathering tests. Such is the popularity of these accelerated tests that many accelerated weathering chambers are now commercially available (SolarLight, 2009).

Many standards have been developed to guide the design of both natural and accelerated weathering tests for polymers. ASTM F2657 – 07 (ASTM, 2009) was developed as a standard test method for outdoor weathering exposure of crosslinked polyethylene tubing. Also, a number of standards have been developed for accelerated weathering of polymers. An example of this is ASTM G-151 (ASTM, 2009), standard practice for exposing non-metallic materials in accelerated test devices that use laboratory light sources. These standard tests seek to establish a correlation between exposure times and depreciation in a readily observable material property.

In view of the increasing importance of PE as a commodity polymer, polymer processors are getting more competitive with their polymer blends and compounds. This recent drive to capture new markets and create innovative product offerings has led to the increasing importance of the weathering performance of the polymer. In this work, a simple laboratory-based accelerated weathering test, using a concentrated light source and intermittent dousing of samples in cold water was developed. It is hoped that this will be a considerable time saver in product R&D and quality control.

2. Experimental Method

PE compounded with varying proportions of performance enhancing fillers was prepared. The fillers used include: Fine clay and ammonium chloride for flame resistance, carbon black at 2 % weight of PE. Carbon black is compounded with polymers as a protective agent against photo oxidation (Halliwell, 1992).

The samples all have dimensions of 25 mm by 25 mm and a thickness of 3 mm. The test setup comprised of an exposure rack and a 500 W halogen floodlight. A thermometer was placed against the back face of the rack to measure the back face temperature. The floodlight was positioned 30 cm away from the exposure rack and served as the radiant energy source. See Figure 1 for details of the experimental setup. Samples were exposed to the radiation from the light source for 10 min. and were then dipped in cold water in a temperature range of between 4 °C and 10 °C. The progressive degradation of the samples caused by this radiation was observed at 30 min. interval over a time period of 300 min.

Index of Degradation (*i*): The index of degradation denoted by *i* is a pre-determined material property that deteriorates in the course of the test. Monitoring of the index of degradation was based on visual examination of the samples for progressive changes in the material property.

For the virgin PE, the indices of degradation are:

(i) Warping (WP): This is the distortion of the shape of the sample
(ii) Loss of Flexibility (LF): Samples were subjected gentle bending during the course of the test and the degree of stiffening was noted.
(iii) Loss of Gloss (LG): This is the gradual disappearance of the glossy texture on the PE surface

For PE compounded with fillers, the indices of degradation are:

(i) Warping (WP): Same as for PE
(ii) Surface Pitting (SP): Appearance of small pits or cavities of the polymer surface
(iii) Surface Erosion (SE): Gradual breaking off or loss of surface material

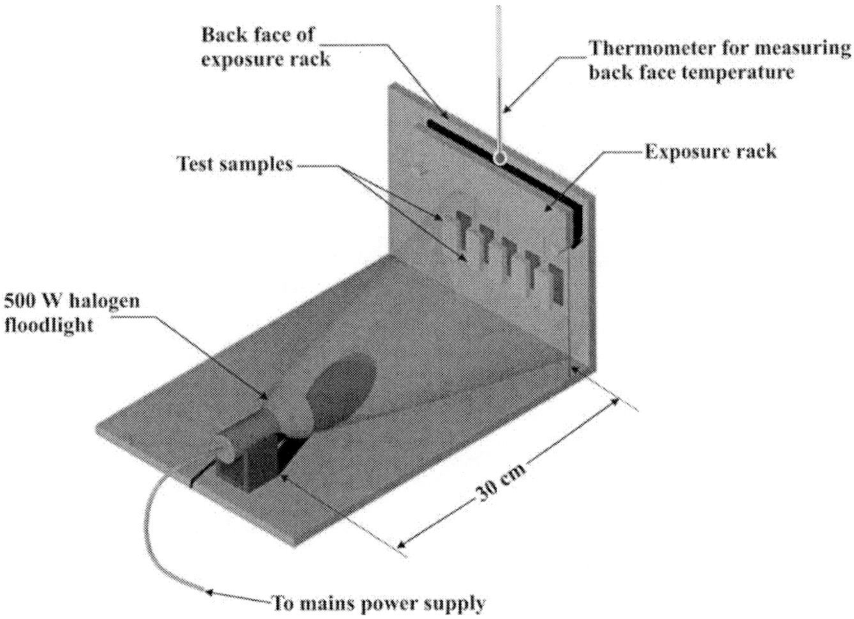

Figure 1. Experimental setup of the accelerated weathering test.

Figure 2 provides illustrations of the different indices of degradation.

The test was carried out in a time period of 300 min and the changes in the degradation indices was monitored at 30 min intervals. A rating system was used to assign quantitative values to the degradation indices. The rating system is given below.

10 – No degradation
8 – Very slight degradation
6 – Reasonably noticeable degradation
4 – Very noticeable degradation
2 – Severe degradation
0 – Complete failure

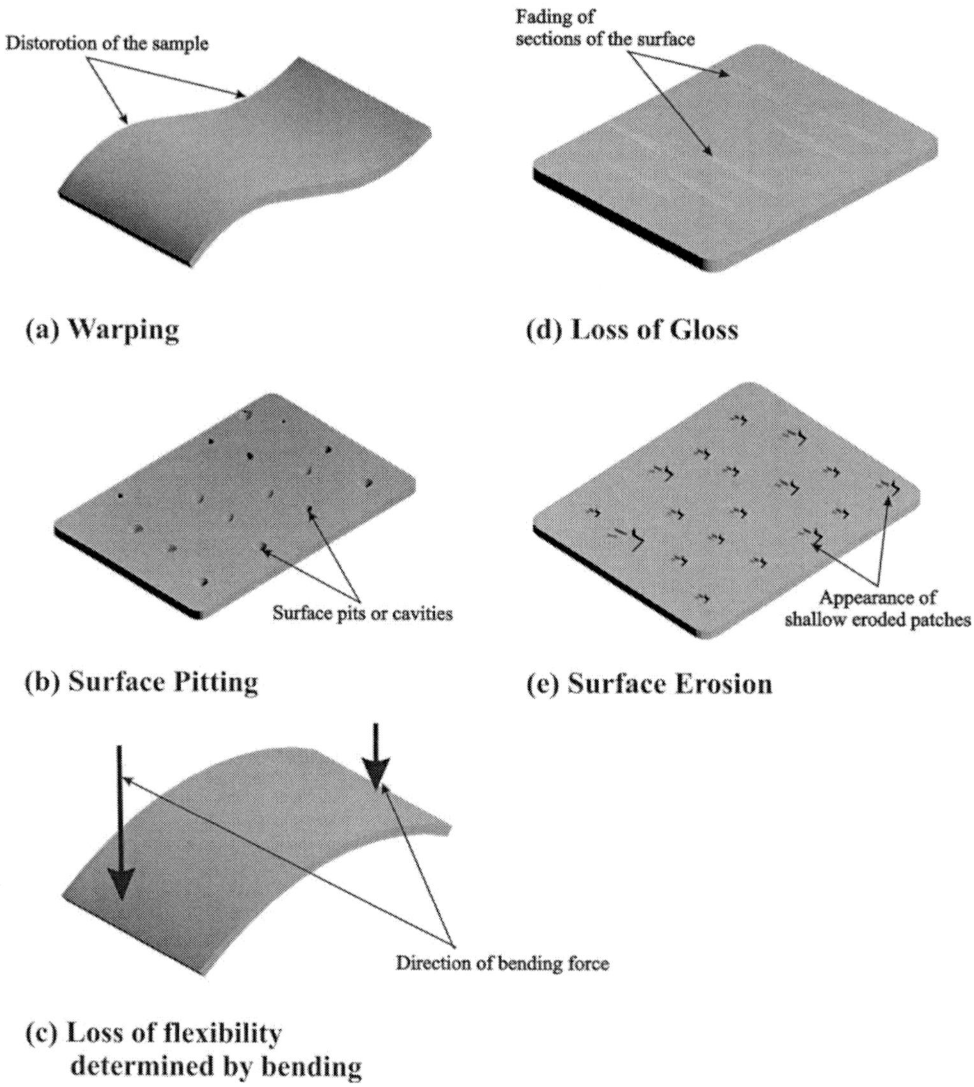

Figure 2. Illustration of the different indices of degradation.

3. Results

The result of the weathering test is presented in Table 1.

Table 1. Results of the accelerated weathering test

Specimen	Degradation Index (*i*)	Time (min.)										
		0	**30**	**60**	**90**	**120**	**150**	**180**	**210**	**240**	**270**	**300**
A	[a]WP	10	10	10	10	10	10	8	6	4	2	2
	[b]LF	10	8	8	6	4	4	4	4	2	2	2
	[c]LG	10	8	8	8	8	6	6	6	6	6	6
	i-total	**30**	**26**	**26**	**24**	**22**	**20**	**18**	**16**	**12**	**10**	**10**
B	WP	10	6	6	4	4	4	4	4	4	4	4
	[d]SP	10	8	8	6	6	6	6	6	6	6	6
	[e]SE	10	8	8	8	8	8	6	4	4	4	4
	i-total	**30**	**22**	**22**	**18**	**18**	**18**	**16**	**14**	**14**	**14**	**14**
C	WP	10	8	8	8	8	8	6	4	4	4	4
	SP	10	8	8	8	8	6	6	6	6	6	6
	SE	10	8	8	8	8	8	8	8	8	8	8
	i-total	**30**	**24**	**24**	**24**	**24**	**22**	**20**	**18**	**18**	**18**	**18**
D	WP	10	8	8	8	8	8	6	4	4	4	4
	SP	10	8	8	8	8	8	8	8	8	6	6
	SE	10	8	8	8	8	8	8	8	8	8	8
	i-total	**30**	**24**	**24**	**24**	**24**	**24**	**22**	**20**	**20**	**18**	**18**
E	WP	10	8	8	8	8	8	6	4	4	4	4
	SP	10	8	8	8	6	6	6	6	6	6	6
	SE	10	8	8	8	8	8	8	8	8	8	8
	i-total	**30**	**24**	**24**	**24**	**22**	**22**	**20**	**18**	**18**	**18**	**18**

[a]WP = warping
[b]LF = Loss of Flexibility
[c]LG = Loss of Gloss
[d]SP = Surface Pitting
[e]SE = Surface Erosion.

Control Sample A (Virgin PE)

The first 60 min. of exposure produced very slight deterioration of the indices. The exposed sample showed good dimensional stability and good flexibility. As the exposure time is increased to 120 min., there was a sharp deterioration of the degradation indices. This was reflected in the value of *i-total* dropping from 28 at 60 min to 20 at 120 min. After 180 min., the first slight change in dimensional stability was observed. The value of *i* for warping dropped slightly for the first time from 10 to 8. Also observed during this interval was a further decrease in the value of *i* for loss of gloss from 8 to 6 and *i*-value for flexibility from 6 to 4. The *i*-value for flexibility also deteriorated further from 6 to 4. At this time interval, the test rack back-face temperature is about 80 °C. All the crystallites present in the polymer microstructure would have melted and structural shape will be maintained mainly by covalent crosslinks. The proportion of crystallites form in a thermoplastic polymer like PE depends partly on the cooling rate from the glass transition temperature (T_g). Thus rapid cooling from this temperature reduces the proportion of crystallite. The degree of covalent crosssIinks however increases due to the radiant energy. The reduction in crystallinity at T_g, increases the tendency to warp because the crystallites are responsible for dimensional stability. The corresponding increase in cross-links creates a network structure and marked the onset of embrittlement and a rapid loss of gloss.

210 – 300 min.: During this time range, there is a highly accelerated rate of deterioration of the degradation indicies. The value of *i-total* falls from 18 at 180 min to 10 at 300 min. The accelerated deterioration could be attributed to the sharp decline is the proportion of crystallites and the increased degree of crosslinking. At this point, the value of the flexibility index had reduced to 2. From the rating system, this depicted severe degradation.

Samples B, C, D and E

The first 60 min.: Samples C, D and E were rated with i-values of 8 (very slight degradation) in all the indices. The exception in this time period was sample B with an i-value rated at 6 for the warping index. The samples were all compression moulded at low compaction pressure. The samples visibly shrunk during this time interval due to the test temperature the presence of void caused by their method of forming. The shrinking surface thus caused the rapid appearance of pits and warping of the samples.

120 min.: By this time the shrinking of the samples had stabilized and the i-value for warping for the samples had assumed constant values of 4 for sample B and 8 for samples C, D, and E. Samples C, D and E also assumed constant i-values of 8 for the other indices while sample B experience some deterioration of these indices. Sample B contained more of the PE and less of the fillers. This indicated that the sample relatively less resistant to the degradative conditions of the test.

180 min.: After 180 min., samples B, C and D exhibited slight decrease in the i-value for the warping index. This was then followed by steady deterioration of the other degradation indices. The values of *i-total* observed in all the compounded sample at 300 min was much higher than for virgin PE. Steady deterioration behaviour observed beyond 180 min indicated an improved resistance to photo-degradation in these samples virgin PE.

210 – 300 min.: Between 180 and 300 min., almost all the samples displayed a constant deterioration of properties. This was reflected in the largely constant values of the degradation indices during this period. After a slight degradation in surface erosion values at 180 min., sample B displayed constant deterioration behaviour.

Weathering changes the physical and optical properties of polymers relative to initial observed properties. The most prominent changes are loss of mechanical properties. Typical changes to mechanical properties caused by photo degradation include: reduced tensile strength, increased embrittlement and possible appearance of surface cracks and chalking (ZEUS Technical White Paper, 2005). These observed property changes in naturally photo-degraded polymer formed the basis for the development of the degradation indices used in this test. The degradation indices, Surface Erosion (SE) and Loss of Gloss (LG) were based on the observed chalking or flaking off of surface material, while Surface Pitting (SP) was based on the appearance of surface cracks. Loss of Flexibility (LF) was developed from the progressive embrittlement of the polymer while Warping was based on the reduction in tensile strength.

Hydrolysis can cause chain scission and consequent failure in several polymers (Salamone, 1999). Although this is a major problem with polymers at high temperature, it can proceed slowly at ambient temperature with rain water as the source of water.

More quantitative property changes like weight loss and direct tensile strength measurements were not considered adequate to monitor the progression of the weathering. The accelerated weathering test included a step involving immersion of the samples in water. Adsorbed water in the samples would give erroneous weight loss readings. An attempt to dry the samples using an oven would introduce an unwanted variable, due to the oven heat, into the test. Natural air-drying on the other hand would take considerable time to accomplished and hence unduly prolong the test. Measurement of reduction in tensile strength would involve subjecting samples to a destructive tensile test. A large number of samples will thus be required at every interval of the test. An even larger number will consequently be needed for the entire test and these will require a much larger exposure area.

The positioning of the floodlight 30 cm from the samples was sufficient to generate an exposure rack back-face temperature of 80 °C necessary to catalyse the degradation. Test temperature plays a prominent role in catalyzing accelerated weathering of polymer. Suits and Hsuan (2003) suggested that the faster degradation obtainable in Xenon arc light laboratory weatherometers may be mainly contributed by the elevated incubation temperature rather than the irradiance. Hence, the accelerated deterioration of the degradation indices thus observed may be mainly as a result of the relatively high back-face temperature.

The test was further accelerated by the thermal cycling of sample temperature due o the periodic dipping in cold water. This caused the rapid manifestation of surface cracks.

The UV component of the radiant light source causes crosslinking reactions to occur in the polymer chains. This leads to chain scission and the formation of covalent crosslinks between adjacent chains. The presence of these crosslinks makes the polymer progressively stiffer and brittle.

4. Conclusion

The increasing use of PE for outdoor applications makes it pertinent to develop quick and practicable weathering tests. The accelerated weathering test described in this work has been able to compare the relative weathering performance of virgin PE and PE compounded with some fillers. From this test, the relative effectiveness of the fillers in protecting the polymer under simulated weathering conditions was determined. The simplicity of this test will be useful in R&D of new PE-based plastic structures destined for outdoor usage. Also, the test may be used in quick quality control assessment of compounded polymer batches and plastic products.

REFERENCES

[1] ASTM International (2009). ASTM F2657 - 07 Standard Test Method for Outdoor Weathering Exposure of Crosslinked Polyethylene (PEX) Tubing. Source: http://www.astm.org/Standards/F2657.htm. Accessed: 20th August 2009

[2] ASTM International (2009). ASTM G151 - 06 Standard Practice for Exposing Nonmetallic Materials in Accelerated Test Devices that Use Laboratory Light Sources. Source: http://www.astm.org/Standards/G151.htm. Accessed: 20th August 2009

[3] Basfar. AA; Idriss-Ali, KM. (2006). Natural weathering tests for films of various formulations of low density polyethylene (LDPE) and linear low density polyethylene (LLDPE). *Polymer degradation and stability*. 91(3):437-443.

[4] Hall, C. (1996). Polymers, In: *Civil Engineering Materials*, 5th Ed., Jackson N, Dhir RK, Macmillan, London.

[5] Halliwell, SM (1992). *Weathering of Polymers*. iSmithers Rapra Publishing, p.19.

[6] Kamal, MR; Huang, B (1992). Natural and artificial weathering of polymers. In: Hamid SH et al. (eds) *Handbook of Polymer Degradation*. Marcel Dekker, New York, pp. 127-168.

[7] Kumar, A; Gupta, RK (1998). *Fundamentals of Polymers*. McGraw-Hill, Singapore, p 38.

[8] Ranby, B; Rabek, JF (1975). *Photo-degradation, Photo-oxidation and Photo-stabilization of Polymers*, Wiley, London.

[9] Salamone, JC (1999). *Concise polymeric materials encyclopedia*. CRC Press, p365.

[10] Singh, B; Sharma, N (2008). Mechanistic Implications of Polymer Degradation. *Polymer Degradation and Stability*, 93(3):561-584.

[11] SolarLight Inc., (2009). *Testing Services*. Source: http://www.solarlight.com/services/services/html. Date Accessed: 16th August 2009.

[12] ZEUS Industrial Products (2005). *Weathering of Plastics*. ZEUS Technical White Paper. pp 1-7

In: Advances in Polymer Research
Editors: Losif Sidorov, Fedor Nikitin, pp. 39-54
ISBN: 978-1-61942-223-0

GRAFTING OF POLY(LAURYL METHACRYLATE) FROM THE SURFACE OF LAYERED SILICATES USING OLIGOMERIC AZO-INITIATORS

*Vikas Mittal**
Institute of Chemical and Bio-engineering, Department of Chemistry and Applied Biosciences, ETH Zurich, 8093 Zurich, Switzerland.

ABSTRACT

To improve the dispersion of filler in the polymer matrices to achieve exfoliated nanocomposites, a number of different methods have been used. Grafting of polymers from the surface of the filler is an attractive approach among these methods. To achieve polymer grafting, it is required to immobilize an initiator on the surface of the filler platelets. This initiator can then be used to graft polymer brushes from the surface of the filler. Initiators for this purpose can either be monocationic or bicationic, and offer their specific limitations and advantages. The use of bicationic initiators leads to generation of less amount of free polymer than the monocationic initiator but has also higher potential of quick termination of the radicals owing to their close proximity with each other. The use of living nitroxide mediated polymerization has also been demonstrated to delay the uncontrolled termination of the radicals which leads to extensive as well as controlled grafting of polymers from the filler surface.

Keywords: nanocomposites, filler, cation exchange, initiator, bicationic, monocationic, living polymerization.

1. INTRODUCTION

Polymer layered silicate nanocomposites have gained tremendous research interest in the recent years owing to enhancement of the properties of polymers at very low filler volume fractions [1-7]. To synthesize these nanocomposites, layered silicate aluminosilicate clays like montmorillonite are generally used as fillers [8,9]. These aluminosilicate filler particles have platy morphology and have different cation exchange capacities i.e. different extents of cations present on the surface of their platelets. Owing to the presence of cations on the surface of the platelets, the platelets are attracted to each other and form thick tactoids with

* BASF SE, Polymer Research, 67056 Ludwigshafen, Germany. vikas.mittal@chem.ethz.ch

the platelets held together by strong van der Waals forces. These platelets do not lead to the required enhancement of the properties and to achieve maximum extent to reinforcement from the fillers, it is required to delaminate these tactoids into single platelets so that the filler is dispersed in the polymer matrices at nanoscale. The delamination of the thick tactoids is not straight forward and depends on different factors like

- Surface modification of the filler platelets
- Compatibility of the organically modified filler platelets with the polymer matrices
- Mode of synthetic process used for the generation of nanocomposites. Shear forces used to mix the polymer and filler together should be optimum in addition to the chemical compatibilization of the filler and polymer matrix in order to achieve an optimum result.

Surface modification of filler platelets is achieved in order to compatibilize them with the polymer matrices so as to accelerate the delamination of filler at nanoscale. It is generally achieved by exchanging the surface cations with the long chain alkyl ammonium ions which enhance the hydrophobicity and organophilicity of the filler platelets thus improving their chances of homogenous mixing in the polymer matrices. Apart from that, the surface modification by cation exchange also leads to increase in the inter-distance between the platelets which is also supportive of the delamination of the filler in the polymer matrix by the action of shear [10,11]. Commonly used alkyl ammonium modifications include trimethyloctadecylammonium, benzylhexadecyldimethylammonium and dimethyldioctadecyl -ammonium etc [10,11]. However, it has been observed that these modifications are only helpful in achieving delaminated nanocomposites when polar polymer matrices are used for the generation of nanocomposites. As the filler interlayers are partially polar even after the surface modification and these can compatibilize better with the polar polymer matrices, therefore, maximum gains in the properties have been reported in the case of polar polymers [1-7]. The non-polar polymers like polypropylene or polyethylene cannot compatibilize with the partially polar filler platelets owing to absence of any positive interactions between them. Thus, the delamination of filler platelets in these matrices is a more serious challenge.

Various methods have been employed in the recent years to achieve optimum surface modification of the filler so as to make the filler platelets also dispersible in non-polar matrices. These include among others [12]

- Exchange of high chain density surface modifications on the filler surface [13]
- Exchange of long alkyl ammonium chains on the filler surface [13,14]
- Grafting of polymer chains either from or to the filler surface [15-21]
- Physical adsorption of the long chain organic molecules or polymer chains on the surface of already modified montmorillonites [22-25]
- Esterification reactions of the long chain alcohols or acids with the carboxyl or hydroxyl groups ionically bound to the surface etc [26-31].

Although the use of all of the above mentioned techniques has been reported to be beneficial for the synthesis of nanocomposites with non-polar matrices, but the grafting of polymeric or oligomeric chains on the surface of the filler platelets is very attractive owing to its flexibility as a number of different polymers can be grafted on the filler surface. These polymer grafted platelets can then be mixed efficiently with the bulk polymer matrices thus achieving delaminated polymer nanocomposites. The grafting of the polymer chains on the filler surface can also be achieved in two different ways: grafting to the surface or grafting from the surface. Grafting of polymer chains 'to' the surface approach focuses on chemical immobilization of end-functionalized polymer chains to the surface by reacting the ends with the reactive groups already immobilized on the surface. It can also be achieved either by attaching a monomer on the surfaces and subsequently polymerizing the monomer in the presence of externally added monomer and initiator. In both the cases, as the preformed long chains approach the surfaces, therefore, the steric hindrance concerns may still hinder the generation of high density grafts on the surface. The generated grafts may also not be very uniform in terms of molecular weight and chain lengths etc. owing to the uncontrolled nature of the process which is generally achieved by non-living conventional free radical polymerization.

In the grafting from the surface approach, an initiator is covalently bound to the surface which is subsequently polymerized with the external monomer without the external addition of any further initiator. This approach has enjoyed more success because of its ability to generate densely packed polymer functionalities in the absence of any kinetic hindrance along with the advantage of covalently bound polymer chains on the surface.

As mentioned above, the grafting from surface approach requires the immobilization of an initiator on the surface of the filler platelets. The initiator for such a purpose needs to perform two functions: it should acts as free radical initiator as well as it should be able to bind ionically on the filler surface during the surface modification step. The initiator can also either attach by both of its ends to the filler surface or only one end may be used for the immobilization, whereas the second end may be free. Thus, to achieve this purpose, synthesis of either monocationic or bicationic initiators is required. The use of monocationic and bicationic initiators has their own advantages and limitations. The use of bicationic initiator on the surface of the filler reduces the amount of free polymer formed during the polymerization as both the end of the initiator molecules are bound to the filler surface. But the use of such initiator also suffers from the possibility of quick termination of the radicals generated after the dissociation of initiator molecules owing to the close proximity of these radicals. The use of monocationic initiator has less possibility of quick termination of the generated radicals as one end being free can diffuse away from the radical on the filler surface. But as one end of the initiator is free, it also leads to the generation of a large amount of free polymer in solution. Thus, but the initiator systems are on one hand helpful in grafting polymers form the polymer surface, but on the other hand, their specific disadvantages, as a result, the choice of initiator is dependant on the nature of the system at hand. The length of initiator molecule is also an important factor as too short chain length may lead to unwanted interaction of these radicals with the filler surface or masking of the radicals by other means. Therefore, the chain length of the initiator molecules should be long enough to provide easy access of the radicals for the grafting reaction. The amount of surface coverage of the filler with the initiator is also a serious concern as the complete coverage of the filler surface with the initiator can lead to the generation of a large number of radicals on the surface which may start to terminate with each other owing to the presences of radicals too near to each other. A mixed brush of the initiator molecules with non-reactive surface modification is desirable to be formed on the filler surface. The coverage of the filler surface with initiator on the other hand should not be too less as it may also limit the efficiency of grafting reaction. The living means of polymerization are important in controlling the molecular characteristics of the polymer chain grafted from the surface, as these methods reduce the termination of the radicals thus allowing longer life times to the radicals.

2. Synthesis of Oligomeric Initiators and Subsequent Filler Processing

Synthesis of Asymmetric Monocationic Initiator

The synthesis of asymmetric monocationic initiator has been reported earlier in references [12,16,20]. Figure 1 also reports the synthetic steps for the generation of the asymmetric monocationic initiator. For this, required amounts of 4,4'-azobis(4-cyanovaleric acid), butanol and dimethylamino pyridine (DMAP) were dissolved in dimethylformamide (DMF). The solution was cooled slowly in an ice bath and dicyclohexylcarbodiimide (DCC) was then slowly added to the solution. After the addition of DCC, the reaction solution was allowed to warm to room temperature and the reaction was run overnight. The reaction led to the generation of urea, which was filtered and the filtrate was poured in cold water. The filtrate was extracted with dichloromethane (CH_2Cl_2), followed by washing with saturated sodium bicarbonate. The resultant organic material was dried over sodium sulphate. The solvent was subsequently evaporated and the thick organic material was poured in cold hexane. The crystals of ester were generated in hexane when the solution was kept in fridge.

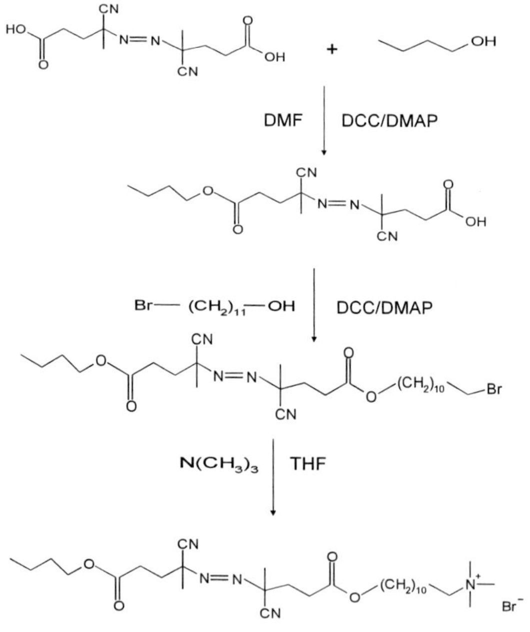

Figure 1. Reaction scheme for the synthesis of asymmetrical monocationic initiator for achieving polymerization from the clay surface [12,16,20]. Reproduced from reference 12 with permission from Taylor and Francis.

The generated ester was further reacted with 11-bromo-1-undecanol by using the similar reaction conditions using the DMAP/DCC in DMF. The working p if the reaction solution in the similar way as described above led to the recovery of diester. The diester required the synthesis of ammonium group at one end of the molecule so as to attach the initiator molecules to the filler surface. This group was generated by reacting the diester with excess of trimethyl amine in tetrahydrofuran (THF). The generated ammonium salt was collected and after washing with THF was stored under low temperature. The overall yield of the synthetic processes for the generation of asymmetrical monocationic initiator was 65%.

Synthesis of Symmetric Bicationic Initiator

Symmetric bicationic initiator was synthesized by following the process reported in the literature [20,32]. Figure 2 reports the synthetic steps for the generation of the symmetric bicationic initiator [20]. For the synthesis, required amounts of 4,4'-azobis(4-cyanovaleric acid), 1-bromo-11-undecanol, DMAP were mixed with dimethylformamide in a round bottom flask. The reagents were stirred at room temperature to get a homogenous solution. After the complete dissolution as achieved, the contents were cooled in an ice water bath. DCC was subsequently added and was allowed to dissolve fully. The reaction was kept for 5 min at 0 °C and subsequently was warmed to room temperature. Similarly in the case of asymmetric initiator, the reaction was allowed to run at room temperature for 24 h. The urea was similarly filtered and the reaction was worked up in water and methylene chloride followed by washing with saturated sodium bicarbonate and drying over sodium sulfate. The evaporation of the solvent from the organic solution led to the generation of viscous solution which was poured in cold methanol. It resulted in the generation of the crystals of the dibromo ester product. The crystals were washed with excess methanol, and were dissolved in THF for the reaction to generate terminal ammonium groups. Slow addition of an excess of trimethyl amine to the THF solution of diester was achieved and the reaction was kept under stirring at room temperature for two days. The solvent was then evaporated and the crystals of the bicationic salt were precipitated by the addition of ether. Overall yield of the synthesis processes for the generation of symmetrical bicationic initiator was observed to be 80%.

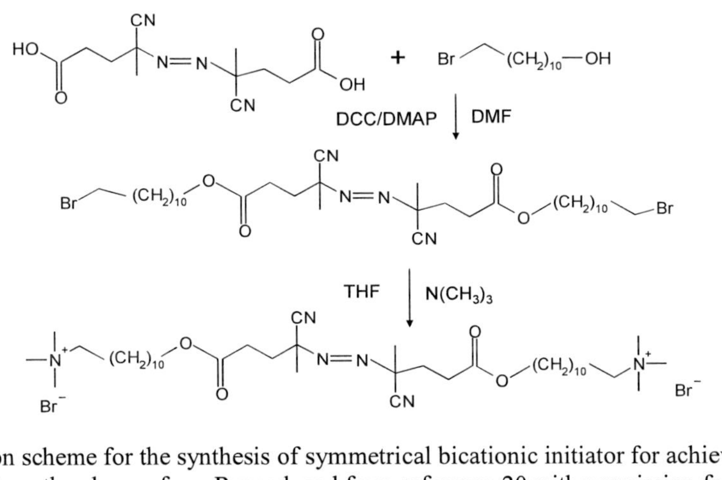

Figure 2. Reaction scheme for the synthesis of symmetrical bicationic initiator for achieving polymerization from the clay surface. Reproduced from reference 20 with permission from Elsevier.

Cation Exchange of the Initiator on the Filler Surface and Polymer Grafting

The generated initiators were exchanged on the filler surface by cation exchange with the inorganic cations present on the filler surface. The amount of the initiator was also varied in order to study the effect of the amount of surface coverage of the filler surface with the initiator on the generated polymer grafts. Dimethyldidodecylammonium bromide was chosen as the non-reactive modification when the monocationic initiator was exchanged on the filler surface. On the other hand, dihexadecyldimethylammonium bromide was used as the non-reactive surface modification for the case of bicationic initiator. The length of the non-reactive surface modification should also be carefully chosen as it should be significantly longer than the initiator chain length, otherwise it may mask the radicals formed after the dissociation of the initiator molecules. It should also be noted that the bicationic initiator an either exchange on the two different platelets through its two cationic terminal groups or both the cationic groups may also be immobilized on the same platelet.

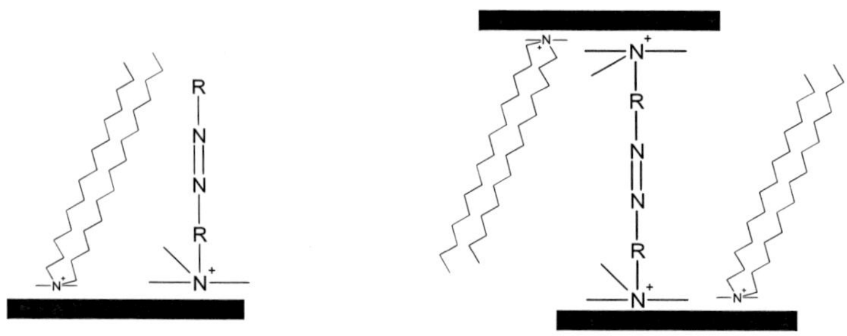

(a) (b)

Figure 3. Representation of the partial modification of the filler surface with (a) monocationic initiator and (b) bicationic initiator.

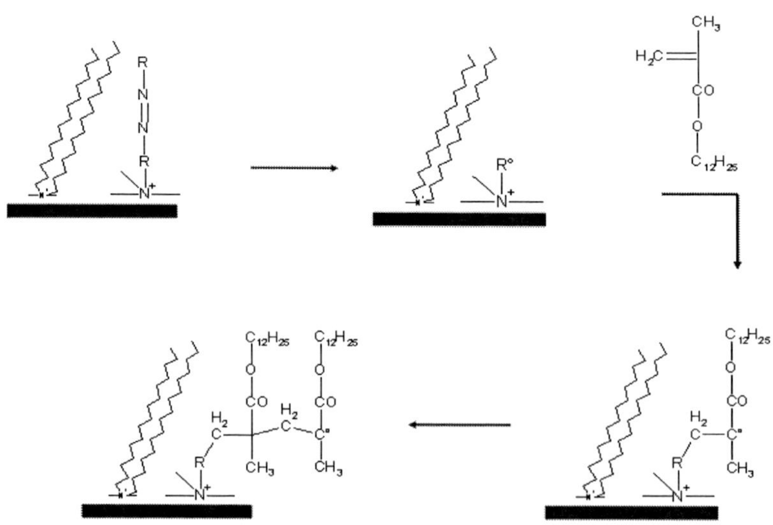

Figure 4. Representation of the polymerization from the surface (for the polymerization of lauryl methacrylate) using the montmorillonite partially modified with asymmetrical monocationic initiator. Reproduced from reference 12 with permission from Taylor and Francis.

Cation exchange with bicationic initiator was achieved in a quantity corresponding to 30% of the CEC of filler and with dihexadecyldimethylammonium bromide corresponding to 70% of the CEC. Similarly, exchange was also achieved with bicationic initiator in a quantity corresponding to 70% of the CEC of filler and with dihexadecyldimethylammonium bromide corresponding to 30% of the CEC. In the case of monocationic initiator, the initiator was exchanged on the surface in an amount corresponding to 30% of the CEC of filler and didodecyldimethylammonium bromide corresponding to 70% of the CEC. Figure 3 shows the schematic of the partial modification of the filler surface with monocationic and bicationic initiators.

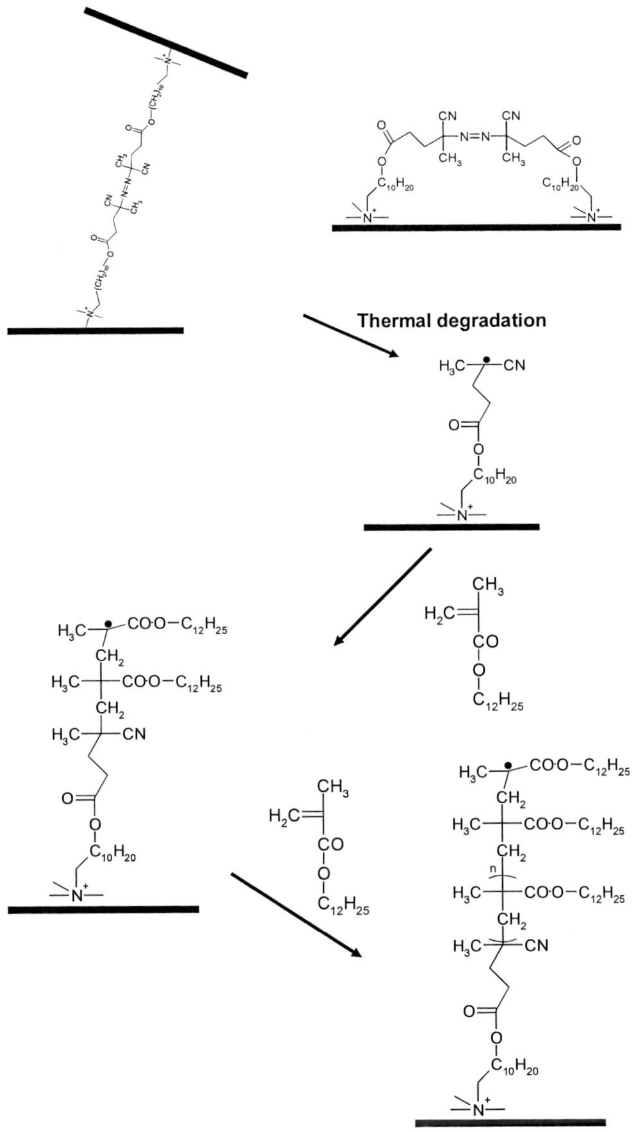

Figure 5. Representation of the polymerization 'from' the surface approach with bicationic initiator bound to the clay surface. It should be noted that the bicationic initiator may be attached either to one platelet or to two platelets. Reproduced from reference 33 with permission.

In order to ascertain if the immobilization of the initiators on the surface of the fillers was successful in achieving the grafting of the polymer chains form the surface of filler platelets, the free radical polymerization of externally added lauryl methacrylate was achieved. The polymerization trials were carried out in toluene as the filler could be efficiently dispersed in toluene and the monomer as well as generate polymer were readily soluble in toluene. The initiator modified filler was dispersed in toluene by sonication as the first step. Required amount of lauryl methacrylate was then added to this dispersion and the dispersion was placed in the oil bath preheated at the polymerization temperature. After the polymerization, the reaction mixture was poured in large excess of ethanol which allows the reacted clay and the non-bound polymer to precipitate. Ethanol mixed with toluene was then decanted and the thick gel was added with isopropyl acetate and then heated at 70 °C. Isopropyl acetates was added to solubilize the non-bound polymer. The solvent was again decanted and the whole process was repeated 4-5 times followed by the evaporation of solvent. Figure 4 and 5 demonstrate the schematic representation of the polymerization reactions using monocationic and bicationic initiators respectively [12,33]. The living polymerization trials were also carried out using nitroxide mediated polymerization. SG1 nitroxide (N-tert-butyl-1-diethylphosphono-2,2′-dimethylpropyl nitroxide) was used for this purpose. An advantage of the nitroxide mediated polymerization is the ease of applicability to the conventional free radical polymerization. The growing chains get end capped with the nitroxide agent thus getting reversibly terminated. These reversible terminated chains become active again after some time and add further monomers before becoming inactive again. This way the life of the radical is significantly enhanced which allows immense control on the termination reactions as well as on the molecular properties of the grafted chains. The reactions using living polymerization trials were carried out in bulk conditions at 120°C. Figure 6 depicts the process of nitroxide mediated polymerization for the polymer grafting from the filler surface.

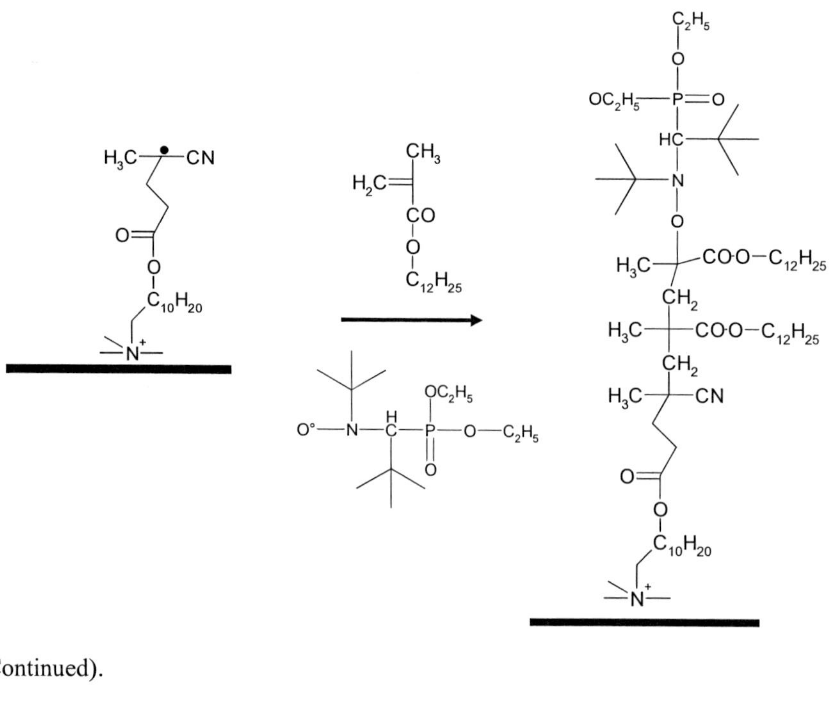

Figure 6. (Continued).

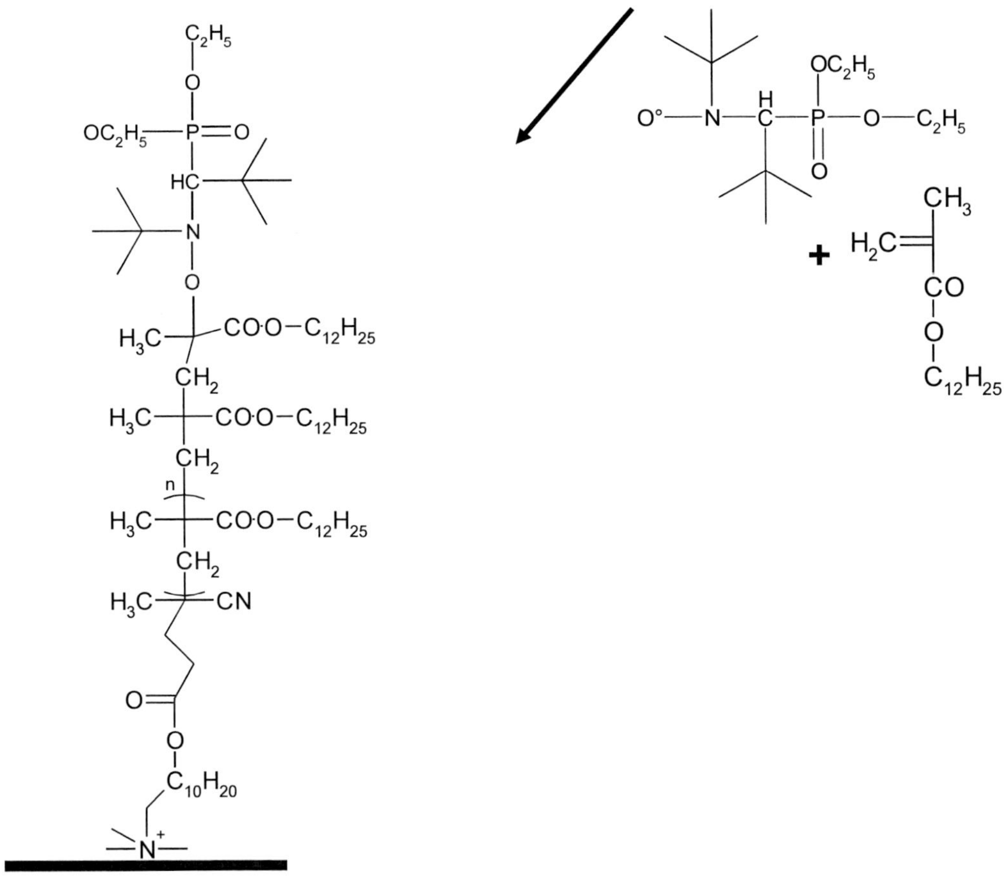

Figure 6. Schematic representation of controlled polymer grafting on the clay surface using nitroxide mediated polymerization of lauryl methacrylate. Reproduced from reference 33 with permission.

3. POLYMERIZATION

3.1. Polymerization Trials with Asymmetric Monocationic Initiator

As mentioned above, the use of asymmetric monocationic initiator for the polymer grafting has less termination of the radicals as out of the two generated radicals by initiator dissociation, one is mobile and can diffuse away from the immobilized radical faster. To further reduce the possibility of termination, use of living polymerization techniques was also tried. Use of living polymerization techniques like nitroxide mediated polymerization and atom transfer radical polymerization have been reported for the synthesis of either controlled graft or block copolymers from the surface [20,34-40]. The initiator structure of the initiator was controlled in order to contain an oligomeric chain in the molecule so as to achieve higher basal plane spacing of the filler when modified with the initiator which would subsequently help to swell the filler better in the solvent and monomer. Figure 7 shows the X-ray diffractograms of clay modified by using asymmetrical monocationic initiator in an amount corresponding to 30% of CEC and dimethyldidodecylammonium in an amount corresponding to 70% of CEC (curve I), the clay of curve I when reacted with lauryl methacrylate in non-living conditions (curve II) and the clay of curve I when reacted with lauryl methacrylate in living conditions (curve III) [12]. The clay before the ion exchange reaction had a basal plane spacing of roughly 1 nm which increased to 1.9 nm after exchange with the initiator molecules indicating that the filler platelets could be successfully exchanged with the initiator molecules. The clay after polymerization in non-living conditions did not show any improvement in the basal plane spacing indicating that the attached initiator molecules did not lead to any grafting of polymer from the surface of the filler. When the filler platelets were subjected to polymerization in the living conditions using the SG1 nitroxide, the basal plane spacing was observed to shift to 3.8 nm indicating that the use of living polymerization conditions was beneficial in eliminating the termination reactions of the radicals leading to the attachment of extensive amounts of polymer chains on the filler surface. The use of living conditions can also be used o control the amount of grafted polymer by controlling the amount of initially added monomer for grafting reaction.

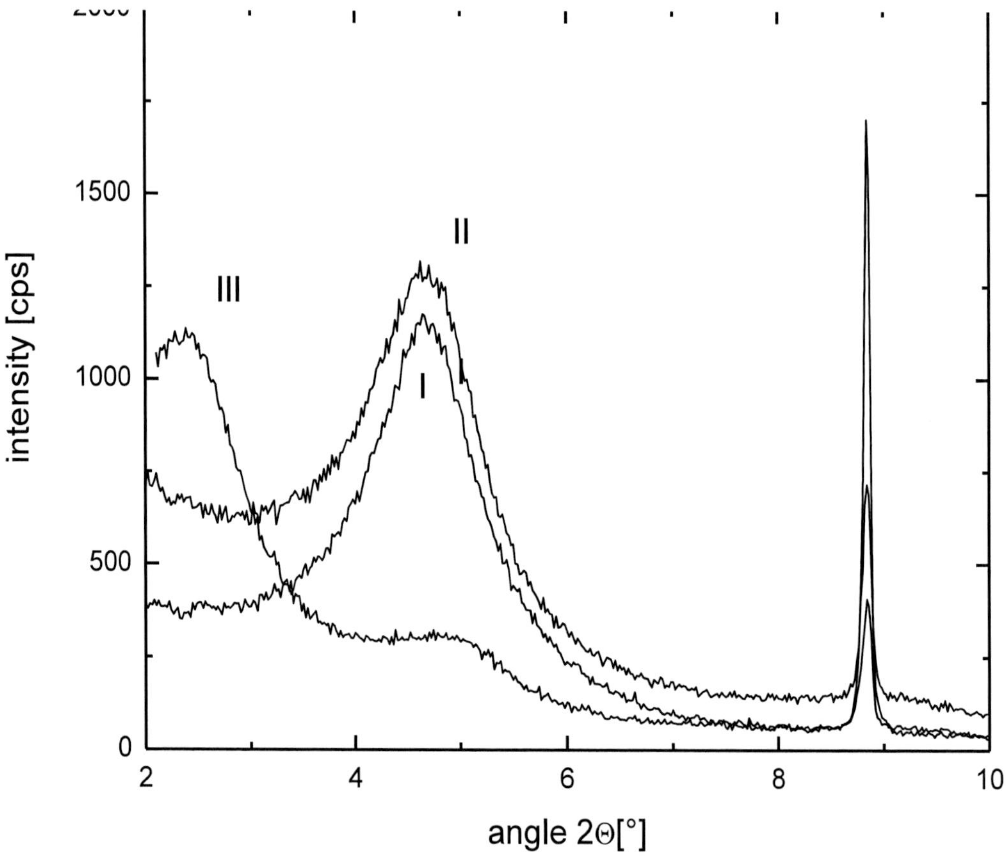

Figure 7. X-ray diffractograms of clay (I) modified by using asymmetrical monocationic initiator in an amount corresponding to 30% of CEC and dimethyldidodecylammonium in an amount corresponding to 70% of CEC, (II) the clay of I when reacted with lauryl methacrylate in non-living conditions and (III) the clay of I when reacted with lauryl methacrylate in living conditions. The peak at 8.84° corresponds to mica platelets used as an internal standard. Reproduced from reference 12 with permission from Taylor and Francis.

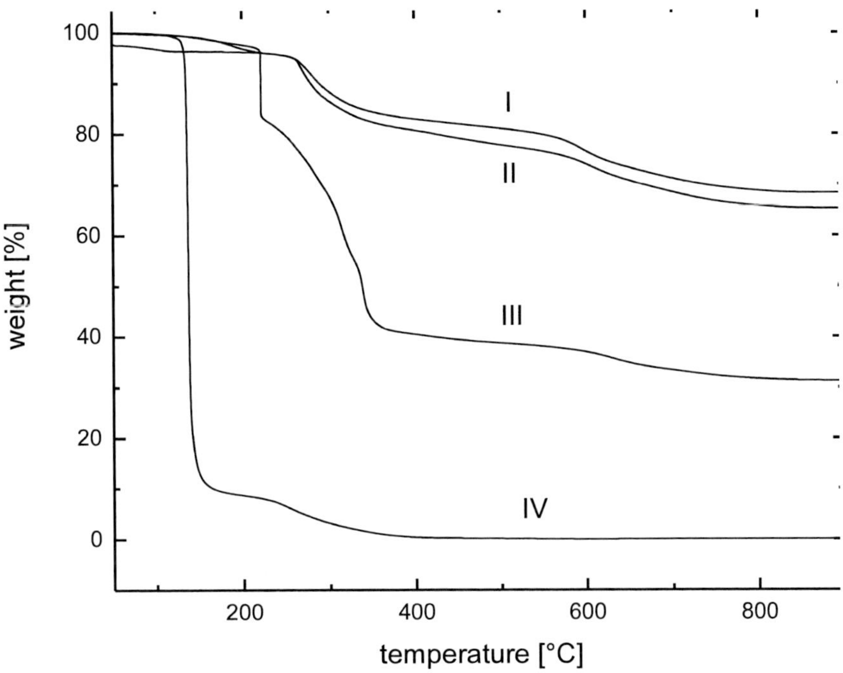

Figure 8. TGA thermograms of clay (I) modified by using asymmetrical monocationic initiator in an amount corresponding to 30% of CEC and dimethyldidodecylammonium in an amount corresponding to 70% of CEC, (II) the clay of I when reacted with lauryl methacrylate in non-living conditions and (III) the clay of I when reacted with lauryl methacrylate in living conditions. The TGA thermogram of pure lauryl methacrylate is shown in curve IV. Reproduced from reference 12 with permission from Taylor and Francis.

Figure 8 shows the TGA thermograms of clay modified by using asymmetrical monocationic initiator in an amount corresponding to 30% of CEC and dimethyldidodecylammonium in an amount corresponding to 70% of CEC (curve I), the clay of curve I when reacted with lauryl methacrylate in non-living conditions (curve II) and the clay of curve I when reacted with lauryl methacrylate in living conditions (curve III) [12]. The TGA thermogram of pure lauryl methacrylate is shown in curve IV for comparison. When the polymerization was performed in non-living conditions, the amount of weight loss was observed to be similar to the filler before polymerization indicating no or minimal attachment of polymer chains from the filler surface, thus, also confirming the findings from X-ray diffraction. When the living polymerization conditions were used for the grafting reaction, the weight loss was much higher indicating that the grafting of large amount of polymer chains from the filler surface and again confirming the X-ray diffraction studies. It is also important that the free polymer formed during the polymerization is washed off efficiently so that the grafted polymer is corresponded rightly owing to the use of initiator and processing conditions. The thermogravimetric analysis of the pure polymer is for this reason is also plotted in Figure 8. The pure polymer degraded at lower temperature than the filler after polymerization confirming that the organic matter was attached to the filler surface and not present as free polymer.

3.2. Polymerization Trials with Symmetric Bicationic Initiator

Figure 9 [20] shows the infra red spectroscopy analysis of bicationic initiator (curve I), clay modified by using bicationic initiator in an amount corresponding to 30% of the CEC and dihexadecyldimethylammonium in an amount corresponding to 70% of the CEC (curve II) and clay modified by using bicationic initiator in an amount corresponding to 70% of the CEC and dihexadecyldimethylammonium in an amount corresponding to 30% of the CEC (curve III). The spectra of the initiator showed the characteristic peak of C=O ester stretching at 1732 cm^{-1}. After exchange of the initiator on the surface of the filler, the presence of C=O stretching peaks at 1725 cm^{-1} similar to the pure initiator was observed which confirms the immobilization of the initiator on the clay surface.

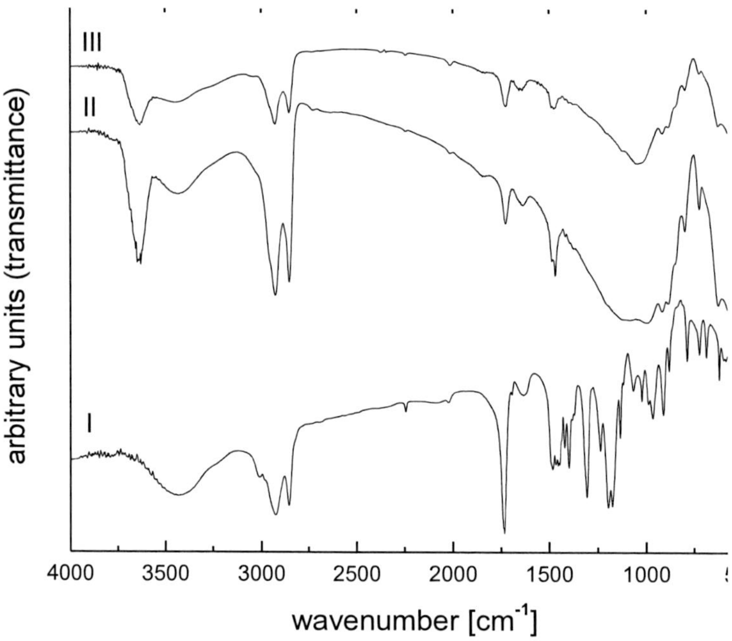

Figure 9. Infra red spectroscopy analysis of bicationic initiator (curve I), clay modified by using bicationic initiator in an amount corresponding to 30% of the CEC and dihexadecyldimethylammonium in an amount corresponding to 70% of the CEC (curve II) and clay modified by using bicationic initiator in an amount corresponding to 70% of the CEC and dihexadecyldimethylammonium in an amount corresponding to 30% of the CEC (curve III). Reproduced from reference 20 with permission from Elsevier.

Figure 10 [20] shows the X-ray diffractograms of the clay modified by using bicationic initiator in an amount corresponding to 30% of the CEC and dihexadecyldimethylammonium in an amount corresponding to 70% of the CEC (curve I), clay of I when reacted in non-living conditions at 60°C (curve II) and clay of I when polymerized in living conditions at 120°C (clay III). When the initiator was exchanged on the surface of the filler in an amount corresponding to 30% of the CEC, 001 basal plane spacing of 2.2 nm was observed. The clay reacted with lauryl methacrylate in non-living conditions at 60°C for 72 h showed a minimal increase of 0.2 nm in the basal plane spacing indicating small extent of polymer grafting from the clay surface. This may be the result of quick termination of the generated radicals after the dissociation of the initiator. Slight increase in the weight loss was also observed in thermogravimetric analysis which confirmed the findings of X-ray diffraction analysis. The clay was also reacted at 120°C in living conditions, the basal spacing increased to 3.4 nm indicating that the use of living conditions was successful in delaying the quick termination of the radicals. The increase in basal plane spacing when using the living polymerization conditions was somewhat lesser than when monocationic initiator was exchanged on the clay surface in an amount corresponding to 30% of the CEC.

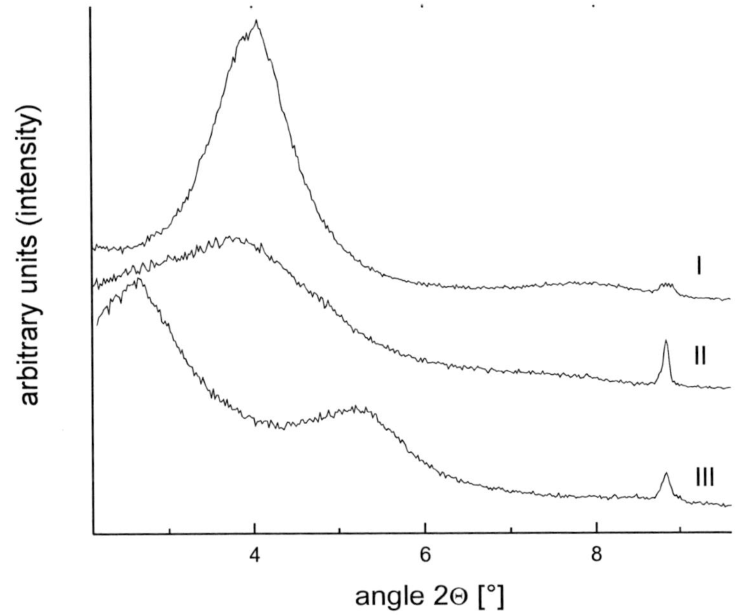

Figure 10. X-ray diffractograms of the clay modified by using bicationic initiator in an amount corresponding to 30% of the CEC and dihexadecyldimethylammonium in an amount corresponding to 70% of the CEC (curve I), clay of I when reacted in non-living conditions at 60°C (curve II) and clay of I when polymerized in living conditions at 120°C (clay III). The peak at 8.84 2θ is the 001 basal plane spacing of mica platelets used as internal standard. Reproduced from reference 20 with permission from Elsevier.

To analyze the effect of the increased extent of filler surface coverage with initiator, bicationic initiator was exchanged on the filler surface in an amount corresponding to 70% of the CEC. Figure 11 shows the X-ray diffractograms of clay modified by using bicationic initiator in an amount corresponding to 70% of the CEC and dihexadecyldimethylammonium in an amount corresponding to 30% of the CEC (curve I), clay of I when reacted in non-living conditions at 60°C (curve II) and clay of I when polymerized in living conditions at 120°C (curve III). The clay with the initiator exchanged on the surface in an amount corresponding to 70% of the CEC had a basal plane spacing of 1.9 nm. The basal plane spacing remained unchanged when the filler was reacted at 60°C in non-living conditions. It indicated that probably the number of radicals generated on the surface of the filler platelets was too high and quickly terminated with each other. The filler when reacted under living conditions at 120°C showed an increased basal plane spacing of 3.8 nm. The higher extent of the polymer grafting when the amount of initiator molecules on the surface was increased indicated that an optimal amount of initiator molecules is required to be present on the filler surface. It should also be noted that the in the case of monocationic initiator, an increased basal plane spacing of 3.8 nm was observed even with 30% coverage of the filler surface with initiator. It thus indicates that the different systems have different optimal amounts of initiator on the filler surface to achieve higher extent of polymer grafting. It has also been earlier reported that polymerization occurred only outside the interlayers when the filler surface did not contain enough amount of monomer bound to it [24], however, the dynamics of the polymerization using initiator bound montmorillonite can be completely different from the monomer bound montmorillonites.

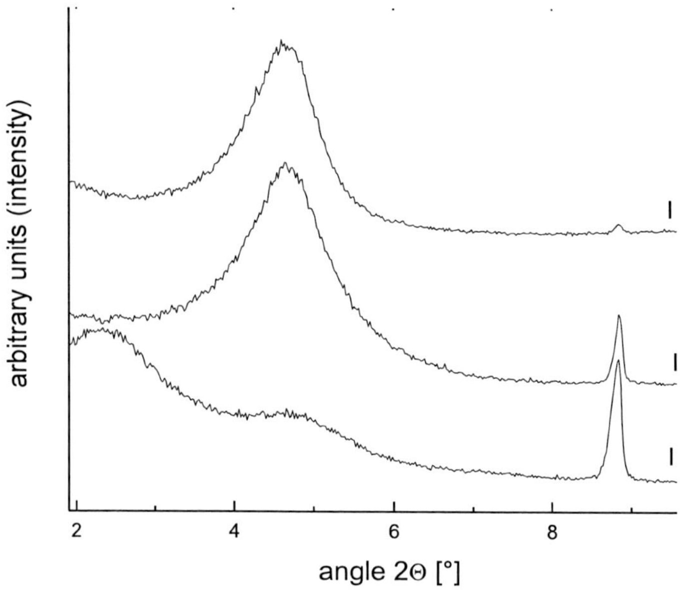

Figure 11. X-ray diffractograms of clay modified by using bicationic initiator in an amount corresponding to 70% of the CEC and dihexadecyldimethylammonium in an amount corresponding to 30% of the CEC (curve I), clay of I when reacted in non-living conditions at 60°C (curve II) and clay of I when polymerized in living conditions at 120°C (curve III). The peak at 8.84 2θ is the 001 basal plane spacing of mica platelets used as internal standard. Reproduced from reference 20 with permission from Elsevier.

REFERENCES

[1] Usuki, A., Kojima, Y., Kawasumi, M., Okada, A., Fukushima, Y., Kurauchi, T., and Kamigaito, O. *J. Mater. Res.* 8 (1993) 1179.

[2] Gilman, J. W., Jackson, C. L., Morgan, A. B., Harris, R., Manias, E., Giannelis, E. P. Wuthenow, M., Hilton, D. and Phillips, S. H. *Chem. Mater.* 12 (2000) 1866.

[3] Giannelis, E. P. *Adv. Mater.* 8 (1996) 29.

[4] Usuki, A., Kawasumi, M., Kojima, Y., Okada, A., Kurauchi, T. and Kamigaito, O. *J. Mater. Res.* 8 (1993) 1174.

[5] LeBaron, P. C., Wang, Z. and Pinavaia, T. *J. Appl. Clay Sci.* 15 (1999) 11.

[6] Osman, M. A., Mittal, V., Morbidelli, M. and Suter, U. W. *Macromolecules* 36 (2003) 9851.
[7] Alexandre, M. Ph. Dubois, *Mater. Sci. Eng. R* 28 (2000) 1.
[8] Bailey, S. W., in: Bailey, S. W. (Ed.), *Reviews in Mineralogy*, Virginia Polytechnic Institute and State University, Blacksburg, 1984.
[9] Bailey, S. W., in: G. W. Brindley, G. Brown (Eds.), *Crystal Structure of Clay Minerals and their X-ray Identification*, Mineralogical Society, London, 1980.
[10] Theng, B. K. G., *The chemistry of clay-organic reactions,* Adam Hilger, London, 1974.
[11] Lagaly, G. and Beneke, K. *Colloid Polym. Sci.* 269 (1991) 1198.
[12] Mittal, V., Phil. Mag. *in print.*
[13] Osman, M. A., Mittal, V. and Suter, U. W., *Macromol. Chem. Phys.* 207 (2007) 68.
[14] Osman, M. A., Rupp, J. E. P. and Suter, U. W., *J. Mater. Chem.* 15 (2005) 1298.
[15] Moet, A. S. and Akelah, A. *Mater. Lett.* 18 (1993) 97.
[16] Fan, X., Xia, C., Fulghum, T., Park, M. K., Locklin, J. and Advincula, R. C. *Langmuir* 19 (2003) 916.
[17] Meier, L. P., Shelden, R. A., Caseri, W. R. and Suter, U. W., *Macromolecules* 27 (1994) 1637.
[18] Velten, U. Shelden, R. A., Caseri, W. R. Suter, U. W. and Li, Y. *Macromolecules* 32 (1999) 3590.
[19] Prucker, O. and Ruhe, J. *Macromolecules* 31 (1998) 602.
[20] Mittal, V., *J. Colloid Interface Sci.* 314 (2007) 141.
[21] Uthirakumar, P., Kim, C. J., Nahm, K. S., Hahn, Y. B. and Lee, Y. S. *Colloids Surf. A: Physicochem. Eng. Aspects* 247 (2004) 69.
[22] Zhou, Q., Frost, R. L., He, H., Xi, Y. and Liu, H. *J. Colloid Interface Sci.* 307 (2007) 357.
[23] Zhou, Q., He, H., Frost, R. L. and Xi, Y. *J. Phys. Chem.* C 111 (2007) 7487.
[24] Greenland, D. J., Laby, R. H. and Quirk, J. P. *Trans. Faraday Soc.* 61 (1965) 2024.
[25] Mittal, V. and Herle, V. *J. Colloid Interface Sci.* 308 (2008) 295.
[26] Guiomar, A. J., Guthrie, J. T. and Evans, S. D. *Langmuir* 15(1999) 1198.
[27] Pan, S., Castner, D. G. and Ratner, B. D. *Langmuir* 14 (1998) 3545.
[28] Lofas, S. and Johnsson, B. *J. Chem. Soc., Chem. Commun.* (1990) 1526.
[29] Persson, H. H. J., Caseri, W. R. and Suter, U. W., Langmuir 17 (2001) 3643.
[30] Wilner, I., Riklin, A., Shoham, B., Rivenson, D. and Katz, E. *Adv. Mater.* 5 (1993) 912.
[31] Mittal, V., *J. Colloid Interface Sci.* 315 (2007) 135.
[32] Fan, X., Xia, C. and Advincula, R. C. *Langmuir* 19 (2003) 4381.
[33] Mittal, V., Advances in Grafting of Polymer Chains 'to' and 'from' the Layered-Silicate Clay Surface, *Encyclopedia of Polymer Composites,* Editors: Mikhail Lechkov and Sergej Prandzheva, Nova Science Publishers, New York, USA, 2009, pp. 76-99.
[34] Hawker, C. J., Barclay, G. G., Orellana, A., Dao, J. and Devonport, W. *Macromolecules* 29 (1996) 5245.
[35] Zhao, H., Argoti, S. D., Farrell, B. P. and Shipp, D. A. *J. Polym. Sci.,* Part A: *Polym. Chem.* 42 (2004) 916.
[36] Georges, M. K., Lukkarila, J. L. and Szkurhan, A. R. *Macromolecules* 37 (2004) 1297.
[37] Grimaldi, S., Finet, J. P., Moigne, F. L., Zeghdaoui, A., Tordo, P., Benoit, D. Fontanille, M. and Gnanou, Y. *Macromolecules* 33 (2000) 1141.

[38] Bottcher, H., Hallensleben, M. L., Nuss, S., Wurm, H., Bauer, J. and Behrens, P. *J. Mater. Chem.* 12 (2002) 1351.
[39] Farcet, C., Lansalot, M., Charleux, B., Pirri, R. and Vairon, J. P. *Macromolecules* 33 (2000) 8559.
[40] Pyun, J. and Matyjaszewski, K. *Chem. Mater.* 13 (2001) 3436.

In: Advances in Polymer Research
Editors: Losif Sidorov, Fedor Nikitin, pp. 55-71
ISBN: 978-1-61942-223-0

THE INFLUENCE OF HIGH APPLICATION RATES OF POLYACRYLAMIDE ON EUBACTERIAL STRUCTURAL DIVERSITY IN AN AGRICULTURAL SOIL

***James A. Entry*[1]*, DeEtta Mills**[*2]*, Krish Jayachandran*[3] *and R. E. Sojka*[1]**

[1] USDA Agricultural Research Service, Northwest Irrigation and Soils Research Laboratory, Kimberly, ID.

[2] Department of Biological Sciences, Florida International University, University Park, Miami, FL.

[3] Department of Earth and Environmental, Florida International University, University Park, Miami, FL.

ABSTRACT

Anionic polyacrylamide (PAM) is a linear, water soluble anionic polymer. The formulation used in this study has a molecular weight of 12 to 15 Mg mole^{-1} and is widely used for erosion control and water quality protection. There has been an issue whether this formulation could possibly have negative effects on soil microbial diversity by altering microbial binding to soil particles or to one another and thus restricting their mobility and role in C and nutrient cycling To explore this possibility, we conducted an eight year study annually applying ultra high rates (more than two orders of magnitude higher than recommended application rates) of PAM to soil and then monitored impacts on soil eubacterial diversity. In July and August, we measured active soil bacterial and fungal biomass and microbial diversity in soils receiving 0 (control), 2691 and 5382 kg active ingredient (ai) PAM ha^{-1}. Active microbial biomass in soil was 27-48% greater in the untreated control than soil treated with 2691 or 5382 kg ai PAM ha $^{-1}$. Active bacterial biomass in soil was 20-30 % greater in the control treatment than in soil treated with 2691 or 5382 kg ai PAM ha $^{-1}$ in August, but not July. Active fungal biomass in soils was 30-50% greater in the control treatment than soil treated with 2691 or 5382 kg ai PAM ha $^{-1}$ in July, but not August. Amplicon length heterogeneity polymerase chain reaction (LH-PCR) DNA profiling was used to access the eubacterial diversity, richness, and evenness in an agricultural soil that received 0 (control), 2691 and 5382 kg active ingredient (ai) PAM ha^{-1}. LH-PCR profiles showed no discernible change in the soil microbial diversity, richness, and evenness due to either of the PAM treatments at any sampling time. We found that although soil receiving these massive PAM application rates and prolonged exposure may reduce active bacterial and fungal biomass, those

[*] Phone: (305) 348-2201, fax: (305) 348-1986.

reductions were not consistent, varying with sampling date. Furthermore these massive application rates did not substantially or consistently affect eubacterial structural diversity, richness or evenness in this agricultural soil.

INTRODUCTION

Polyacrylamide (PAM) is a generic term, referring to a broad class of compounds. There are hundreds of PAM formulations that vary in polymer chain length and number and kinds of functional group substitutions. In some chain segments PAM amide functional groups are substituted with groups containing sodium ions or protons (Sojka et al., 2007). These protons freely dissociate in water, providing negative charge sites in soils. These formulations are linear water soluble anionic polymers with typical molecular weights of 12 to 15 Mg $mole^{-1}$ (Entry et al., 2002). They are commercially available polymers that accelerate separation of solids from aqueous suspensions in sewage sludge dewatering, mine slurry conditioning, paper manufacture, clarification of refined sugar and fruit juices, and as a thickening agent in animal feed preparations. Anionic PAM molecules are used in agriculture for erosion control and runoff water quality protection to avoid the negative environmental and safety drawbacks of cationic and non-ionic formulations. PAM formulations used for erosion control are restricted to formulations containing less than 0.05% free unreacted acrylamide, which is the same purity standard for potable water treatment use. Anionic, moderate molecular-weight PAM products for erosion control are either fine granular forms or concentrated liquid emulsions of PAM and mineral spirits (Sojka et al., 2007). The latter category contains a surfactant to help disperse the PAM when mixed with water. Granular materials and emulsified concentrates require substantial turbulence or agitation and high flow rate at the point of addition to water in order to dissolve PAM (Sojka et al., 1998c).

Water soluble anionic PAM was found to be an effective in controlling erosion and increasing infiltration in furrow irrigation water (Sojka and Lentz, 1997; Sojka et al., 1998 a, b). Coulombic and Van der Waals forces attract soil particles to the PAM molecule (Orts et al., 1999; 2000) enhancing particle cohesion and increasing resistance to shear-induced detachment and prohibiting particle transport in runoff. The few particles that detach are quickly flocculated by PAM, settling them out of the transport stream. Strong positively charged ions are needed in water when using anionic PAM for infiltration and erosion control (Orts et al., 2002; Wallace and Wallace, 1996). These ions act as a bridge between anionic soil surfaces and the anionic PAM macromolecules. For example, modest amounts of Ca^{++} in the water shrink the electrical double layer surrounding soil particles and bridge the anionic surfaces of soil particles and PAM molecules, promoting flocculation (Orts, et al., 2002; Wallace and Wallace, 1996). The Ca^{++} ion has a double charge with small hydrated radius that favors flocculation, whereas, Na^{+} ions have a large hydrated radius which is less effective for ion bridging and can leading to dispersion rather than flocculation of solids at higher Na^{+} concentrations. Water low in electrolytes or with high Na^{+} concentrations can be easily amended through addition of $Ca(NO_3)_2$ fertilizer or gypsum, i.e. $Ca(SO_4)_2$. PAM delivery in irrigation water is extremely efficient, because it needs only to stabilize the thin veneer of soil directly active in the erosion process. In furrow irrigation, PAM conditions the treated soil (about 25% of the field surface area) to only a few millimeters depth; thus only 1-2 kg ha^{-1} of PAM per irrigation is necessary to prevent erosion (Sojka et al., 2007).

The ability of PAM to flocculate nutrients and microorganisms, removing them from waste water, has been investigated (Entry et al., 2002). Sojka and Entry (2000) found that PAM-treatment reduced algae, active bacterial active fungal and microbial biomass relative to the control treatment after water traveled 1 meter in furrows at 7.5 and 15.5 liter min^{-1}. After water traveled 40 meters at 7.5, 15.5, and 22.5 liter min^{-1}, PAM-treatment reduced algae, bacterial , fungal biomass and overall microbial biomass relative to the control treatment. Entry and Sojka (2000) found that PAM+$Al(SO_4)_3$, and PAM+CaO mixtures reduced populations of total and fecal coliform bacteria and fecal streptococci in cattle, fish, and swine wastewater leachate and surface runoff by approximately one hundred to one thousand fold compared to no treatment. PAM+$Al(SO_4)_3$, and PAM+CaO removed large amounts of NH_4^+, PO_4^{-3} and total P in cattle and swine wastewater leachate and surface runoff (Entry and Sojka, 2001). PAM+$Al(SO_4)_3$, and PAM+CaO did not remove significant amounts of NH_4^+, $PO_4^{-3}$and total P from fish wastewater leachate or surface runoff. There were substantially lower concentrations of NO_3^-, NH_4^+, PO_4^{-3}, total P, K, Ca and Mg in fish wastewater than cattle or swine wastewater. If nutrient concentrations in wastewater are low, there is a smaller chance that they will contact binding sites on the PAM molecule or Al^{+3} and Ca^{+2}. Therefore PAM+$Al(SO_4)_3$, and PAM+CaO may be unable to remove significant fractions of total nutrients from water when nutrient concentrations are extremely low.

Entry and Sojka (2000) found that PAM+$Al(SO_4)_3$, and PAM+CaO mixtures reduced populations of total fecal coliform bacteria and fecal streptococci in cattle, fish, and swine wastewater leachate and surface runoff by approximately one hundred to one thousand fold compared to no treatment. PAM reduced populations of total coliform and fecal coliform bacteria in swine manure leachate from columns containing four different soil types ranging from sand to clay by at least ten fold compared to soil columns without PAM. In the same study, PAM+$Al(SO_4)_3$, and PAM+CaO treatments reduced populations of total coliform and fecal coliform bacteria from 10 to 100 fold compared to the control treatment in water running off agricultural files amended with three different manure sources. PAM, PAM+ $Al(SO_4)_3$, and PAM+CaO seem to be indiscriminate as to the type and species of microorganisms removed from water. The PAM mixtures substantially reduced the numbers of pathogenic bacteria in runoff prior to entering public water systems. PAM+$Al(SO_4)_3$, and PAM+CaO compounds also removed significant amounts of NH_4^+, PO_4^{-3} and total P in cattle and swine wastewater leachate and surface runoff (Entry and Sojka, 2001). There were substantially lower concentrations of NO_3^-, NH_4^+, PO_4^{-3}, total P, K, Ca and Mg in fish waste than cattle or swine wastewater. If nutrient concentrations in wastewater are low, there is less chance that they will contact binding sites on the PAM molecule or Al^{+3} and Ca^{+2}. Anionic PAM should preferentially bind bacteria that have a negative charge. Positively charged bacteria will most likely be bound to the anionic PAM molecule through ion bridging (Sojka et al., 2007).

On the same plots used in this study, Sojka et al., (2006) found that fatty acid methyl ester (FAME) profiles showed no discernible change in the soil microbial community due to either of the PAM treatments at any sampling time. In contrast, seaonsal analysis of metabolic characteristics using Biolog® GN plates yielded separation of the non-amended control soils from those plots receiving the high PAM application rate in spring but not summer (Sojka et al., 2006). Comparisons of the three sampling times by both the fatty acid and Biolog® analyses indicated that the microbial metabolic potential present in spring were different from those sampled in summer, which was associated with temperature and increased rhizoshpere

C associated with cropplants. In spring, microbial communities that received a high PAM application generally responded more to a wider range of carbohydrates and fewer amino and carboxylic acids than those from the non-amended control soils. The objective of this study was to determine the influence of very high rates of soil-applied anionic PAM on soil microbial biomass and eubacterial structural diversity.

MATERIALS AND METHODS

Study Site

The study was conducted at the USDA Agricultural Research Service's Northwest Irrigation and Soils Research Laboratory in Kimberly, Idaho. The soil in the test area was Portneuf silt loam (coarse-silty, mixed, superactive mesic Durinodic Xeric Haplocalcid), with 10-21% clay and 60-75% silt, and organic matter content of approximately 13 g kg^{-1}. Saturated paste extract electrical conductivity (EC) of this soil ranges from 0.7 to 1.3 dS m^{-1}, with exchangeable sodium percentage (ESP) of 1.4 to 1.7 and pH of 7.6-8.0 with a $CaCO_3$ equivalent of 2-8%. Slope on this site was 1.5%. Soil physical and chemical parameters on these treatments are described in detail in Sojka et al. (2006). Plots were 40 m long x 4 m wide.

Experimental Design

We took a total of 54 samples. There were 3 PAM treatments (5382 kg PAM ha^{-1}, 2691 kg PAM ha^{-1} and a control x 3 replicated plots x 2 sampling times (July and August) x 3 separate samples taken from each plot during each sampling period. The experimental design was a randomized complete block with three replications.

Treatments were: 5382 kg PAM ha^{-1}, 2691 kg PAM ha^{-1} and a control (no PAM applied) with 3 replications (plots) of each treatment. We took three samples in July and August, 2001 from each plot.

Crop Growth, Fertilizer and Pesticide Application

Bean, (*Phaseolus vulgaris* L), wheat (*Triticum aestivum* L.), corn (*Zea Mays* L.) and potato (*Solanum tuberosum* L.) were grown the first year (1995) in various parts of the study area and all plots received 146 kg ha^{-1} N and 156 kg ha^{-1} P as well as the herbicides Ethalfluralin, Cycloate, EPTC, Metribuzin and Metolachlor. In the second and third years only bean was grown and all plots received 146 kg ha^{-1} N and the herbicide Ethalflurin. In the fourth and fifth years only corn was grown and all plots received 168 kg ha^{-1} N and herbicide EPTC. In the sixth year only bean was grown and all plots received 90 kg ha^{-1} P and herbicides Ethalflurin , Cycloate, EPTC, Metribuzin and Metolachlor.

Polyacrylamide Application

The polyacrylamide copolymer used was a dry granular material having an approximate molecular weight of 12-15 Mg $mole^{-1}$, with an 18% negative charge density (provided by CYTEC Industries of Wayne, NJ and marketed under the trade name Superfloc 836A). We incorporated 897 kg PAM ha^{-1} into the soil in the spring every year for three years to plots receiving 2691 kg PAM applied ha^{-1} from 1998-2000 (Table 1).

We incorporated 897 kg PAM ha^{-1} into the soil in the spring every year for six years to plots receiving 5382 kg PAM applied ha^{-1} from 1995-2000. Prior to each irrigation PAM application in treated plots also involved the spreading of granular PAM on the surface of an approximately 0.1 m^2 area in the furrow, corresponding to the first meter of furrow below inflow spigots. Successive application amounts were 35, 35, and 15 g of material (approximately 28, 28, and 12 g of active ingredient, which is referred to as the patch PAM application method as described in the USDA National Resource Conservation Service conservation protective standard (Sojka et al., 1998c). Since the PAM contains approximately 41.5% C and 17.7 % N PAM treatments received 465 kg C ha^{-1} and 199 kg N ha^{-1} each year while receiving 897 kg PAM (Table 1). Since Superfloc 836A polyacrylamide degrades at approximately 10% yr^{-1} in soil (Entry et al., 2008), we estimated the amount of PAM present in the soil for each year (Table 1).

Water Application

Water was applied as furrow irrigation from a storage pond via spigoted plastic pipe to a conventionally tilled field that had been disked to 10 cm depth in autumn and spring, then roller harrowed following incorporation of fertilizer, herbicides and PAM prior to planting. Furrows, 40 m in length were approximately 10 cm in depth, and were prepared with weighted 75° v-shaped furrow shaping tools.

Furrow spacing in the various crops was 76 cm. Irrigation was on every other furrow only, in wheel-track furrows on the first irrigation, non-wheel furrows in the second irrigation and in wheel-track furrows again on the third irrigation and so forth. Per hectare sediment-loss and infiltration depth were calculated based on inflow and outflow rate measurements, sediment concentration of outflow, and spacing between irrigated furrows, following previously published protocols (Sojka et al., 1992; 1994). Irrigation water electrical conductivity (EC) was 0.5 dS m^{-1} with a sodium adsorption ratio (SAR) of 0.4 to 0.7.

Soil Sampling

Three separate 2.5 cm diameter soil cores were collected from the top 30 cm of mineral soil from three different locations in each plot in June, July and August, 2001. One hundred grams mineral soil for each sample was collected from each plot. Soil samples were thoroughly mixed, sieved to pass a 1 mm opening and split into 5 equal parts. Soils were analyzed for active bacterial and active fungal biomass determination and microbial community parameters as described below.

Table 1. Amount of carbon (C) and nitrogen (N) contained in polyacrylamide (PAM) treatments over a six year period from 1995 to 2000

	------------1995---------			------------1996---------			------------1997----------			------------1998----------			-------------1999--------			-------------2000--------		
Treatment	PAM Applied	C	N	PAM Applied	C	N	PAM Applied	C	N	PAM Applied	C	N	PAM Applied	C	N	PAM Applied	C	N
kg ai PAM ha^{-1}	- kg ha^{-1}-	- kg ha^{-1}-		- kg ha^{-1}-	- kg ha^{-1}-		- kg ha^{-1}-	- kg ha^{-1}-		- kg ha^{-1}-	- kg ha^{-1}-		- kg ha^{-1}-	- kg ha^{-1}-		- kg ha^{-1}-	- kg ha^{-1}-	
No PAM	0	0	0	0	0	0	0	0	0	0	0	0	0	0	0	0	0	0
Low PAM	0	0	0	0	0	0	897	465	199	897	465	199	897	465	199	897	465	199
PAM remaining	0	0	0	0	0	0	0	0	0	897	465	199	1704	465	199	2430	465	199
High PAM	897	465	199	897	465	199	897	465	199	897	465	199	897	465	199	897	465	199
PAM remaining	897	465	199	1704	465	199	2430	465	199	3085	465	199	3673	465	199	4230	465	199

* The High PAM treatment received a total of 5382 kg ai PAM ha^{-1} over six years. The Low PAM treatment received a total of 2691 kg ai PAM ha^{-1} over six years.

Soil was stored in air and moisture-tight plastic bags at $4^{O}C$ at moisture conditions similar to those in the field. Soil was prepared for microbial analysis within 24 hr of collection to minimize effects of storage on microbial activity (West et al. 1986).

Microbial Biomass Analysis

Active and total bacteria and fungi numbers and biomass in soil were determined for each treatment using methods described by Ingham and Klein (1984). Active fungi were estimated by taking a 1.0 g soil sample which was diluted in 9 ml of a phosphate buffer (pH 6.0) and shaken at approximately 120 rpm for 5 min. A 1 ml aliquot was removed and stained with 1 ml of a 20 μg ml^{-1} fluorescein diacetate (FDA) solution in a 0.2 M phosphate buffer for 3 min. One ml of 1.5% agar in a 0.1 M phosphate buffer (pH 9.5) was added to the FDA suspension. The sample was mixed and an aliquot placed on a microscope slide containing a cavity of known volume (Ingham and Klein, 1984). Immediately after preparation, slides were examined for FDA-stained hyphal length by epifluorescent microscopy. Total fungal biomass was estimated by measuring the length and diameter of hyphae in 3 to 60 fields with phase - contrast microscopy. Three slides were evaluated from each sample and ten fields per slide were evaluated with phase contrast microscopy for total (stained and unstained) hyphal length, and three transects per field were evaluated for FDA-stained (active) hyphal length at X 160 total magnification.

Iodonitrotetrazolium (INT) stain was used for counting active bacteria (Stamatiadis et al., 1990). A 1 ml sample of initial soil suspension was diluted to a final dilution of 0.2 mg soil in 4 ml buffer. The suspension was incubated with 4 ml of filtered INT buffer for 60 min in the dark at 20 °C. Total bacteria per ml of water were estimated from the mean number of bacteria (fluorescent and non-fluorescent bacteria), their average diameter and length per field. Three slides were evaluated for each sample and ten fields per slide were evaluated using epifluorescent oil-immersion microscopy to determine numbers and size of fluorescent (stained) and total bacteria (Lodge and Ingham, 1991).

Bacterial biomass was computed from the numbers of active and total bacteria and active and total fungal biomass was determined from hyphal length. Bacterial biomass was computed from the number of soil bacteria per gram of soil by assuming that the bacterial spheres were 1 μm in diameter (Jenkinson and Ladd, 1981). Active and total fungal biomass were computed by assuming average hyphal diameter to be 1 μm in diameter and then multiplying by the length of observed hyphae (Jenkinson and Ladd, 1981). A carbon to volume conversion factor of 120 μg C mm^{-3} was used for bacteria and fungi, assuming 1.1 g cm^{-3} wet density, 20% dry matter content, and 0.41 carbon content in the bacterium or fungus (Jenkinson and Ladd, 1981).

DNA extraction and PCR conditions

Whole community genomic DNA was extracted from replicate (3) frozen soil samples with slight modifications to the FastDNA®SPIN Kit for Soil (QBiogene, Vista, CA) (Mills et al, 2003). After placing 500 mg of homogenized soil in the MULTIMIX 2 tubes with the appropriate buffers and reagents, samples were processed in the FastPrep® instrument (QBiogene, Vista, CA) for 20 sec at setting 5.5.

Table 2. LH-PCR primer sequences. (Based on *E.coli* numbering of the 16S rRNA gene)

Primer name (hypervariable regions)	Sequence (5' to 3')	5' modification
27 F (V1+V2)	AGA GTT TGA TCM TGG CTC AG	6-FAM
355R (V1+V2)	GCT GCC TCC CGT AGG AGT	none
P1F	GCG GCG TGC CTA ATA CAT GC	6-FAM
P1R	TTC CCC ACG CGT TAC TCA CC	none
338F (V3)	ACT CCT ACG GGA GGC AGC AG	HEX
518R	ATT ACC GCG GCT GCT GG	None

*Sequence primer M=C:A, Y= C:T, K = G:T, W=A:T (possible nucleotide substitutions).

Samples were incubated for 45 minutes at 70°C (to increase DNA yield and aid in cell lysis), placed back into the FastPrep® instrument and then processed for an additional 10 seconds at 5.5. High molecular weight (HMW) DNA was quantified using a Bio-Rad fluorometer (Bio-Rad, Richmond, CA) and 10 ng of HMW DNA was used as the working stock in subsequent PCR reactions. Only the forward primer of each primer set was fluorescently labeled (PE Biosystems, Foster City, CA) and paired with individual non-fluorescent reverse primers for the V1, V3 and V1+V2 hypervariable regions (Table 2).

Positive controls used DNA from pure laboratory cultured isolates. Negative controls used only diethylpyrocarbonate-treated (DEPC) water. Final concentrations of PCR reaction mixtures were: 1X PCR buffer, 0.25 mM MgCl2, 0.250 mM dNTPs (Boehringer, Mannheim, GmbH), 0.5 μM forward and reverse primers, 0.25U AmpliTaq Gold polymerase (PE Biosystems, Foster City, CA), 0.1% (w/v) bovine serum albumin (BSA), fraction V (ICN Biomedicals, Inc, Aurora, OH), 10 ng HMW DNA, and DEPC treated water to make up the final volume.

The PTC-100 Programmable Thermal Cycler (M.J. Research, Inc., Watertown, MA) was programmed for an initial denaturing step at 94°C for 11 minutes followed by 25 cycles of 94°C for one minute, 55°C for one minute, and 72°C for one minute with a final extension step at 72°C for 10 minutes. LH-PCR products were loaded directly onto the polyacrylamide gels without further purification.

Electrophoresis

All samples were denatured in a 5:1:1 mixture of deionized formamide (98%, Sigma, St. Louis, MO), Blue Dextran-EDTA loading dye, and GeneScan 500 ROX internal standard (PE Biosystems, Foster City, CA), heated to 94°C for 3 minutes, chilled on ice until loaded on a 36 cm WTR 5% Long Ranger polyacrylamide gel (Bio-Rad, Richmond, CA) and then run on a ABI® 377 DNA sequencing instrument. Gels were run for 3.5 hours using standard electrophoresis run parameters.

Standardization and Analyses of Profiles

Microbial community profiles were collected and analyzed using the ABI Prism™ GeneScan®, ABI Prism™ Genotyper® software (PE Biosystems, Foster City, CA) and diversity, evenness and similarity indices were calculated using custom macros written for Microsoft Excel (Microsoft Corp., Seattle, WA). In GeneScan®, analysis parameters were set to the local Southern size calling, no peak correction and a minimum noise threshold of 50 fluorescent units. Standardized binning criteria were used to identify the reproducible peaks in each profile. The parameters were that each analyzed amplicon had to appear in at least three of the nine replicate profiles and, the relative intensity ratio was equal to or greater than 1%. The reason for setting these binning parameters was to eliminate the error introduced by the collection and analysis software (Dunbar et al., 1999). The GeneScan® software analyzes the amplicons to the nearest 0.01 base pair. By rounding data to the nearest integer, however, two peaks that may actually represent the same amplicon can erroneously be called as two separate fragments due to slight variations in migration during electrophoresis runs.

Standardization between Sample Profiles

In addition to using the same concentration of quantified HMW DNA in each PCR reaction, standardization of peaks between samples was performed in the following manner. Total fluorescence intensities (i.e., heights) were calculated for each profile as described in Dunbar, et al (2001). For example, if the total peak intensities for two profiles were 25,000 and 50,000, respectively, each peak in the latter profile would be standardized by a correction factor of 0.5. Relative intensity ratios were calculated using the corrected data by dividing each individual peak height by the total peak height of each electropherogram. This procedure was repeated for each comparison between sample profiles.

Statistical Analysis

Bacterial, fungal and microbial biomass were subjected to a one way analysis of variance (ANOVA) for a completely randomized design (Snedecor and Cochran, 1994; Kirk, 1995). Residuals were normally distributed with constant variance. Statistical Analysis Software programs (SAS Institute Inc., Cary, NC) were used to conduct the analysis of variance. Significance of treatment means was determined at $p \leq 0.05$ with the Least Square Means test. Raw data from the 16S rRNA hypervariable domains were aligned by amplicon size for each sub-site, averaged and the averaged data was used for all other calculations. The Shannon-Weaver diversity index, phylotype richness, and evenness parameters were calculated as described elsewhere (Dunbar et al., 1999). The data matrices were converted to binary data (presence/absence) and similarity indices were calculated using the Sørensøn's Index (pairwise similarity values) (2). SPSS Version 13.0 for Windows (SPSS Inc., Chicago,

IL) was used to compare indices. ANOVA with the Bonferonni post-hoc test was used to assess the replicate sub-plots within the test samples. The mean indices for each site were subsequently tested for significance using sample t-tests.

RESULTS

Active bacterial, fungal and microbial biomass was not consistently affected by high PAM additions. In August, active bacterial biomass in soil was 20-30% greater in the control treatment (not receiving PAM) than soil treated with 2691 or 5382 kg ai PAM ha^{-1} but this difference was not observed in July (Table 4). There was no difference in soil active bacterial biomass between the 2691 or 5382 kg ai PAM ha^{-1} treatments regardless of sampling time. Active fungal biomass was 30-50% greater in the control treatment than soil treated with 2691 or 5382 kg ai PAM ha^{-1} in July but, again, not in August (Table 3). There was no difference in soil active fungal biomass between the 2691 or 5382 kg ai AM ha^{-1} treatments regardless of sampling time. Active microbial biomass in soil was 27-48% greater in the control treatment than soil treated with 2691 or 5382 kg ai PAM ha^{-1}.

Table 3. Richness, diversity, and evenness indices for three concatenated hypervariable 16S rRNA domains: V1, V1+V2, and V3

Month	July			August		
Treatment	Control	High PAM	Low PAM	Control	High PAM	Low PAM
Richness (S)	33	39	35	31	34	34
H_{max}	3.50	3.66	3.56	3.43	3.53	3.53
Diversity (H)	2.72	2.76	2.68	2.73	2.72	2.77
Evenness (E)	0.78	0.75	0.75	0.80	0.77	0.78

[a] Richness (S) = #of peaks in each sample. Diversity (H) = - p_i) $\ln(p_i)$ where p_i is the relative ratio of individual peak heights; Evenness (E) = H/H_{max} where $H_{max} = \ln(S)$.

Table 4.Active bacterial biomass, active fungal biomass, and active microbial biomass after in soils treated with 2691and 5382 kg ai PAM ha^{-1}

Month	Treatment	Active Bacterial Biomass	Active Fungal Biomass	Active Microbial Biomass
		---------------------- mg C g^{-1} soil----------------------		
July	Control (0 kg ai PAM applied ha^{-1})	5.31 b	10.64 a	15.95 a
	2691 kg ai PAM applied ha^{-1}	4.86 b	6.64 b	11.51 b
	5382 kg ai PAM applied ha^{-1}	5.39 b	5.32 b	10.71 b
August	Control (0 kg ai PAM applied ha^{-1})	9.13 a	6.28 b	15.42 a
	2691 kg ai PAM applied ha^{-1}	7.20 b	6.93 b	12.54 b
	5382 kg ai PAM applied ha^{-1}	6.33 b	4.70 b	11.03 b

[a] In each column, values followed by the same letter are not significantly different as determined by the Least Square Means Test ($P \leq 0.05$; n = 27).

We found no differences in eubacterial diversity, H_{max}, richness, and evenness indices for V1, V1+V2, and V3 hypervariable 16S rRNA domains in July or August (Table 3). In July, the control soil had a higher peak in the V1 region at the base pair (bp) 81 amplicon length than soil treated with 2691 or 5382 kg ai PAM ha^{-1} (Figure 1). Soil treated with 2691 or 5382 kg ai PAM ha^{-1} had higher peaks in the V3 region at bp 195 than the control soil (Figure 2). The control soil had higher peaks in the V1+V2 region at bp 315 and bp 343 than soil treated with 2691 or 5382 kg ai PAM ha^{-1} (Figure 3). Soil treated with 2691 or 5382 kg ai PAM ha^{-1} had higher peaks in the V1+V2 region at the bp 317 than the control soil.

In August, the control and soil treated with 2691 kg ai PAM ha^{-1} had a higher peak in the V1 region at bp 69 than soil treated with 5382 kg ai PAM ha^{-1} (Figure 4). Soil treated with 5382 kg ai PAM ha^{-1} had a higher peak in the V1 region at the bp 71 than the control and soil treated with 2691 kg ai PAM ha^{-1}. In August, there was no difference in any bp in the V3 region regardless of PAM treatment (Figure 5). Soil treated with 2691 or 5382 kg ai PAM ha^{-1} had higher peaks in the V1+V2 region at the bp 315 than the control soil (Figure 6).

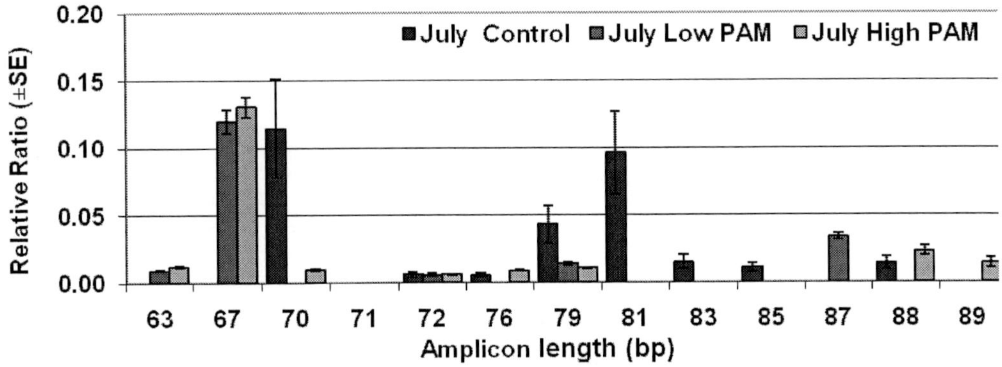

Figure 1. Comparison of amplicon lengths in the V1 region of eubacterial 16SrRNA genes in July from soils receiving no PAM (control), low PAM (2691 kg ai PAM ha-1) and high PAM (5382 kg ai PAM applied ha-1) each year from 1998-2000.

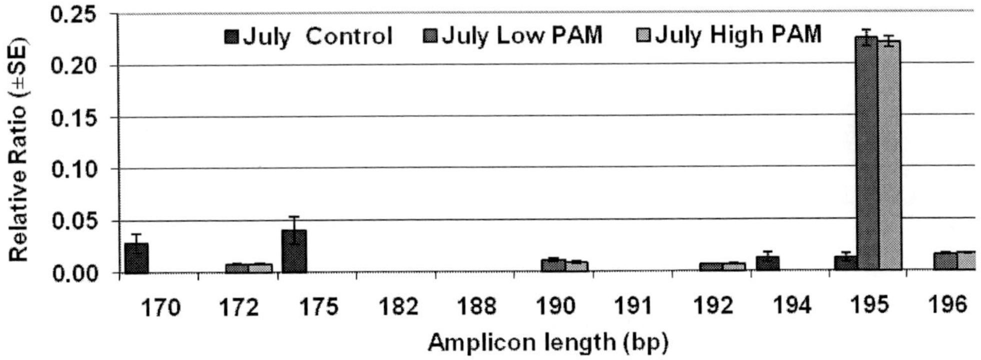

Figure 2. Comparison of amplicon lengths in the V3 region of eubacterial 16SrRNA genes from soils in July receiving no PAM (control), low PAM (2691 kg ai PAM ha^{-1}) and high PAM (5382 kg ai PAM applied ha^{-1}) each year from 1998-2000.

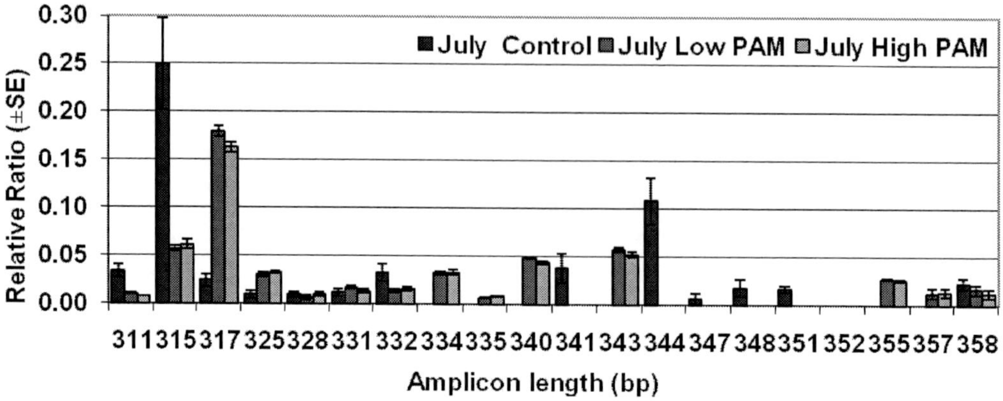

Figure 3. Comparison of amplicon lengths in the V1+V2 region of eubacterial 16SrRNA genes from soils in July receiving no PAM (control), low PAM (2691 kg ai PAM ha^{-1}) and high PAM (5382 kg ai PAM applied ha^{-1}) each year from 1998-2000.

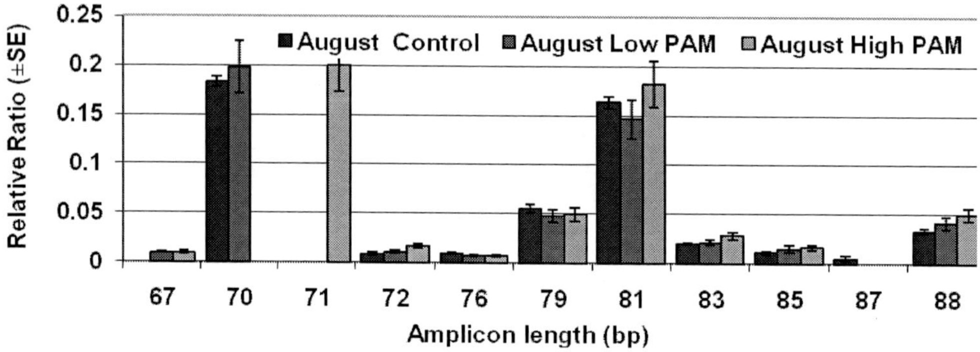

Figure 4. Comparison of amplicon lengths in the V1 region of eubacterial 16SrRNA genes in August from soils receiving no PAM (control), low PAM (2691 kg ai PAM ha-1) and high PAM (5382 kg ai PAM applied ha-1) each year from 1998-2000.

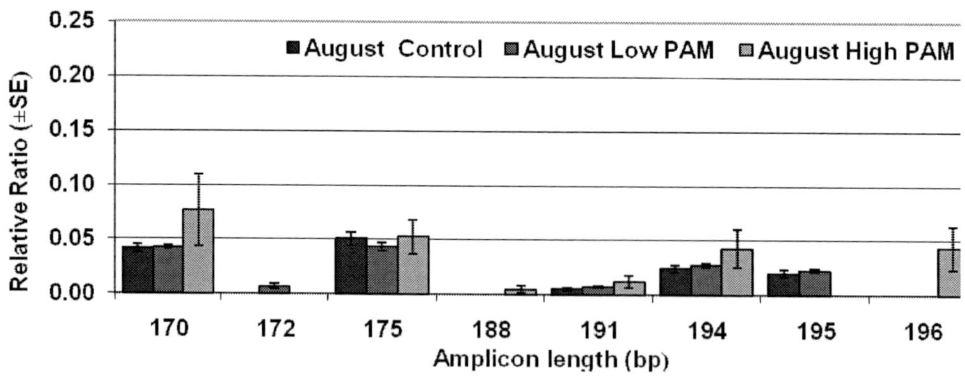

Figure 5. Comparison of amplicon lengths in the V3 region of eubacterial 16SrRNA genes in August from soils receiving no PAM (control), low PAM (2691 kg ai PAM ha^{-1}) and high PAM (5382 kg ai PAM applied ha^{-1}) each year from 1998-2000.

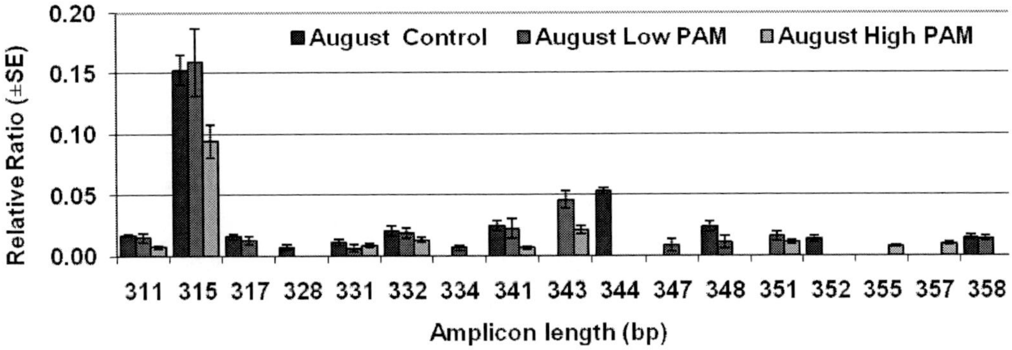

Figure 6. Comparison of amplicon lengths in the V1+V2 region of eubacterial 16SrRNA genes in August from soils receiving no PAM (control), low PAM (897 kg ai PAM ha^{-1}) and high PAM (2691 kg ai PAM applied ha^{-1}) each year from 1998-2000.

Discussion

Eubacterial community diversity, H_{max}, richness, and evenness indices for V1, V1+V2, and V3 hypervariable 16S rRNA domains did not differ among PAM treatments in July or August. We found small differences in a total of seven amplicons in the V1, V1+V2, and V3 hypervariable 16S rRNA domains. Our results are consistent with Sojka et al. (2006) who found that whole soil FAME and Biolog® profiles showed that the soil microbial community did not change in response to the PAM application rates. Analysis of nutritional characteristics using Biolog® GN plates, showed that there was an apparent separation of the non-amended control soils from those plots receiving the high PAM application rate in June, but not in July or August (Sojka et al., 2006). In contrast, comparisons of sampling times by both the FAME and Biolog analyses indicated that the microbial communities present in June were different from those sampled in July and August, both taxonomically and metabolically (Sojka et al., 2006).

PAM could increase microbial biomass because of the addition of the easily degradable organic material, or possibly due to N enrichment (discussed below). One might expect PAM to slowly limit microbial activity by binding microorganisms to soil particles or to one another, restricting their mobility and access to carbon and nutrients or reducing N-fixing bacteria via nitrogen additions. Considering the extremely high rates of PAM application in this study, the additional N contained in the commercial polymer formulations may possibly affect soil microbial community rather than the addition of PAM itself (Sojka et al., 2006). There have been mixed, and sometimes conflicting, reports regarding the effects of PAM application to agricultural soils on bacterial biomass (Nadler and Steinberger, 1993; Steinberger et al., 1993; Kay-Shoemake et al., 1998a; 1998b). PAM-treated soils had larger populations of culturable bacteria when planted to potatoes as compared to those planted to beans (Kay-Shoemake et al., 1998a). Negative impacts have not been documented for aquatic macrofauna, edaphic microorganisms, or crop species for the anionic PAMs when applied at recommended concentrations and rates (Kay-Shoemake et al., 1998a; 1998b). Total and fecal coliform bacteria in soil and water were not affected by 10 kg PAM ha^{-1} (Spackman et al., 2003). PAM addition to agricultural soils increased nitrification of added urea (Kay-

Shoemake et al., 2000b), but no other significant impacts of PAM-application on fertilizer fate have been noted. Watwood and Kay-Shoemake (2000) found that sorptive dynamics of the common pesticides, 2,4-D and atrazine, are not dramatically altered by PAM treatment, although the authors found slight changes in desorption and degradation rates. Although PAM additions to field soils may reflect changes in microbial carbon utilization patterns, these effects are masked by impacts of other field variables, such as crop cover type or nutrient status (Watwood and Kay-Shoemake, 2000). PAM effects on biota are greatly buffered due to adsorption and deactivation associated with the suspended impurities when introduced into waters containing sediments, humic acids or other impurities, (Goodrich et al., 1991; Buchholz, 1992). Studies reporting increases or decreases in bacterial numbers for PAM-treated soils indicate that these effects are likely site, season or cultural practice specific and may interact with other important variables, such as nutrient levels, crop cover type or herbicide regimes (Sojka et al., 2006).

Results from three separate methods, LH-PCR in this study, FAME and Biolog® (Sojka et al. (2006) found massive PAM additions to this soil show that the soil microbial community did not change appreciably in response to these application rates. However, microorganisms are degrading PAM at approximately 9.8% yr^{-1} in this soil (Entry et al., 2008). The authors found that when they incorporated 897 kg PAM ha^{-1} into this soil ever year for 6 years to plots (i.e. receiving 5382 kg ai PAM ha^{-1} from 1995-2000), the PAM degraded at approximately 4.5% yr^{-1} for the first 6 years.

After 12 years, in 2006, the authors found 1466 kg ai PAM remaining, indicating that PAM degraded at approximately 9.8% yr^{-1} (Entry et al., 2008). The slow degradation rates of PAM at high application rates may have been the result of PAM clumping and therefore not being readily accessible to soil microorganisms. The 5382 and 2691 kg ai PAM ha^{-1} application rate was more than two orders of magnitude higher than application rates ever likely to be encountered in actual field agricultural scenarios, since PAM is traditionally applied at 1 -2 kg ai PAM ha^{-1} per irrigation or 10 -12 kg ai PAM ha^{-1} yr^{-1} (Sojka et al., 2007; 1998a; 1998b; Lentz et al., 2000; 1998).

The Superfloc A836 used in this study has been shown to be nontoxic to most organisms (Barvenik, 1994; Seybold, 1994; Bologna et al., 1999), contains <0.05% acrylamide monomer (AMD), which is a neurotoxin. PAM products with AMD concentrations <0.05% are considered safe under prescribed use (Barvenik, 1994). AMD has a half life in tens of hours in soil and water due to rapid breakdown by microorganisms (Lande et al, 1979; Shanker et al., 1990).

Bologna et al. (1999) showed that AMD is not absorbed by plant tissues and apparently breaks down rapidly even when injected directly into living plant tissue. PAM does not revert to AMD in the natural environment upon degradation (MacWilliams, 1978). The PAM used in this study degrades approximately 10% per year as a result of physical, chemical, biological and photochemical processes and reactions (Entry et al., 2008; Tolstikh et al., 1992). Because the overall microbial community and functional effects of such massive amounts of PAM addition to soil for prolonged periods was relatively minor, it is unlikely that PAM, used according to NRCS rate and quality recommendations for soil conditioning to prevent erosion and protect runoff water quality, is unlikely to pose a threat to the soil microbial community.

REFERENCES

Barvenik, F.W. 1994. Polyacrylamide characteristics related to soil applications. *Soil Sci.* 158, 235-243.

Bologna, L.S., F.F. Andrawes, F.W. Barvenik, R.D. Lentz, R.E. Sojka. 1999. Analysis of residual acrylamide in field crops. *J. Chromat. Sci.* 37, 240-244.

Buchholz, F.L. 1992. Polyacrylamides and polyacrylic acids. In: B. Elvers, S. Hawkins and G. Schulz (Eds.) Ullmann's *Encyclopedia of Industrial Chemistry*. Vol. A21.VCH Weinheim, Germany. pp. 143-146.

Dunbar, J., S. Takala, S. M. Barns, J. A. Davis, and C. R. Kuske. 1999. Levels of bacterial community diversity in four arid soils compared by cultivation and 16S rRNA gene cloning. *Appl. Environ. Microbiol.* 65, 1662-1669.

Dunbar, J., L. O. Ticknor, and C. R. Kuske. 2001. Phylogenetic specificity and reproducibility and new method for analysis of terminal restriction fragment profiles of 16S rRNA genes from bacterial communities. *Appl. Environ. Microbiol.* 67, 190-197.

Entry, J.A., R.E. Sojka and B.J. Hicks. 2008. Carbon and nitrogen stable isotope ratios measurement of anionic polyacrylamide degradation in soil. *Geoderma*. 145. 8-16.

Entry, J.A., I. Phillips, H. Stratton, and R.E. Sojka. 2003. Polyacrylamide, polyacrylamide+ Al(SO4)3 and polyacrylamide +CaO remove microorganisms and nutrients from animal wastewater. *Environ. Pollut.* 121, 453-462.

Entry, J.A., R.E. Sojka, M.E. Watwood, and C. Ross. 2002. Environmental and agricultural applications of polyacrylamide preparations. *Environ. Pollut.* 120, 191-200.

Goodrich, M.S., L.H. Dulak, M.A. Freidman, and J.J. Lech. 1991. Acute and longterm toxicity of water-soluble cationic polymers to rainbow trout (Oncorhynchus mykus) and the modification of toxicity by humic acid. *Environ. Toxicol. Chem.* 10, 509-551.

Ingham, E.R. and D.A Klein. 1984. Soil fungi relationships between hyphal activity and staining with fluorescein diacetate. Soil Biol. Biochem. 16, 273-278.

Jenkinson, D.S. and Ladd, J.M. 1981. Microbial biomass in soil: Measurement and turnover. In: E.A. Paul and J.N. Ladd (Eds.) *Soil biochemistry*. Vol 5 Marcel Dekker, New York pp. 415-471.

Kay-Shoemake, J.L., M.E. Watwood, R.D. Lentz, and R.E. Sojka. 1998a. Polyacrylamide as an organic nitrogen source for soil microorganisms with potential impact on inorganic soil nitrogen in agricultural soil. *Soil Biol. Biochem*. 30, 1045-1052.

Kay-Shoemake, J.L., M.E. Watwood, R.D. Lentz, and R.E. Sojka. 1998b. Polyacrylamide as a substrate for microbial amidase. *Soil Biol. Biochem*. 30, 1647-1654.

Kay-Shoemake, J.L., M.E. Watwood, R.E. Sojka, and R.D. Lentz. 2000a. Soil amidase activity in polyacrylamide (PAM) treated soils and potential activity toward common amide containing agrochemicals. *Biol. Fertil. Soils* 31, 183-186.

Kay-Shoemake, J.L., M.E. Watwood, L. Kilpatrick, L. and K. Harris. 2000b. Exchangeable ammonium and nitrate from different nitrogen fertilizer preparations in polyacrylamide-treated and untreated agricultural soils. *Biol. Fertil. Soils* 31, 245-248.

Kirk, R.E. 1995. Experimental design: Procedures for the behavioral sciences. 2nd ed. Brooks Cole Publishing Co, Belmont, CA.

Lande, S.S., S.J. Bosch, and P.H. Howard. 1979. Degradation and leaching of acrylamide in soil. *J. Environ. Qual.* 8, 133-137.

Lentz, R.D., Sojka, R.E. and Robbins, C.W. 1998. Reducing phosphorus losses from surface - irrigated fields: emerging polyacrylamide technology. J. Environ. Qual. 27, 305-312.

Lentz, R.D., Sojka, R.E. and. Ross, C.W 2000. Polymer charge and molecular weight effects on treated irrigation furrow processes. Int. J. Sed. Res. 1, 17-30.

Lodge, D.J. and Ingham, E.R. 1991. A comparison of agar film techniques for Estimating fungal biovolumes in litter and soil. *Agric. Ecosyst. Environ*. 5, 31-37.

MacWilliams, D.C. 1978. Acrylamides. In: I. Kirk and D.F. Othmer (Eds.). *Encyclopedia of Chemical Technology*, 3rd. Ed., Vol. 1. Wiley, New York, pp. 298-311.

Mills, D. K., K. Fitzgerald, C.D. Litchfield and P.M. Gillvet. 2003. A comparison of DNA profiling techniques for monitoring nutrient impact on microbial community composition during bioremediation of petroleum-contaminated soils. *J. Microbiol. Methods*. 54, 57-74.

Nadler, A. and Y. Steinberger. 1993. Trends in structure, plant growth and microorganism interrelations in the soil. *Soil Sci*. 115, 114-122.

Orts, W.J., R.E. Sojka, G.M. Glenn, and R.A. Gross. 1999. Preventing soil erosion with polymer additives. *Polymer News*. 24, 406-413.

Orts, W.J., R.E. Sojka, and G.M. Glenn. 2000. Biopolymer additives to reduce erosion-induced soil losses during irrigation. *Industrial Crops and Products*. 11, 19-29.

Orts, W.J., R.E. Sojka, and G.M. Glenn. 2002. Polymer additives in irrigation water to reduce erosion and better manage water infiltration. *Agro Food Industry Hi-Tech*. 13, 37-41.

Seybold, C.A. 1994. Polyacrylamide review: Soil conditioning and environmental fate. *Comm. Soil Sci. Plant Anal*. 25, 2171-2185.

Shanker, R., C. Ramakrishna, and P.K. Seth. 1990. Microbial degradation of acrylamide monomer. *Arch. Microbiol*. 154, 192-198.

Snedecor, W.G. and W.G. Cochran, 1994. *Statistical methods*. Iowa State University Press, Ames, Iowa. 354 p.

Spackman, R., J.A. Entry, R.E. Sojka, and J.W. Ellsworth. 2003. Polyacrylamide for coliform bacteria removal from agricultural waste water in irrigated forage production. *J. Soil Water Cons*. 58, 276-283.

Sojka, R.E. D.L. Bjorneberg, J.A. Entry, R.D. Lentz, and W.J. Orts. 2007. Polyacrylamide (PAM) in agriculture and environmental land management. *Advan. Agron*. 92, 76-141.

Sojka, R.E., J. A. Entry, and J.J. Furhmann. 2006. The influence of high application rates of polyacrylamide on microbial metabolic potential in an agricultural soil. *Appl. Soil Ecol*. 108,405-412.

Sojka, R. E. and J.A. Entry. 2000. Influence of polyacrylamide application to soil on movement of microorganisms in runoff water. *Environ. Pollut*. 108, 405-412.

Sojka, R.E., R.D. Lentz, C.W. Ross, T. Trout, D.L. Bjorneberg, and J.K. Aase. 1998a. Polyacrylamide effects on infiltration in irrigated agriculture. *J. Soil. Water. Cons*. 54, 325-331.

Sojka, R.E., R.D. Lentz, and D.T. Westermann. 1998b. Water and erosion management with multiple applications of polyacrylamide in furrow irrigation. *Soil Sci. Soc. Am. J*. 62, 1672-1680.

Sojka, R.E., R.D. Lentz, D.L.Bjorneberg, and J.K. Aase, 1998c. The PAMphlet: A concise guide for safe and practical use of polyacrylamide (PAM) for irrigation-induced erosion control and infiltration enhancement. USDA-ARS Northwest Irrigation and Soils Research Lab, Kimberly, ID, Station Note #02-98.

Sojka, R.E., D.L. Carter, and M.J. Brown, 1992 Imhoff cone determination of sediment in irrigation runoff. *Soil Sci. Soc. Am. J.* 56, 884-890.

Sojka, R.E., R.D. Lentz, and J.A. Foerster. 1994. Software utilizing Imhoff cone volumes to estimate furrow-irrigation erosion. *J. Soil and Water Conser.* 49, 400-406.

Stadmatiadis, S., J.W. Doran, and E.R. Ingham. 1990. Use of staining inhibitors to separate fungal and bacterial activity in soil. *Soil Biol. Boichem.* 22, 81-88.

Steinberger, Y., S. Sarig, A. Nadler, and G. Barnes. 1993. The effect of synthetic soil conditioners on microbial biomass. *Arid Soil Res. Rehabil.* 7, 303-306.

Tolstikh, L.I., N.I. Akimov, I.A. Golubeva, and I. Shvetsov. 1992. Degradation and stabilization of polyacrylamide in polymer flooding conditions. *Int. J. Polymeric Material.* 17, 177-193.

Watwood, M. and J. Kay-Shoemake. 2000. Impact of polyacrylamide treatment on sorptive dynamics and degradation of 2,4-D and atrazine in agricultural soil. *Journal of Soil Contamination.* 9, 133-147.

Wallace, A. and G.A. Wallace. 1996. Need for solution or exchangeable calcium and/or critical EC level for flocculation of clay by polyacrylamides. In: *Proceedings: Managing Irrigation-Induced Erosion and Infiltration with Polyacrylamide.* (Eds. R.E Sojka and R.D. Lentz.). University of Idaho Misc. Pub. 101-96, pp 59-63.

West. A.W., D.J. Ross, and J.C. Cowling. 1986. Changes in microbial C, N, P, and ATP contents, numbers and respiration on storage of soil. *Soil Biol. Biochem.* 18, 141-148.

In: Advances in Polymer Research
Editors: Losif Sidorov, Fedor Nikitin, pp. 73-107
ISBN: 978-1-61942-223-0

POLYMERIC INTRACORPOREAL DEVICES FOR THE MANAGEMENT OF NEUROLOGICAL DISORDERS

Viness Pillay[*1], *Thiresen Govender*[1], *Yahya E. Choonara*[1], *Valence M. K. Ndesendo*[1], *Lisa C. du Toit*[1], *Girish Modi*[2] *and Dinesh Naidoo*[2]

[1] University of the Witwatersrand, Department of Pharmacy and Pharmacology, Parktown, 2193, Johannesburg, South Africa.

[2] University of the Witwatersrand, School of Neurosciences, Department of Neurology, Parktown, 2193, Johannesburg, South Africa.

ABSTRACT

The complexity of the brain and the membranous blood-brain barrier (BBB) has proved to be a significant limitation to the systemic delivery of pharmaceuticals to the brain rendering them sub-therapeutic and ineffective in the treatment of neurological diseases. Apart from this, lack of innovation in product development to counteract the problem is also a major contributing factor to a poor therapeutic outcome. Various innovative strategies show potential in treating some of the neurological disorders; however drug delivery remains the most popular. To attain therapeutic drug levels in the central nervous system (CNS), large, intolerable systemic doses are generally administered. The success of maintenance therapy in many neurological diseases depends on a number of variables, including the constant release of neurotherapeutics, a reduction in the dosing frequency, a greater antipsychotic drug bioavailability and ultimately improved patient compliance, many of which is not achievable by conventional oral or parenteral formulations.

This article reviews the therapeutic implantable polymeric and transdermal devices employed in an attempt to effectively achieve therapeutic quantities of drug across the BBB over a prolonged period, to improve patient disease prognosis.

Keywords: Membranous blood-brain barrier (BBB), Polymer-based implantable devices, Intracranial implants, Subcutaneous and subdermal implants, Intrathecal implants, Transdermal patches.

---

* Corresponding author: Email: viness.pillay@wits.ac.za, Tel: +27-11-717-2274. Fax: +27-11-642-4355, Fax2Email: +27-86-553-4733.

INTRODUCTION

The brain is considered to be the body's central hub, essential for managing a variety of vital bodily functions and maintaining normality within the healthy body. Higher order neural activity is the consequence of a combination of essential conditions in the brain namely; the efficient liberation of nutrients to the brain, effective waste elimination, neuroprotection from toxic and neuroactive substances circulating within the body, maintenance and repair processes and ultimately homeostasis of the neural micro-environment (Abbott, 2005; Cole and Vassar, 2007; Hsu et al., 2008; Ewing, 2009; Cunnane et al., 2010; Schlegelmilch et al., 2010). Any imbalances in the above-mentioned processes may result in neuropathological, cerebrovascular, inflammatory, carcinogenic or neurodegenerative disorders (Abbot, 2005). Various innovative strategies including, but not limited to, Deep Brain Stimulation (DBS), cell grafts and gene delivery, have shown potential in treating some of the above mentioned disorders, however drug delivery remains the most popular. Apart from common neuropathological and neurodegenerative diseases, another prime example includes the ever-growing AIDS pandemic. During the course of the disease the HI-Virus may infect the brain. The density of the viral load in the peripheral compartments is effectively reduced by highly active anti-retroviral therapy (HAART) however this is not the case in the brain. Owing to the fact that drugs such as azidothymidine, 3TC, or protease inhibitors have minimal penetration into the brain parenchyma, the virus seeks refuge here and continues to multiply, shortening even further the lifespan of the infected individual (Antinori et al., 2003; Kandanearatchi et al., 2003; Pardridge, 2005). Such an occurrence exemplifies the delicate nature of the brain and the vast array of disorders which may affect it. Furthermore, it emphasizes the great need for prolonged drug delivery for CNS disorder in patients on maintenance therapy.

Compliance with medication regimens is a pivotal factor in the treatment of serious psychiatric disorders, and ultimately determines the treatment outcome. Non-compliance to treatment regimens is a major impediment to the long-term treatment and positive therapeutic outcome of a variety of CNS disorders and may result in deterioration in mental capacity, loss of employment and relationships and more importantly relapse and re-hospitalization. A major challenge with continuous relapse is that reduction after the next episode is extremely difficult to achieve (Keith et al., 2004; Addington et al., 2010). Non-compliance may be deliberate or accidental and can be as a result of numerous factors (Centorrino et al., 2001; Irani et al., 2004; Rabin et al., 2008; Bulloch and Pattern, 2009).

Breen and Thornhill (1998) categorized the reasons for non-compliance into three distinct classes namely; medication, patient and provider-related reasons. Intolerable side-effects, ineffectiveness, complexity and cost of treatment regimens are medication-related reasons for non-compliance with medication side-effects being the most significant and likely cause of non-compliance of all three classes (Goodwin and Vieta, 2005; Singh et al., 2006; Birnbaum and Sharif, 2008).

The patient's culture and attitude, symptoms of the disorder and the presence of substance abuse are patient-related factors impacting on compliance. Communication between the practitioner, the patient and the patient's family's coupled with the differences in perceptions of patient treatment outcomes and medication effectiveness is provider-related reasons affecting compliance (Breen and Thornhill, 1998).

Compounding the problems of CNS disorders are the difficulties experienced by caregivers. Much emphasis is placed on the sufferers of CNS disorders and far less on the caregivers and family faced with the responsibility of caring for them. The chronic, debilitating nature of certain CNS-related disorders may be physically, emotionally, psychologically and financially challenging to the caregiver. The burden of care is a phrase applied to the caregiver that describes the hardship and consequences faced by them in their quest to assist those living with CNS disorders. Burden of care also encompasses distressing thoughts and perceptions such as self-blame, embarrassment, shame and guilt, all of which may be debilitating to the caregiver (Awad and Voruganti, 2008). Caring for these patients can prove to be an arduous task. As a result, caregivers are prone to large amounts of stress and the risk of depression and physical susceptibility (Ferrara et al., 2008).

The advent of utilizing polymer-based implantable devices offers many advantages including the ability to accurately control the release of specific quantities of active drugs, macromolecules and proteins over prolonged periods of time as well as the protection of active molecules from degradation prior to release.

The issues of non-compliance and the consequences of missing doses, the daily use of oral psychotropic medication and some of the intolerable peripheral side-effects of most psychotropic medication are virtually eliminated. In addition, polymeric implantable devices decrease the need for having a constant caregiver for CNS patients who are on chronic maintenance therapy. This article reviews the polymer-based implantable and transdermal devices utilized in the treatment of major CNS disorders.

2. POLYMER-BASED IMPLANTABLE DEVICES FOR MAJOR CNS DISORDERS

2.1. Intracranial Implants

2.1.1. Drug Delivery Approaches Employed to Traverse the Blood Brain Barrier

The membranous BBB has always proved to be a significant limitation to the systemic delivery of pharmaceuticals to the brain rendering them sub-therapeutic and ineffective in the treatment of neurological diseases (Abbott, 2005; Gelperina, 2006; Koo et al., 2006; Koziara et al., 2006; Krauze et al., 2006; Kwan *P and* Brodie*, 2006;* Tiwari et al., 2006; Nelson et al., 2010). The failure of majority of neurotherapeutic drugs relies on the fact that these drugs are only transported minimally across the BBB.

Apart from this, lack of innovation in product development to counteract the problem is a major contributing factor to a poor therapeutic outcome.

In order for a small drug molecule to cross the BBB, it must fulfill two general criteria namely: (a) a molecular mass below 400-500 Da and (b) high lipid solubility. It is evident that most drugs do not possess these criteria (Pardridge, 2005). Consequently, the most debilitating and life-threatening diseases are ineffectively and sub-therapeutically treated. BBB is a pharmacological and physiological barrier situated at the capillaries of the brain where a large variety of cells are present (Figure 1). Tight junctions form an integral morphological trait characteristic of the brain micro-vessel endothelial cells that form the BBB.

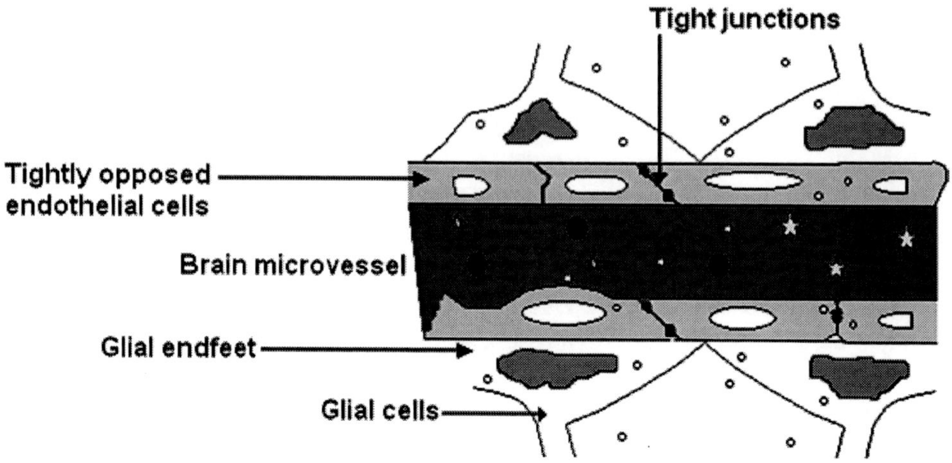

Figure 1. A schematic depicting the morphology of the BBB, [Adapted from Patel, 2007].

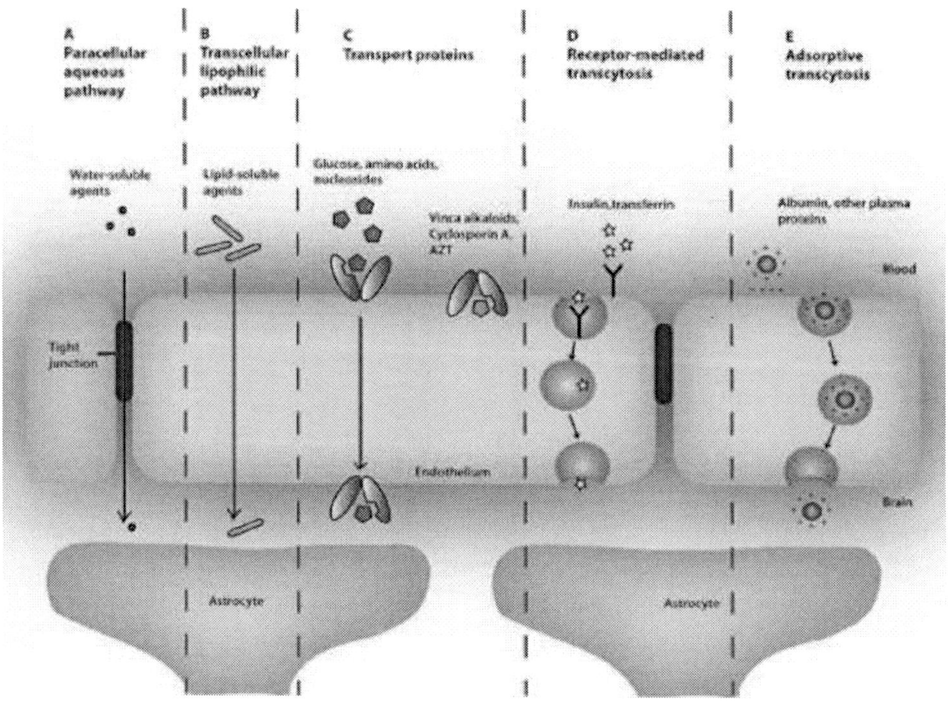

Figure 2. Schematic of the BBB illustrating the various routes of molecular transport across brain capillary endothelial cells: A) water-soluble agents can travel between the endothelial cells via tight junctions, but passage is highly restricted; B) in contrast, lipid-soluble agents easily penetrate the BBB by diffusing across the endothelial cell membranes; C) carrier proteins present in the membrane facilitate crossing of substances such as monosaccharides, amino acids, peptides, nucleosides, choline, and organic cations and AZT; D) receptors specific for transferrin, insulin, insulin-like growth factor, lipoproteins, and leptin are also exposed on the surface of the endothelium to aid in penetration via transcytosis; and E) Positively charged molecules can traverse the BBB through adsorptive transcytosis, [Adapted from Bennewitz and Saltzman W.M., 2009].

This coupled with a weakened pinocytic activity and the absence of fenestrations; effectively limits the passage of certain compounds from the blood into the extra-cellular cerebral environment. While nutrients and essential gases are readily exchanged through the barrier, larger molecules such as immune cells, certain microbes and a large variety of drugs are denied entry through the barrier (Soppimath et al., 2001; Wang et al., 2002; Jaypraker et al., 2009).

Figure 2 is a schematic of the BBB illustrating the various routes of molecular transport across brain capillary endothelial cells. The elucidated pathways can be exploited to enhance the entrance and accumulation of AED's into the brain parenchyma.

A difference between the uptake and efflux processes in the BBB determines the net uptake of a given drug by the brain through the BBB (Figure 3) (Scherrmann, 2002). Specific properties of the endothelial cells and the systemic disposition of the drug are integral factors that regulate the uptake of the drug. Drug binding to blood cells or plasma proteins elucidates the unbound drug fraction that may influence uptake. Protein binding may however limit drug uptake by the brain (Park S. and Sinko, 2005; Banks, 2009). The efflux of drug from the brain extra cellular fluid (BECF) into the blood through the BBB is modulated by similar properties however protein binding in the BECF does not exist due to the absence of proteins. Consequently, the quantities of drug crossing the BBB in both directions are actively controlled by the permeability of the endothelial cells and their ability to metabolize the drug.

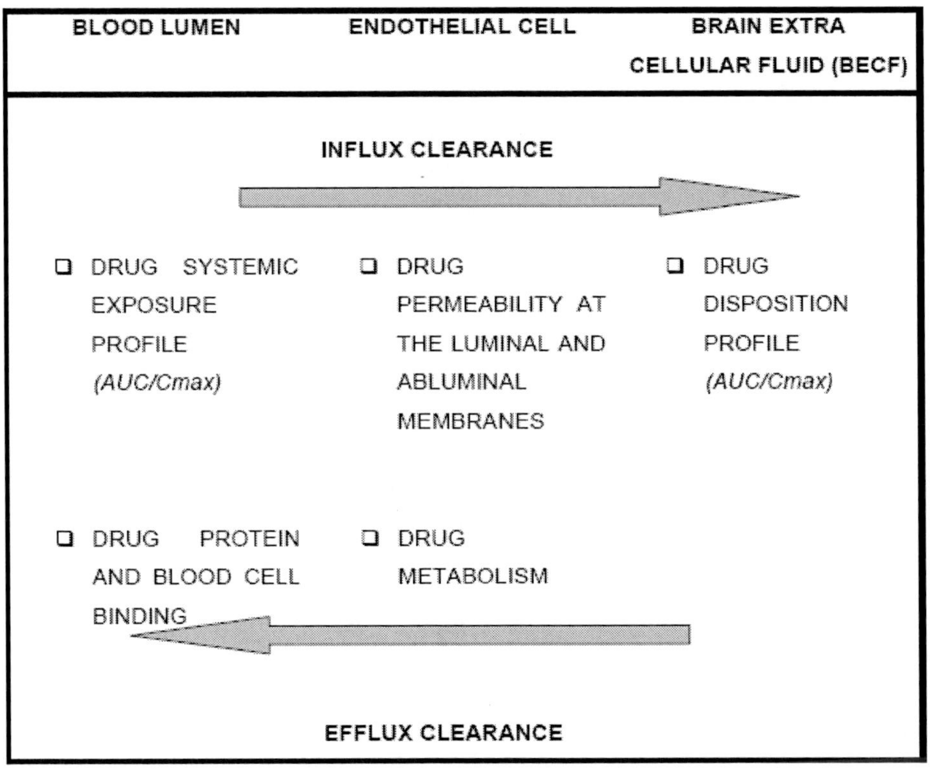

Figure 3. Representation of the net uptake of a drug by the brain through the endothelial cell of the BBB determined by the difference of the overall uptake and efflux processes, [Adapted from Scherrmann, 2002].

A variety of techniques have been adopted in an attempt to obtain therapeutic drug levels across the BBB without having to administer potentially toxic oral or intravenous doses of a specific drug. Included in this is injecting the patient at the carotid artery with a hypertonic solution such as mannitol, essentially shocking the tight junctions of the BBB and subsequently opening them momentarily. During this window period, the drug is administered via the carotid artery (Rapoport and Thompson, 1973; Yamaguchi et al., 1990; Guo-yu et al., 2000; Cudmore, 2002; Bryan, 2004). The major shortcoming of this method is that it is non-selective and during the time of junction opening, substances that are potentially toxic to the brain, may enter (Bryan, 2004). Another method is direct injection into the CNS (Cudmore, 2002). Convection-enhanced delivery deals with the continuous injection of therapeutics under positive pressure through the interstitial space of the parenchyma. This technique, used for optimizing brain disease treatment, allows the passage of larger therapeutic substances across the BBB (Raghavan et al., 2006). Other strategies make use of endothelial cell proteins that would conventionally transport hormones and nutrients to the brain. It is evident that this method could transport a host of molecules, including genes, enzymes and monoclonal antibody drugs, previously too large and incapable of crossing the BBB (Ferber, 2007). Some scientists have used nasal inhalation as a delivery route with direct access to the brain through olfactory and trigeminal nerves, which relay sensory signals from the nose and mouth (Ferber, 2007). Studies have shown that this depends on the molecular weight and lipophilicity of the drug (Perras et al., 1999; Fehm et al., 2000; Illum, 2000; Quay, 2001). Insulin-like growth factor has been shown to bypass the BBB via the intranasal route (Frey, 2001).

Another pivotal area of research involves the use of nanobiotechnology to effectively cross the BBB. Modern therapeutic agents generally have a low permeability across biological membranes, are unstable in the physiological environment, have poor solubilities and essentially exhibit a very low bioavailability. Nanobiotechnology is believed to counteract these negative traits and confer upon the drug an enhanced therapeutic efficacy (Sahoo et al., 2008). Despite the previous methods of drug delivery across the BBB, much focus is on developing novel delivery systems using nano-particles and nano-liposomes (Bryan, 2004). Using a variety of chemically synthesized polymeric nanoparticles, researchers are exploring the possibilities of delivering oligonucleotides, genes, antineoplastic agents and MRI contrast agents (Silva, 2007). According to the structural organization, biodegradable nanoparticles are classified as nanocapsules and nanospheres for which the drug molecules may either be entrapped inside or adsorbed on the surface (Figure 4).

Site-specific drug delivery using polymeric implantable devices circumvents the difficulties posed by the BBB, limits systemic side-effects, decreases serum protein binding and reduces hepatic metabolism and peripheral drug inactivation. In addition, it avoids the need to administer potentially toxic doses of drugs to achieve therapeutic doses in the CNS (Domb, 1995; Begley, 2004).

Varieties of biocompatible synthetic and non- synthetic polymers may be explored for local drug delivery to the brain and include the non-biodegradable ethylene-vinyl acetate copolymer (EVAc), certain polyurethanes and acrylate-based hydrogels as well as the biodegradable poly (lactic-co-glycolic acid) (PLGA), poly (lactic acid) (PLA), poly (hydroxybutyrate), poly (caprolactone), polyanhydrides, cellulose and gelatin (Zhong and Bellamkonda, 2008). Examples of non- synthetic polymers employed for local drug delivery to the brain for the treatment of epilepsy are silk-based polymers.

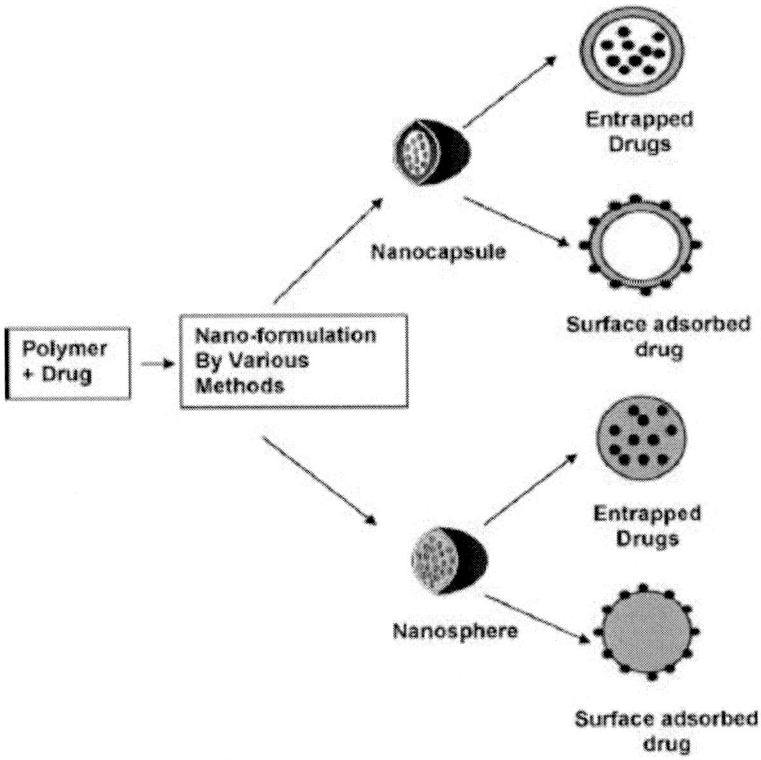

Figure 4. A schematic depicting the type of biodegradable nanoparticles with the drug either entrapped inside or adsorbed on the surface of the nanocapsules or nanospheres, [Adapted from Kumari et al., 2009].

2.1.1.1. Ethylene Vinyl Acetate Copolymers AVAc as a Prototype Biocompatible Polymer Employed for Local Drug Delivery to the Brain

EVAc polymers have recently found a wide application in as an intracerebral delivery vehicle for a variety of neurological conditions (During et al., 1989; Freese et al., 1989; Hoffman et al., 1990; Saltzman et al., 1999; Sanberg et al., 1993). The first polymer-based Adenosine Augmentation Therapy (AAT) approach was performed by Boison and co-workers (1999) using EVAc polymers engineered to release adenosine. Employed was 10%(w/v) solution of purified EVAc in methylene chloride into which 20% adenosine (w/w) was added. This suspension was frozen and then lyophilized to remove the solvent out of the mixture. The remaining mixture was pressure extruded at 50°C into cylindrical adenosine-EVAc rods. Single polymers (1 × 0.4mm) released an amount of around 20 to 50ng adenosine per day into Ringer solution. Individual polymers were implanted into the lateral brain ventricles of rats that had been kindled in the hippocampus. It was observed that, in these rats stimulus-induced epileptic seizures were drastically reduced after polymer implantation. This was demonstrated by a strong reduction of stage 5 seizures for at least 7 days and by suppression or reduction of epileptiform electrical afterdischarges up to 3 days as recorded bilaterally in the hippocampus. Upon diminishing the supply of adenosine, the antiseizure effects gradually decreased and had expired after 14 days. Control implants that were loaded with BSA failed to display any antiepileptic activity or changes in stimulus-evoked EEGs. It was further confirmed that the local release of 20 to 50ng adenosine per day is sufficient to exert a significant antiepileptic effect. This study demonstrates that focal controlled release of adenosine within the brain can suppress epileptic seizures.

2.1.1.2. Silk-Based Polymers as an Alternative to Synthetic Drug Delivery-Systems for the Treatment of Epilepsy

A novel generation of silk-based drug-delivery systems has recently been developed as an alternative to synthetic drug delivery-systems for the treatment of epilepsy (Boison, 2009). Among the several advantages that they offer is that silk fibroin is biocompatible (Altman et al., 2003) and capable of biodegrading slowly (Horan et al., 2005). Noteworthy is that as an important property for sustained release applications, the degradation lifetimes of silk can be regulated by the extent of physical crosslinking for beta sheet formation, thus allowing control of degradation timeframes from weeks to years (Horan et al., 2005). Silk can also be processed under aqueous conditions, thus allowing for the incorporation of labile biomolecules without loss of biological activity (Jin and Kaplan, 2003; Li et al., 2006). Furthermore, the frequent use of silk sutures in brain and nerve tissue confirms the feasibility of implantation of silk biomaterials in the brain (Boison, 2009).

In a study that was conducted by Wilz and co-workers (2008), silk-based polymers were engineered to release target release rates of 0, 40, 200, and 1000ng adenosine per day. These polymers were implanted into the infrahippocampal fissure of rats prior to the onset of kindling. It was demonstrated that by using this approach the focal adenosine release from silk-based polymers is effective in retarding kindling epileptogenesis in a dose-dependent manner (Boison, 2009). Combining polymers designed to release the specified target doses for period limited to 14 days with a kindling-epileptogenesis paradigm extended to 20 days after polymer implantation (i.e. a total of 48 kindling stimulations distributed over 20 days), it was possible to document kindling progression after expiration of the polymers (Boison, 2009. In this study there was a clear dose-dependency in the average number of stimulations needed to elicit the first stage 1 or 2 seizure. These findings suggested a linear dose-response relation. Notably, recipients of polymers releasing a target dose of 1000ng adenosine per day did not display any behavioral seizure during the first 22 stimulations (i.e. stage 0) despite the presence of after-discharges in EEG recordings, whereas control rats had reached stage 3 seizures at that time. As adenosine-release from the polymers began to wear off, seizures gradually developed with progressive intensity. This delayed seizure-development curve was parallel to the curve of control rats, but shifted to the right by 7 days. These findings suggest a potential antiepileptogenic effect of intracerebral adenosine release.

One of the very first attempts to use a controlled release polymeric system implanted in the brain was intended to improve labeling of perivascular meningeal projections in the trigeminal ganglia of the cat model. Elucidating the neuronal connections within the central and peripheral nervous system was made possible through the use of the marker enzyme, Horseradish peroxidase (HRP). When using HRP in the central nervous system, the enzyme was often in aqueous solution and administered either by microiontophoresis or stereotaxic injections. A major shortfall of using HRP in an aqueous solution is the fact that the enzyme diffuses away from the target site once injected ultimately resulting in false positive results. This is a problem when using the enzyme to map out specific pathways in the brain. This study therefore established an efficient means of overcoming the problem which proved to be effective. Incorporating the HRP within a poly (vinyl alcohol) (PVA) casting solution and simply applying this to the cerebral blood vessels, localizes the HRP delivery to a greater extent. Diffusion of HRP was further curbed by modification of PVA using pure polymer (Mayberg et al., 1981). The ease of implantation led to many researchers exploring the possibility of polymer-based neurotherapy and site-specific drug delivery to the brain for the treatment of various CNS disorders.

2.1.2. Application of Polymer-Based Implantable Devices in Parkinson's Disease

2.1.2.1. Biocompatible Dopamine-Loaded Polymeric Matrix Device

The proposal of using monolithic polymeric brain implants for the treatment of various CNS disorders long exists. The development of a biocompatible dopamine-loaded polymeric matrix device for intracerebral implantation bears testament to this. The device was developed to optimize CNS disorder therapy and to overcome the difficulties presented with drug delivery to the brain. The device comprised dopamine as the active drug dispersed within a variety of EVAc matrix configurations. Initially, it was noted that the *in vitro* cumulative dopamine release from the conventional configurations was found to depend on the original concentration of dopamine matricized within the EVAc network. Modification of the initial systems by coating them with an impermeable layer of EVAc and then forming a void or aperture in the coating resulted in constant, linear and desirable dopamine release characteristics. This optimized formulation was then subjected to *in vivo* analysis in the rat model. Stable dopamine levels in the brain extra-cellular fluid were reached twenty days post implantation and it remained stable for the duration of the experiment which lasted sixty five days. All rats survived the study and histopathological analysis revealed that the striatal morphology was normal (During et al., 1989; Freese et al., 1989). Despite the promising results of the study, a major concern regarding such a formulation is the fact that EVAc is a non-biodegradable polymer. Apart from implanting the device, additional surgery will be required to remove the device. Furthermore, the sterility of an EVAc device may be questioned if left non-degraded and bathing in the CSF.

2.1.2.2. Glial Cell Line Derived Neurotrophic Factor (GDNF)

The symptoms experienced by the patients suffering from Parkinson's disease (PD) are primarily due to the loss of axons in the striatum which is a result of degeneration and deterioration of nerve cells in the substantia nigra. Early attempts to repair striatal dopaminergic neurons by grafting fetal cells were limited, primarily due to ethical concerns, mortality of the grafted cells and the disposability of donor tissues. Glial cell line derived neurotrophic factor (GDNF) represents the relatively recent identification of neurotrophic factors capable of regenerating and protecting the affected dopaminergic neurons. Despite the benefits of GDNF, its use clinically is limited due to the complexity involved with delivering proteins to brain tissue and the fact that its receptor is located in several areas of the brain causing neurological side-effects. Pre-clinical trials have been conducted and include; the use of non-biodegradable polymeric devices, intraparenchymatous injection, fibrin glue preparation, the grafting of genetically-engineered cells, *in vivo* gene transfer and intracerebro-ventricular infusion. These studies led to a trial using the intracerebro-ventricular infusion method however results depicted a lack of niagrostriatal neuron regeneration and the occurrence of neurological side-effects thus necessitating an alternate means of delivery (Unsicker, 1994; Tomac et al., 1995; Cheng et al., Choi-Lundberg et al., 1997; Rosenblad et al., 1998; Hoane et al., 1999; Kordower et al., 1999; Aoi et al., 2000; Nakao et al., 2000; Tornqvis et al., 2000).

2.1.2.3. Biodegradable PLGA Microspheres Containing GDNF

In a study by Jollivet and co-workers (2004), biodegradable PLGA microspheres containing GDNF were implanted in the rat brain striatum. The Sprague-Dawley rats utilized represented a partial and progressive model of PD. Microspheres were prepared by specialized water in oil in water (W/O/W) emulsion-solvent evaporation technique. Resultant PLGA microspheres produced were smooth, regular and contained a GDNF entrapment of 84%. Prior to stereotaxic implantation into the striatum, microspheres were dispersed in a suspending agent comprising distilled water, Tween 80 and mannitol. Each rat was implanted twice with 1.5mg microspheres in 10μl suspending agent. *In vivo* release kinetics confirmed that GDNF can be released for up to 2 months. In addition, implantable microspheres were well tolerated and most importantly demonstrated the ability to stimulate sprouting of the preserved dopaminergic fibers with synaptogenesis. This was accompanied by a significant improvement in behavior and function. This study provided a unique and promising advance in the treatment of PD. A major shortfall to the study was the fact that the microspheres were not optimized allowing for a smaller volume for implantation. In addition, there is a need for complementary studies with retrograde labeling (Jollivet et al., 2004). *In vivo* GDNF release could have been improved through incorporation of the microspheres within a polymeric matrix however; the relative ease of microsphere implantation may have been compromised and could have resulted in the need for sophisticated neurosurgery. Furthermore, incorporating the microspheres in a suspending agent could cause premature release of GDNF by diffusion.

2.1.2.4. Polymeric Scaffold for Site-Specific Delivery of Dopamine

A novel comprehensive *in vitro* and *in vivo* study by Pillay and co-workers (2009) illustrated he combined benefits of polymeric nanotechnology and polymer scaffold science to augment the site-specific delivery of dopamine for the possible treatment of PD. The study encompassed the design, biometric simulation and optimization of dopamine-loaded cellulose acetate phthalate nanoparticles matricized within a structured intracranial binary crosslinked alginate scaffold. Nanoparticles were prepared by adapted oil in water emulsification-diffusion process and displayed a drug entrapment efficiency of approximately 63% after optimization using a box-benhken experimental design. The alginate scaffolds, also optimized by a box-benhken experimental design, were prepared by crosslinking an alginate in water solution with a primary crosslinking solution, molding the resultant gel-like mass, lyophilizing and suspending the lyophilized structures in a secondary crosslinking solution. Integration of the nanoparticles within the scaffold was completed through a lyo-fusion process. Upon completion of extensive and successful *in vitro* testing, the device was stereotactically implanted into the frontal lobe parenchyma of the Sprague-Dawley rat model (Figure 5). Blood and cerebrospinal fluid (CSF) samples were taken at predetermined intervals and subjected to Ultra Performance Liquid Chromatography analysis. Results from this study illustrated that the device can provide prolonged, controlled, desirable release of dopamine in the CSF of the Sprague-Dawley rat thus establishing its potential as a candidate for the long-term management of PD (Pillay, 2009). Digital images depicting the surgical procedure employed for stereotactically implanting the polymeric scaffold device into the frontal lobe parenchyma of the Sprague-Dawley rat model are presented in Figure 5.

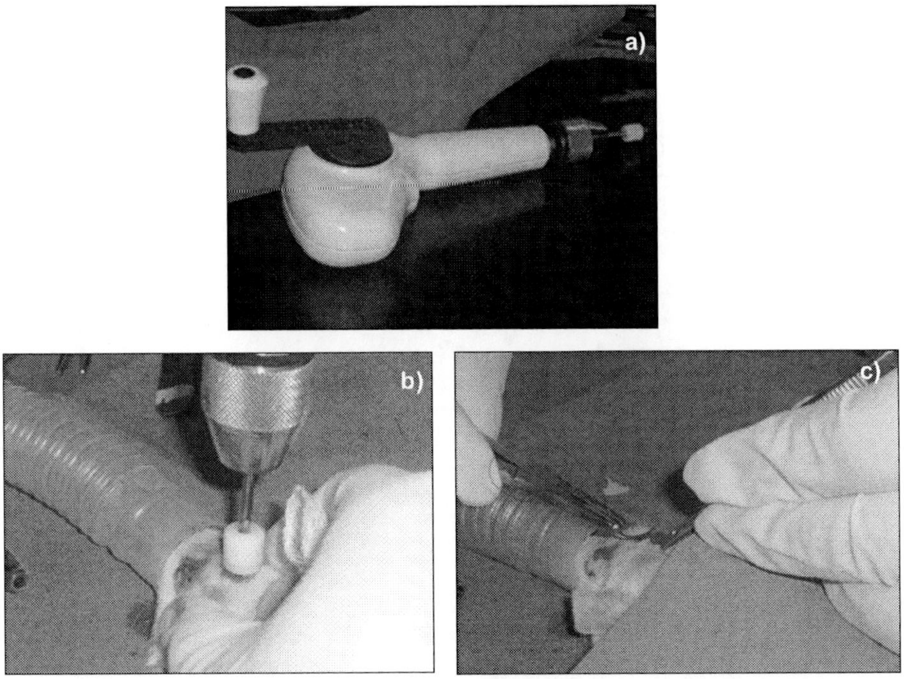

Figure 5. Digital images depicting the surgical procedure employed for stereotactically implanting the polymeric scaffold device into the frontal lobe parenchyma of the Sprague-Dawley rat model: a) manual drill used for incision; b) actual perforation of the skull; and c) suturing of the incision [Source: Pillay, 2009].

2.1.3. Application of Polymer-Based Implantable Devices in Alzheimer's Disease

2.1.3.1. Bethanechol-Loaded Microspheres for Implantation into Denervated Hippocampal Region of the Brain

The impairment in learning and memory associated with Alzheimer's disease (AD) is hypothesized to be as a result of a deficiency of acetylcholine in the cortex of the brain. The treatment outcomes of conventional therapy for AD have generally been poor. This, together with the severe recurrent side-effects associated with conventional therapies could be due to a lack of local specificity within the brain as conventional dosage forms are unable to deliver therapeutic agents to specific sub-regions of the brain persistently. To counteract this challenge, bethanechol-loaded microspheres were synthesized. These biodegradable microspheres comprised a 50: 50 copolymer blend of poly-bis-p-(carboxyphenoxy)propane anhydride and sebacic acid (PCPP-SA) and were implanted in the denervated hippocampal region of the rat brain to assess local administration effects. The effectiveness of the technology was determined by a performance score achieved by rats when completing a maze exercise. Results revealed that hippocampal implants were well-tolerated by all rats. Furthermore, the rats implanted with the bethanechol-loaded microspheres generally exhibited significant and desirable improvement post-implantation, although individual performances did vary within this improved treatment group. Desirable treatment outcomes occurred throughout the duration of the experiment and the study thus established a novel method of localizing neurotransmitter delivery which proved effective in reversing lesion-induced spatial memory deficits in the rat brain (Howard et al., 1989). Figure 6 is a schematic depicting the microsphere implantation technique employed.

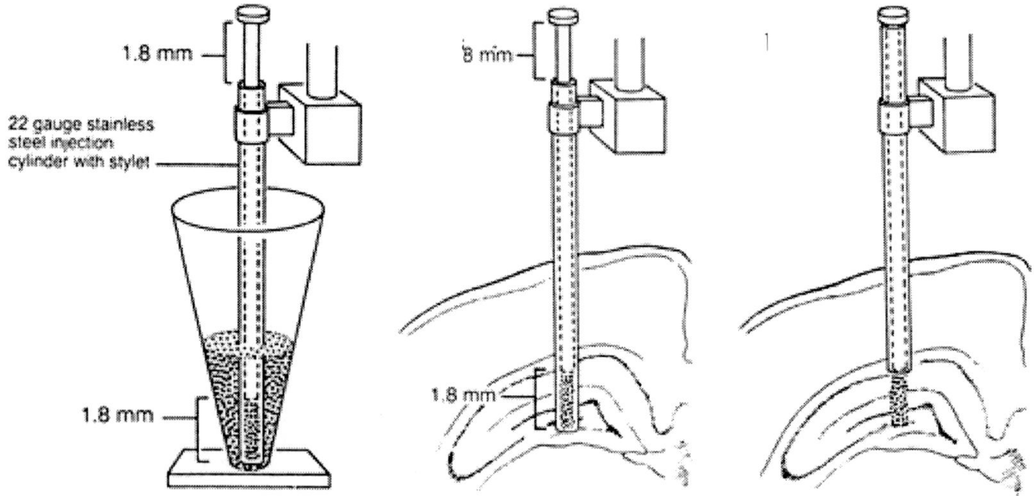

Figure 6. A schematic depicting the microsphere implantation technique: a) microspheres are impacted into a No. 22 stainless steel cylinder to a depth of 1.8 mm; b) The implantation assembly is stereotactically placed into the hippocampal formation; and c) the implantation stylette is held stationary as the stainless steel cylinder is slowly raised, leaving a polymer implant column within the hippocampus [Adapted from Howard et al., 1989].

2.1.3.2. A Biocompatible EVAc Matrix Incorporating Nerve Growth Factor (NGF) for the Potential Site-Specific Treatment of AD

Powell and co-workers (1989) synthesized a biocompatible EVAc matrix incorporating nerve growth factor (NGF) for the potential site-specific treatment of AD. The first stage in the process of synthesizing these matrices involved the production of solid particles containing NGF and ficoll. Briefly, the NGF was reconstituted with phosphate-buffered saline (PBS) and bovine serum albumin (BSA). The solution underwent quick-freezing and lyophilization thereby resulting in a homogenous powder which was then crushed to obtain a uniform particle size. Purified EVAc was dissolved in methylene chloride and the NGF/ficoll powder was added to this solution to obtain a 30-50% loading. The resultant solution was vortexed, producing a homogenous suspension which was cast into a cooled level glass mould.

This resulted in the formation of solidified slabs that were subjected to methylene removal. Resulted slabs were cut into uniform disks and were placed in 10ml PBS at 40°C. BSA was added to the PBS to stabilize the NGF following its release from the polymeric matrix. Despite the limitations of EVAc as discussed in section 2.1.2.1, the study demonstrated a controlled release of NGF over a period of one month and the stimulation of neurite sprouting in NGF sensitized PC-12 cells thus confirming its potential for the treatment of AD (Powell et al., 1989).

2.1.4. Application of Polymer-Based Implantable Devices in Huntington's Disease

Huntington's disease (HD) is a hereditary neurodegenerative disease that is more often than not fatal. The fact that genetic screening may be used to predict persons destined to acquire the disease, makes treatment strategies and formulation approaches for HD very interesting.

It has been demonstrated that several neurotrophic factors such as NGF can prevent the neurodegeneration and the behavioral characteristics associated quinolinic acid (QA)-induced HD however; various limitations with the delivery of these proteins do occur including the ability to attain controlled and sustained delivery of these proteins. Menei and co-workers (1999) successfully synthesized NGF-loaded PLGA microspheres to evaluate its release kinetics and its effects against QA-induced excitotoxic damage. Smooth, regular, porous PLGA microspheres were produced by a W/O/W emulsion-solvent evaporation technique.

In vitro release kinetics was evaluated utilizing radio-labeled NGF, PC-12 cell bioassay and immunozymatic bioassay. This was followed by *in vivo* evaluation following sterotaxic implantation in the rat striatum. Microspheres were implanted a week prior to infusion with QA. *In vitro* results showed that approximately half of the encapsulated NGF was released in a sustained manner over a period of five weeks. Implanted microspheres appeared to contain NGF after two and a half months, degraded after three months and were effective in reducing QA-induced lesions by 40% when compared to control models.

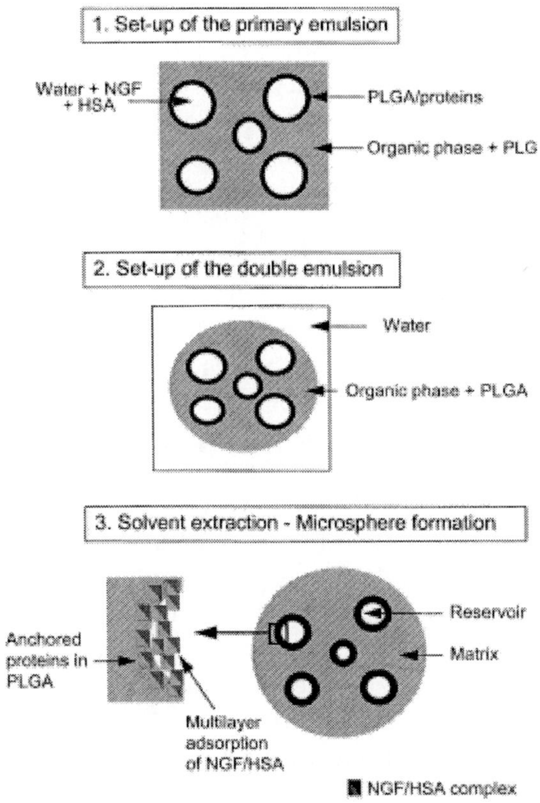

Figure 7. A schematic depicting the formation and morphology of PLGA/NGF-loaded microspheres, [Adapted from Benoita et al., 2000].

The study confirmed that biodegradable NGF-loaded polymeric microspheres implanted locally are neuroprotective against excitotoxic lesions and thus may be effective in the treatment of HD (Menei et al., 2000). Figure 7 is a schematic depicting the formation and morphology of PLGA/NGF-loaded microspheres.

2.1.5. Application of Polymer-Based Implantable Devices in Epilepsy

2.1.5.1. Gamma-Aminobutyric Acid (GABA)-Loaded and Unloaded Polymeric Matrices

Intracerebral implantable devices have been used for the delivery of anti-epileptic drugs (AEDs) and agents known to be beneficial in the treatment of epilepsy. Gamma-aminobutyric acid (GABA)-loaded and unloaded polymeric matrices were successfully implanted in the substantia nigra of the rat brain to evaluate its anti-epileptic potential following electrical stimulation. On the second day following implantation, rats implanted with GABA-loaded matrices experienced only focal limbic seizures and were devoid of major generalized seizures of whereas rats implanted with unloaded matrices experienced generalized seizures. However, GABA levels did drop a week after implantation and thus the anti-epileptic effects of the implant diminished (Kokaia et al., 1994). Kubek and co-workers (1998) successfully developed a polyanhydride-based micro-disk implant containing thyrotropin-releasing hormone (TRH) intended for the treatment of epilepsy. TRH is not a traditional AED but is an endogenous hormone known to have anticonvulsant effects in animal seizure models and intractable epileptic patients. Its use however, is limited by its minimal site-specific bioavailability and short duration of action. The micro-disk was sterotactically implanted in the amygdala of the rat kindling model of temporal lobe epilepsy. The implant was successful in extensively suppressing the kindling expression. In addition, anticonvulsive effects were still observed fifty days after implantation thus verifying that the device is a potential long-term epilepsy treatment modality (Kubek et al., 1998).

2.1.5.2. Ethylene-Vinyl Acetate Copolymer (EVAC) Implants Containing Phenytoin

Phenytoin is a commonly used AED that has been investigated for use intracerebrally. EVAc implants containing phenytoin have been implanted in the cortex of the rat model. Seizures, induced by the delivery of cobalt chloride to the cortex, were significantly reduced in rats receiving phenytoin from the implant when compared to a control group (Tamargo et al., 2002). In another study, phenytoin was loaded in a titanium dioxide (TiO_2) reservoir and implanted in the temporal lobe of Wistar rats. The device was deemed biocompatible and safe as no histological damage was observed. The study, however, did not assess the extent of anticonvulsive effects produced by the device (Lopez et al., 2007).

In a related study, a nanostructured sol-gel TiO_2 reservoir was implanted into the amygdala of kindled rats. The reservoir contained valproic acid and twelve months after implantation, it remained undamaged and showed no evidence of injury to surrounding neuronal tissue (Lopez et al., 2007). The use of intracerebral AED delivery in general is highly beneficial considering the spontaneous onset of seizures and the intolerable systemic side-effects of AEDs. Prolonged therapy is ideal as issues of AED non-compliance are abolished and epileptic patients are kept in maintenance and safeguarded from the consequences of seizures in general.

2.1.5.3. Stem Cells as Direct Transplantation Tools for Epilepsy Therapy

Recently, stem cells, stem cell-derived neural precursors, neuronal cell lines and fetal hippocampal neurons have received much attention as direct transplantation tools for epilepsy therapy (Boison, 2007; Raedt et al., 2007; Shetty and Hattiangady, 2007; Loscher et al., 2008; Suter and Krause, 2008; Boison, 2009). It has been shown that with the use of stem cells, therapeutic strategies are possible and may synergistically augment each other, **as depicted in** Figure 8.

The first strategy is based on reconstitution of hippocampal circuitry by intrahippocampal transplantation of fetal hippocampal neurons. In an elegant set of experiments Shetty and co-workers (2007) have demonstrated that this approach is indeed feasible and can provide therapeutic benefit in post status epilepticus models of epilepsy. Noteworthy is that, fetal hippocampal cell grafts exhibited robust long-term survival (> six months) and integration in animal models of epilepsy (Rao et al., 2006). While functional integration of graft-derived cells is needed for these restorative approaches, stem cell- or neural progenitor-derived brain-implants can be engineered to release therapeutically active molecules with the aim to provide therapeutic benefit by paracrine mechanisms. Thus, a combination of functional integration and paracrine drug delivery may provide synergistic benefits.

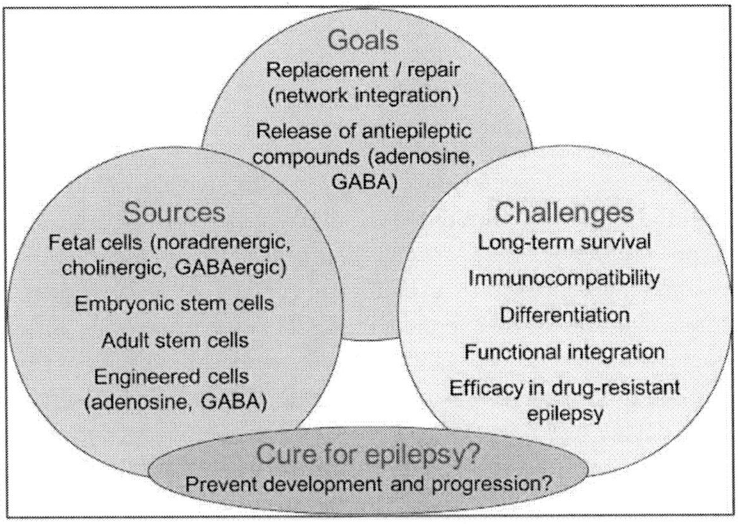

Figure 8. Goals and challenges of stem cell therapies for epilepsy.

2.1.6. Application of Polymer-Based Implantable Devices in Brain Tumors and Carcinogenesis

2.1.6.1. Gliadel® Technology for the Localized Treatment of Glioblastoma Multiforme

Gliadel® is an innovative technology, approved by the United States Food and Drug Authority (FDA) and other regulatory agencies worldwide, for the localized treatment of glioblastoma multiforme. Despite aggressive therapy with surgery, chemotherapy and radiation, the prognosis for primary malignant gliomas is poor, as the average survival time is below a year. Furthermore, recurrence of tumors is a common problem. The Gliadel®

polymeric device is a PCPP-SA copolymer blend and houses the drug carmustine (BCNU) (Figure 9). Up to eight Gliadel® wafers may be utilized following removal of the tumor and results appear to be promising. Pre-clinical testing demonstrated a localized inflammatory reaction which was not linked to any neurological deficits or pathological changes in the relevant animal model. The localized inflammation was transient and resolved within a period of five weeks.

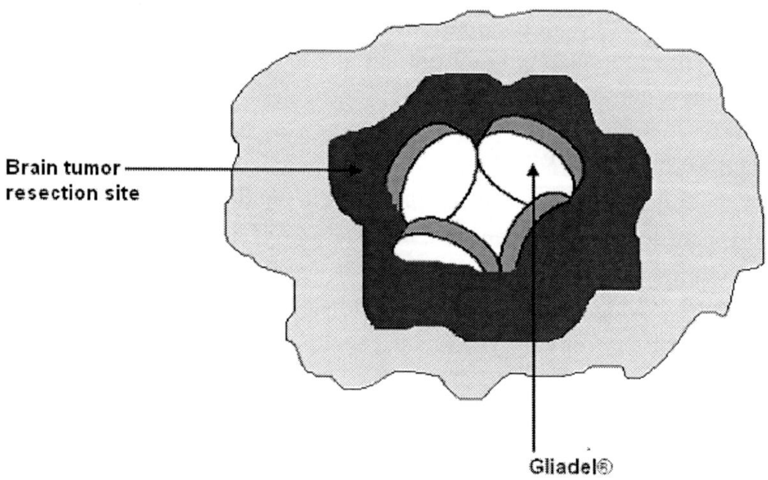

Figure 9. Implantation of Gliadel® at tumor resection site [Adapted from Jain et al., 2005].

Biodistribution studies indicated that the device exhibited desirable drug concentrations at the site of action for prolonged periods. Metabolic distribution and elimination studies were also performed on the device and results confirmed the biodegradable nature of the copolymer blend and its potential to be used for the specific delivery of drug to certain target organs. Furthermore, the device is excreted from the body principally through the kidneys whereby the water-soluble BCNU and sebacic acid are rapidly removed and the carboxyphenoxypropane is gradually excreted due to its insoluble nature. Toxicity studies established that the device was non-toxic and biocompatible. Lastly, efficacy testing revealed a 5.4-7.3-fold increase in median survival as compared to a 2.4-fold attained by conventional systemic therapy. Clinical trials confirmed its safety and efficacy and hence it was granted FDA approval in 1996 (Domb et al., 1994 and 1995; Brem et al., 1995; Guerin et al., 2004; Jain et al., 2005).

2.1.6.2. Amsacrine-Loaded Polymeric Device for Interstitial Malignant Glioma Chemotherapy

An amsacrine-loaded polymeric rod is an additional matrix type implantable delivery system investigated for use in interstitial malignant glioma chemotherapy. This is an amsacrine-loaded EVAc rod synthesized by an extrusion method. *In vitro* release studies displayed a burst release of amsacrine followed by a stable sustained linear release for a minimum of six months. Further *in vitro* studies utilizing rat glioma (RG2) in cell culture depicted that the device was effective in killing these cells in a dose-dependent manner. *In vivo* studies employing a tumor-implanted rat model demonstrated potent anti-tumor activity as well as an increased survival when compared to control models. In addition, the device proved to be biocompatible as no evidence of toxicity was observed (Wahlberg et al., 1996).

Combination therapy with a biomodulator drug, radiosensitizer and external beam radiotherapy (EBR) has been investigated for the treatment of intracranial glioma in the rat model. Tumors, induced by sterotaxic implantation of cultured C6 cells into the rat brain, were treated with a polyanhydride-based, bis(*p*-carboxyphenoxy)propane-sebacic acid (CPP-SA), implantable polymeric rod. The C6 cell line is a clonal line derived from a rat glioblastoma induced by administration of *N*-nitrosomethylurea in Winstar rats. Once implanted, the C6 cells produce intracerebral tumors that closely resemble the traits of spontaneous gliomas. Bromodeoxyuridine (BrdUrd), a radiosensitizing agent and *N*-(phosphonacetyl)-L-aspartic acid (PALA), a biomodulator drug were incorporated into the polyanhydride-based implantable rod. Implantation of the BrdUrd-PALA-loaded polymeric rod was at the site of C6 cell implantation twelve days following tumor implantation. EBR treatment was given as 10-MeV X-rays following anesthesia. The efficacy of the combination therapy was evaluated on the basis of survival of the rats and the "triple therapy" was shown to be highly effective with clinical potential (Li et al., 2004). Regardless of the promising results of the study, treatment involves two procedures namely; implantation of the polymeric rod followed by anesthesia and EBR. This can be highly technical and may prove to be impractical.

2.1.6.3. Drug-Loaded PLGA Microspheres as an Effective Anti-Cancer Agent against Malignant Glioma

The use of carboplatin as an effective anti-cancer agent against malignant glioma is limited due to the significantly high systemic dose required to achieve therapeutic levels at the tumor. This is primarily due to the inability of the drug to traverse the BBB. Chen and co-workers (1997) successfully synthesized carboplatin-loaded PLGA microspheres that were delivered intracerebrally in the rat model. There were no signs of neurotoxicity or systemic toxicity associated with the microspheres although implantation did cause a transient, localized inflammatory reaction that was well tolerated. *In vivo* release studies demonstrated apparent zero-order kinetics over a period of one month thus confirming a novel method of prolonged effective carboplatin delivery for the treatment of malignant glioma (Chen et al., 1997).

In another study, Gavin and co-workers (2005) ascertained that Poly(D,L-Lactide) microspheres carboplatin can be easily prepared by the emulsification-spray drying technique. The microspheres obtained (Figure 10) shown a well controlled delivery of carboplatin *in vitro*. Both drug-free and carboplatin-containing microspheres displayed skin biocompatibility. Furthermore, they didn't induce any local toxic reaction (Gavin et al., 2005). These results indicate that biodegradable microspheres may have a potential application as an efficient system for the local delivery of carboplatin in supporting surgical resection of tumors.

A pilot study by Menei and co-workers (1999) demonstrated the use of biodegradable lactic and glycolic acid copolymer microspheres for the local and sustained delivery of 5-Fluorouracil (5-FU) for the radiosensitization of glioblastoma following surgical resection. The study included eight patients with newly diagnosed glioblastoma. Following resection, patients were implanted with 5-FU microspheres in the wall of the surgical bed. EBR was given to patients approximately seven days post implantation.

Figure 10. SEM depicting spray-dried Poly (D,L-Lactide) microspheres containing carboplatin for post surgical chemotherapy against malignant glioma (magnification 575x362), [Source: Gavin et al., 2005]. Spray-Dried Poly(D,L-Lactide) Microspheres Containing Carboplatin for pt060117_figure1.jpg 575 × 362 - Spray-Dried Poly(D,L-Lactide) Microspheres Containing Carboplatin for.

Results depicted sustained concentrations of 5-FU in the CSF and relatively low concentrations in the blood. Furthermore, limited systemic toxicities were observed and the advent of using implantable microspheres for the treatment of glioblastoma was very promising (Menei et al., 1999).

A Phase 1 clinical study by Menei and co-workers (2004) involved implantation of the microspheres by sterotaxy, into deeply situated and inoperable tumors. The microspheres were generally well tolerated except in four of ten patients, who experienced worsening of pre-existing symptoms. There were no signs of edema or hematological complications and the system was therefore deemed efficient for the treatment of certain inoperable and deep brain tumors (Menei et al., 2004). Roullin and co-workers (2004) determined the effectiveness radiosensitizing 5-FU-loaded PLGA microspheres implanted sterotactically in Sprague-Dawley rats bearing C6 tumors. The study proved that the controlled release of 5-FU from these PLGA microspheres is a highly effective approach for radiosensitizing gliomas (Roullin et al., 2004). A related study utilizing NGF-loaded PLGA microspheres focused on the intraseptal implantation of the microspheres in the region of axotomized septal cholinergic neurons. To evaluate the microspheres *in vivo* activity, both NGF-loaded and unloaded PLGA microspheres were sterotactically implanted in the rat brain. A third group of non-treated axotomized rats were also evaluated. Both NGF-loaded and unloaded microspheres displayed an identical non-specific glial cell reaction post implantation. Furthermore, the microspheres were biocompatible as there was no evidence of neuronal toxicity. The survival of axotomized cholinergic neurons was significantly enhanced following NGF-loaded microsphere implantation whereas unloaded microspheres offered no neuroprotection. Non-treated rats displayed a noticeable loss of cholinergic neurons on the lesion side when compared to the contra-lateral side. The study thus confirmed the neuroprotective capabilities of NGF and an effective implantable polymeric carrier of the neurotrophic factor in PLGA (Pèan et al., 2000).

The use of implantable microspheres in general can be enhanced and optimized in the formulation process. Administration of microspheres which are not matricized, leads to the delivery of a finite quantity of active delivered all at once. Incorporating the microspheres in a biocompatible matrix further controls the release of active and release rates ranging from months to years can be achieved. This is beneficial for patients suffering from neurodegenerative diseases and psychiatric conditions where prolonged therapy is required. Furthermore, microspheres are protected from degradation and it is theoretically easier to achieve delivery at a desired region of the brain and at the very site of implantation as microspheres are not allowed to disperse to multiple regions of the brain.

2.1.6.4. Paclitaxel-Loaded PLGA Discs for the Treatment of Malignant Glioma

Biodegradable microfiber discs and sheets are additional technologies investigated for the treatment of malignant glioma *in vitro* and *in vivo*. The PLGA discs and sheets containing paclitaxel were produced by a specialized electrospinning process. The fibrous nature of the discs and sheets are advantageous to drug release in that it displays a greater surface area to volume ratio. *In vitro* release studies depicted a sustained release of paclitaxel for over eighty days after an initial minor burst release. *In vitro* cell apoptosis testing revealed that the PLGA discs/sheets were superior to conventional paclitaxel delivery in terms of cytotoxic effects and drug sustainability. *In vivo* studies involved the implantation of C6 glioma cells subcutaneously in the BALB/c nude male mice model followed by implantation of the polymeric matrices after twelve days. Mice treated with the PLGA discs/sheets displayed significant tumor inhibition when compared to controls and mice treated with conventional paclitaxel therapy (Ranganath and Wang, 2008). Earlier related studies by Lee and co-workers (2005) involved the synthesis of a BCNU-loaded PLGA wafer for 9L gliosarcoma chemotherapy. Wafers were produce by vortex-mixing the BCNU and PLGA and then compressing using a Carver Hydraulic Press. The anti-tumor effects of the wafers were investigated *in vitro* against 9L gliosarcoma cells and *in vivo* against a subcutaneous solid tumor model of 9L gliosarcoma. The wafers inhibited the growth of the tumors significantly when compared to BCNU powder and increasing the dose of BCNU within the PLGA wafer resulted in an increased release rate of BCNU as well as significant regression of tumor (Lee et al., 2005). The major shortfall with these studies is that the *in vivo* models used possessed subcutaneous-induced tumors. Whilst desirable tumor inhibition was observed, this does not indicate directly, the effects on surrounding brain tissue or brain function as the environments differ. Furthermore, it is difficult to make the assumption that the same behavior will be observed intracranially.

2.1.6.5. Therapeutic Efficacy Study in an Intracranial Human Glioblastoma Xenograft Model Using Paclitaxel-Loaded Polylactide-Co-Glycolide (PLGA) Nanofiber Discs

In one of the most recently conducted studies, Ranganath and co-workers (2010) managed to evaluate and demonstrate the therapeutic efficacy of local chemotherapy of malignant glioblastoma using 9.1% (w/w) paclitaxel-loaded polylactide-co-glycolide (PLGA) nanofiber discs (F1) and 16.7% (w/w) paclitaxel-loaded PLGA microspheres entrapped in hydrogel matrix (F3), through bioluminescence luciferase imaging. 9.1% (w/w) paclitaxel-loaded polylactide-co-glycolide (PLGA) nanofiber discs (F1) were fabricated and characterized for morphology and size distribution. Along with F1, three other formulations,

9.1% (w/w) paclitaxel-loaded PLGA submicron-fiber discs (F2), 16.7% (w/w) paclitaxel-loaded PLGA microspheres entrapped in hydrogel matrices (F3 and F4) were intracranially implanted in BALB/c mice and the coronal brain sections were analyzed for bio-distribution of paclitaxel on 14, 28 and 42 days post-implantation. BALB/c nude mice with intracranial human glioblastoma (U87 MG-luc2) were used in this study. Animals were randomized to intracranial implantation of F1 and F3 with paclitaxel dose of 10 mg/kg, placebo F1, placebo F3, weekly intratumoral injection of Taxol® (10 mg/kg) or no treatment and the treatment response was analyzed by bioluminescence imaging and histological (HandE, Ki-67) examinations. Enhanced, therapeutic paclitaxel penetration (~1 μM) in the mouse brain up to 5 mm from the implant site even after 42 days post-implantation from F1 and F3 was confirmed and deduced to be diffusion/elimination controlled (Ranganath et al., 2010). Figure 10 shows an increase in bioluminescence signal intensity (tumor progression) in animals with no treatment (sham) and placebo implants over the treatment period (Figure 10a, b and c), where as a decrease (tumor inhibition) is observed in animals treated with Taxol®, F1 and F3 (Figure 11d, e and f).

Histological examination (HandE staining) of U87 MG xenograft after 34 days of treatment showed tumor cell invasion at the tumor–normal brain parenchyma interface in sham and placebo groups as seen in Figure 12a, b and c. Notably, F1 and F3 (Figure 12e and f) had fewer tumor cells. Images of Ki-67 staining showed higher degree of tumor proliferation in sham and placebo groups (Figure 12a, b and c) in comparison to Taxol®, F1 and F3 groups (Figure 12d, e and f) indicating treatment success for the later.

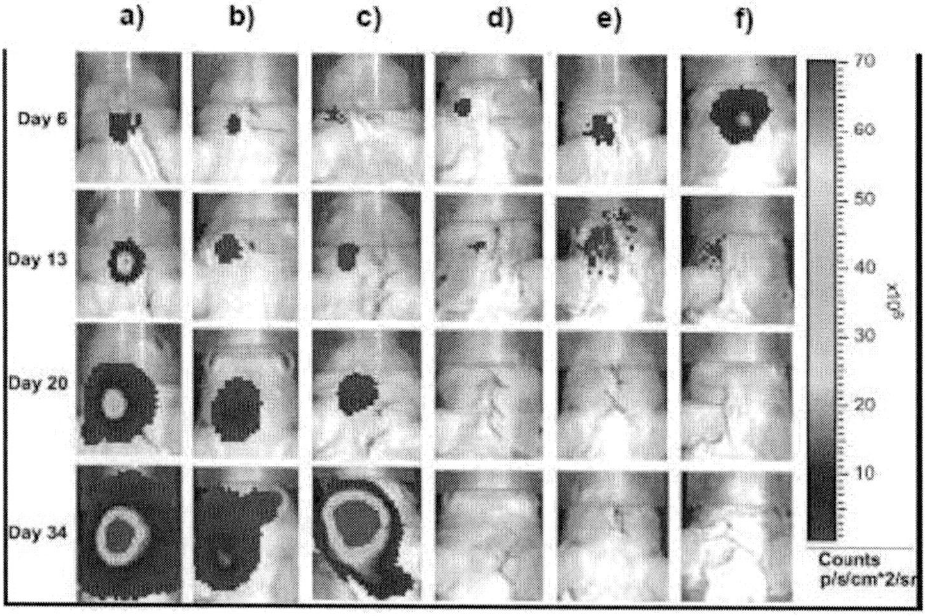

Figure 11. Representative BLI images of intracranial U87 MG-lu2 xenograft in mice from sham, placebo and experimental groups over times as follows: a) sham control; b) placebo 1; c) placebo 3; d) Taxol® control; e) F 1 group; and f) F 3 group. Luciferase expressing tumor cells are displayed in the brain with a red-blue color bar; red color indicating the highest bioluminescence intensity. [Source: Ranganath et al., 2010].

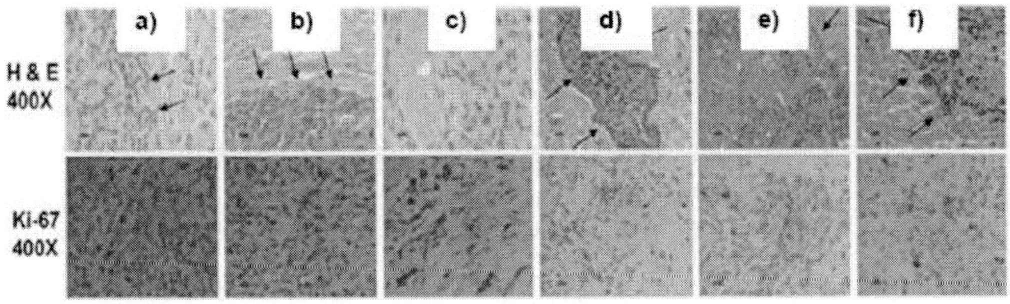

Figure 12. Histological (HandE) examination (top row) and immunostaining of Ki-67 antigen (bottom row) of U87 MG-luc2 xenograft after 34 days of treatment with various groups in mice. Arrows indicate tumor–normal brain parenchyma interface, [Source: Ranganath et al., 2010].

Overall, F1 and F3 (Figure 11e and f) demonstrated significant (≈30 fold) tumor inhibition and significantly low tumor proliferation index after 41 days of treatment in comparison to sham and placebo controls (Figure 11a, b and c). This was an indication that submicron/nanoscale implants were able to demonstrate optimal paclitaxel pharmacokinetics in the brain/tumor with significant tumor inhibition in a glioblastoma xenograft model in mice and hence may be potentially useful in the treatment of highly recurrent malignant glioblastoma.

2.1.7. Application of Polymer-Based Implantable Devices in Cerebral Edema

The use of a dexamethasone-loaded EVAc matrix system as an intracranial or intraperitoneal implant for the reduction of cerebral edema associated with brain tumors has been well studied (Reinhard et al., 1991; Park et al., 2003; Tamilvanan et al., 2008; Saraon et al., 2010). Steroids have been shown to be effective in the treatment of cerebral edema however; their use is limited due to their systemic side-effects. Owing to this, this study has investigated an alternate method of steroid delivery to the brain which involved the local or site-specific delivery of dexamethasone using a polymeric carrier. The Fisher 344 rat model was used and tissue concentration post implantation was quantified by high performance liquid chromatography and selective extraction. Dexamethasone was present in the brain in significant concentrations for up to twenty one days post intracranial implantation. Further research suggested that it was possible to attain drug release for up to several months. The study confirmed that intracranial implantation of dexamethasone-loaded polymeric implants are just as effective as high-dose systemic dexamethasone when treating peritumoral edema and has significant benefits over conventional dosage forms (Reinhard et al., 1991).

2.2. Subcutaneous and Subdermal Implants

Subcutaneous implants have the potential to be an effective means of treating certain psychoses such as schizophrenia (Siegel et al., 2002; Rabin et al., 2008). Surgically implantable haloperidol-loaded biodegradable PLGA pellets have demonstrated stable drug release patterns over a period of five months *in vitro* thereby superseding conventional depot formulations as a more effective means of treating psychotic episodes over prolonged periods of time. Furthermore, *in vivo* results in the rat model depicted an increase in striatal dopamine

(D2) receptor expression thus validating the effectiveness of the drug delivery system (Siegel et al., 2002). A similar study by Rabin and co-workers (2008) utilizing risperidone-loaded PLGA implantable pellets and rods describes drug release patterns for a minimum of two months confirming the value of such a system for the long-term treatment of schizophrenia. Should a patient experience intolerable side-effects, the implants can be immediately removed thereby conferring upon the systems, a degree of reversibility not offered by conventional dosage forms. Furthermore, the biodegradability of the system ensures that no additional surgery will be required to remove the device. Other applications of subcutaneous implants relate to the use of non-biodegradable but biocompatible polymer matrices for the treatment of cancer pain and opioid dependence (US Patent 6126956). Included in this, is the incorporation of hydromorphone within a polyethylene-vinyl acetate matrix which was subsequently coated with methyl methacrylate. The device offers the controlled release of hydromorphone with pseudo-zero order kinetics for prolonged periods thereby providing adequate analgesia for cancer pain and significantly controlling opiate addiction. Hydromorphone levels are kept within therapeutic range, preventing the potentially lethal consequences of fluctuating drug levels. In addition, the device does not cause an immune response and cannot be abused. Gold and co-workers (1993) demonstrated the use of implantable morphine pellets in the rat model to effectively determine the time course for the progression of tolerance, dependence and abstinence (Gold et al., 1993). Related studies display the use of buprenorphine (Probuphine®) (Figure 13) and naltrexone implantable devices for opioid dependence (Iyer et al., 2007; Montoya and Vocci, 2008; Reece, 2009; White et al., 2009).

Naltrexone, an opioid receptor antagonist, proved to be a desirable option when formulating the subcutaneous implant owing to the fact that non-compliance is a major impediment to treatment and furthermore, the drug is marketed as an oral formulation. Probuphine®, intended for the treatment of heroin addiction, comprises the opioid receptor partial agonist buprenorphine within an EVAc matrix in the form of match-stick sized formulations implanted in the subdermal region with the potential to release drug over a period of six months.

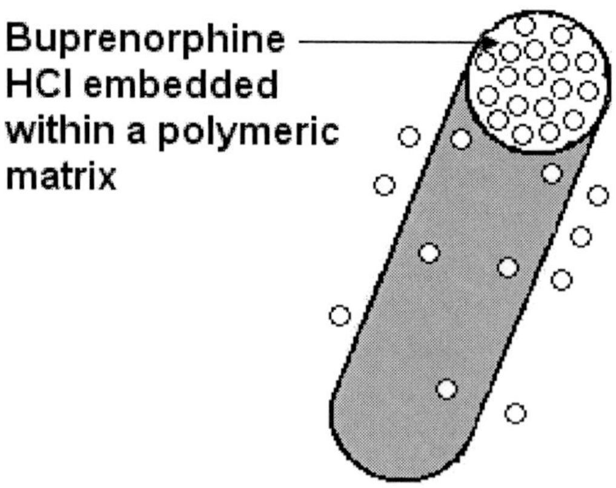

Figure 13. Probuphine® comprising buprenorphine HCl embedded within an EVAc matrix for the potential prolonged treatment of opioid addiction modified from http://www.titanpharm.com/products-candidates-probuphine.php [Accessed November 21, 2010].

This sustained release formulation elicits therapeutic dominance over the conventional daily or every-other day sublingual formulations (White et al., 2009). The non-biodegradable nature of the EVAc may prove disadvantageous however the use of biodegradable polymers may yield toxic metabolites upon degradation thereby rendering the product hazardous. The use of subcutaneous and subdermal implantable devices in general offers a host of benefits when compared to conventional dosage forms. Among these include, a reduced hepatic metabolism and peripheral drug inactivation resulting in a greater bioavailability of the drug when compared to oral dosage forms. Furthermore, drug release over an extensive period provides remission and stability to the patient for prolonged periods when compared to depot formulations. Despite the advantages, systemic toxicities are not evaded as seen with the majority of intracerebral implants.

2.3. Intrathecal Implants

The foundations of intrathecal drug delivery (ITDD) are based on the fact that therapeutic efficacy of strong analgesics can be achieved by the action of the drugs at the dorsal horn of the spinal cord, without the need to administer potentially toxic doses of the drug systemically. Conventional types of ITDD devices include: the percutaneous catheter (tunneled or not tunneled) used with an external pump; the totally implanted catheter with a subcutaneous injection port connected to an external pump; fully implanted fixed rate intrathecal drug delivery systems and fully implanted programmable intrathecal drug delivery systems. Due to cost issues, external pumps are preferred to implantable systems for the management of cancer pain (The British Pain Society, 2008). A novel approach by Jeon and co-workers (2006) investigated the possibility of intrathecally implanting micro-encapsulated xenogenic adrenal medullary chromaffin cells (ACCs) for the purpose of chronic neuropathic pain relief. Neuropathic pain can be a result of injury or dysfunction of the CNS and is generally unresponsive to conventional analgesic or to interventional therapies. Furthermore, it is a major medical and social burden with considerable cost implications. Owing to the fact that the ACCs produce endogenous opioids and catecholamines which act as effective analgesics in the CNS, their antiallodynic effects on chronic neuropathic pain induced by chronic constriction injury (CCI) of the sciatic nerve in the rat model were investigated. ACCs, obtained from bovine adrenal glands, were encapsulated with alginate and poly-L-lysine, rendering them undetectable by the immune system of the body. The major advantage of micro-encapsulation is the fact that high dosage immunosuppressant drugs are not required for ACCs implantation. High dose immunosuppressants are associated with severe systemic side-effects and susceptibility to opportunistic infections. ACCs-loaded microcapsules were produced by first suspending the cells in a sodium alginate and sodium chloride mixture and dropping this suspension into a calcium chloride solution thereby forming ACCs-loaded microcapsules. The microcapsules were coated in poly-L-lysine and finally re-suspended in a sodium alginate solution to form the outer layer of the capsule membrane. Implantation of approximately five hundred ACCs-loaded and unloaded microcapsules occurred a week after CCI in two randomly allocated rat groups. Briefly, implantation involved a L5 posterior laminectomy and an incision in the dura mater. The outcomes of the study suggest that CSF levels of endogenous opioids and catecholamines were significantly higher than controls and the intrathecal implantation of encapsulated ACCs resulted in a decreased mechanical and cold allodynia in a neuropathic pain model (Jeon et al., 2006). The use of microcapsules can also be further optimized to achieve prolonged desired release rates as discussed with microspheres in section 2.1.6. Considering the financial burden associated with treatment of chronic neuropathic pain, this approach would be particularly valuable.

3. Transdermal Patches

Transdermal drug delivery systems (TDDS) are of great benefit when a patient experiences intolerable gastro-intestinal side-effects as well as dysphagia when using conventional oral dosage forms. Furthermore, it may be of particular importance to patients who should not self-medicate with their analgesia and those with cognitive impairment. The skin presents a large surface area for absorption of drug and it is a non-invasive procedure. In addition, first-pass metabolism is avoided and predictable extended duration of activities may be achieved (Sugibayashi and Morimoto, 1994; Naik et al., 2000). Theoretically, TDDS operate via a simplistic mechanism whereby relatively high concentrations of a particular drug are loaded within a patch which is applied to the skin. The high concentration of drug in the patch relative to the lower blood concentration facilitates diffusion of the drug into the blood through the skin maintaining a constant drug concentration in the blood (Chopda, 2006) (Figure 14).

TDDS systems bearing nicotinic agonists, nicotine in particular have been vastly used as an adjunct to smoking cessation and in the treatment of nicotine withdrawal (Balfour and Fagerström, 1996). Furthermore, these agents have been found to treat and alleviate the symptoms of cognitive impairment associated with neurodegenerative diseases, psychoses such as schizophrenia and attention disorders (Decina et al., 1990; Sahakian and Jones., 1991; Jones et al., 1992; Levin, 1992; Adler et al., 1993; Wilson et al., 1995; Conners et al., 1996; Levin and Simon, 1998; Gorell et al., 1999; White and Levin, 1999; Rusted et al., 2000; Rezvani and Levin, 2001; Quik, 2004; Singh et al., 2007). It is believed that the neuroprotective properties of nicotine are a result of stimulation of the α7 nicotinic receptors. Stimulation of these receptors allow for an increased release of dopamine from the striatum thereby conferring protection upon the neurons of the nigrostriatal pathway (Rezvani and Levin, 2001; Singh et al., 2007). This proves to be beneficial when considering the neurodegenerative diseases namely PD and AD. Through unknown mechanisms, transdermal nicotine can improve the symptoms of ADHD.

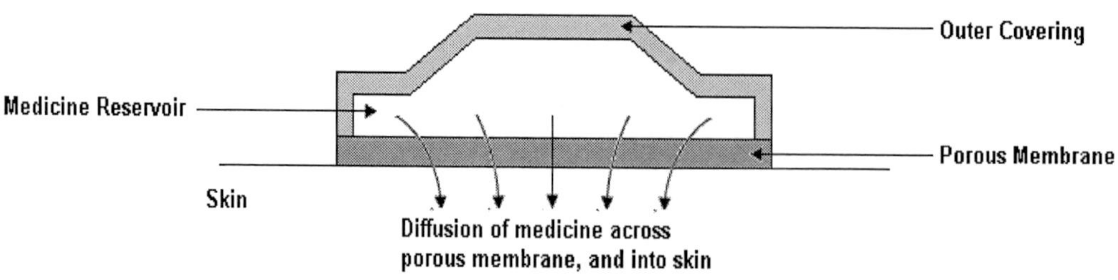

Figure 14. Simplistic representation of the mechanism of action of a typical TDDS, [Adapted from Shah, 2008].

Drugs such as methylphenidate and amphetamine are conventional therapeutics used for ADHD and work by increasing dopamine stimulation. Considering the fact that nicotine causes increased dopamine release, the two are closely related which may be the reason for nicotine's beneficial traits when treating ADHD (Rezvani and Levin, 2001). A similar study by Arnold and co-workers (2007) establishes the use of a novel methylphenidate patch as an effective means of treating ADHD over a prolonged period of time. In this system, the matrix serves as the skin adhesive as well as the reservoir. By manipulating patch size and duration of patch adhesion to the body, it is possible to calculate the optimum dose required for the individual patient (Arnold et al., 2007). A study conducted by Sittl and co-workers (2002) describes the use of a buprenorphine transdermal system as an effective means of treating severe chronic pain associated with cancer and other disorders, without needing to use oral analgesic preparations. In addition, transdermal buprenorphine was associated with an improved duration of sleep. Central and gastro-intestinal side-effects experienced were mild to moderate in nature and were probably associated with the initiation of opioid therapy (Sittl et al., 2002). A novel transdermal system containing the antidepressant drug selegiline, has revealed significant inhibition of central monoamine oxidase A (MAO-A) and B (MAO-B) enzymes. Although the overall antidepressant effects of the system are relatively modest, this system possesses the advantage of bypassing liver and intestinal monoamine oxidase enzymes (Amsterdam, 2003). Commercially available TDDS affecting the CNS include; NeuproR®, a rotigotine-loaded TDDS used to alleviate the symptoms of early-stage idiopathic PD, Nicotinell®, Nicodern® and Prostep® containing nicotine for smoking cessation, Matrifen® and Duragesic® containing the opioid fentanyl for the treatment of moderate to severe pain and Transderm-Scop® housing the drug scopolamine for the treatment of motion sickness. Patented TDDS include a transdermal buprenorphine device for the treatment of pain associated with sickle cell disease (Shah, 2008). Despite the advantages of TDDS, a major shortcoming of the technology is the fact that systemic side-effects are not completely prevented and may still affect patient compliance and disease prognosis especially with disorders of the CNS (Henningfield et al., 2005). Furthermore, the device may not be suitable when using drugs which require rapid titration. A minor disadvantage of the system could possibly be that the patient has complete control and access with regards to removal of the device at any time due to occurrences such as local irritation at the site. Although trivial, such spontaneous actions may negatively affect disease prognosis especially when addressing certain CNS disorders.

4. Advantages and Disadvantages of Polymeric-Controlled Release and Convection-Enhanced Drug Delivery

Local delivery to brain tumors can be accomplished using two strategies namely polymeric-controlled release and convection- enhanced delivery (CED). Each technique strives to address the need for controllable intracranial drug delivery. The two technologies offer unique benefits and suffer distinctive limitations, as listed in Table 1.

The principle disadvantage to controlled release is the restricted drug distribution. The limitations imposed by diffusion within the brain limit drug penetration to a region smaller than the typical invasive area of malignant gliomas.

Table 1. Advantages and disadvantages of polymeric-controlled release and convection-enhanced drug delivery [Adapted from Sawyer et al., 2006]

	Advantages	Disadvantages
Polymeric-controlled release	Sustained drug release Define release kinetics Tunable release properties Low invasiveness Low peak drug release limits tissue damage Biocompatible Localized delivery	Poor drug penetration Drug dosage limited by implant size
Convection-enhanced delivery (CED)	Large drug distribution volume Flexible therapy protocol Consistent drug concentration	Invasive Long infusion times Potential high Delivery intracranial pressures Unpredictable drug distribution

CED's strength lies in the potential for large distribution volume of infused drugs. The long infusion times and unpredictable distribution have also caused an abundance of side effects in clinical trials (Sawyer et al., 2006). Trials using paclitaxel have noticed a high rate of drug related adverse events that include wound dehiscence, inflammation, and edema (Mardor et al., 2005; Popperl et al., 2005). These side effects are attributed to drug backflow along the catheter, and drug localization in the perivascular and subarachnoid spaces. Infusing an encapsulated drug may decrease the incidence of these adverse events.

Conclusions

CNS disorders are relatively common and are generally sub-therapeutically treated. There is thus a great need for prolonged, controlled psychotropic drug release which allows for an extended duration of action, increased patient compliance and a decrease in rates of relapse and care-giver burden. This is, more often than not, unachievable by conventional therapies. Implantable polymeric and transdermal drug delivery systems have been shown to be capable of curbing the difficulties associated with conventional therapies. Despite some of the disadvantages of these systems, the insurmountable benefits outweigh the risks and there is thus potential for polymer-based implantable and transdermal devices being used as first-line therapy when managing CNS disorders.

REFERENCES

Abbot N.J. Physiology of the blood–brain barrier and its consequences for drug transport to the brain, *International Congress Series* 1277, (2005), 3-18.

Addington, J., Piskulic D. and Marshall C. Psychosocial Treatments for Schizophrenia Current Directions in Psychological Science, 19(4) (2O10) 260-263.

Adler L.E., Hoffer L.D., Wiser A. and Freedman R. Normalization of auditory physiology by cigarette smoking in schizophrenic patients, *Am .J. Psychiatry,* **150,** (1993), 1856–1861.

Altman G.H., Diaz F., Jakuba C., Calabro T., Horan RL., Chen J., Lu H., Richmond J., Kaplan D.L. Silk-based biomaterials. *Biomaterials,* 24, (2003), 401–416.

Amsterdam J.D. A double-blind, placebo-controlled trial of the safety and efficacy of selegiline transdermal system without dietary restrictions in patients with major depressive disorder, *Journal of clinical psychiatry*, 64, (2003), 208-214.

Antinori A, Cingolani A, Giancola ML, Forbici F, De Luca A, Perno CF. Clinical implications of HIV-1 drug resistance in the neurological compartment, *Scand J. Infect Dis Suppl,* **106,** (2003), 41–44.

Aoi M., Date I., Tomita S. and Ohmoto T. The effect of intrastriatal single injection of GDNF on the nigrostriatal dopaminergic system in hemiparkinsonian rats: behavioral and histological studies using two different dosages, *Neurosci Res.,* 36(4), (2000), 319–325.

Arnold L.E., Lindsay R.L., Lopez F.A., Jacob S.E., Biederman J., Findling R.L. and Ramadan Y. Treating attention-deficit/hyperactivity disorder with a stimulant transdermal patch: the clinical art, *Pediatrics,* 120(5), (2007), 1100-1106.

Awad A.G. and Voruganti L.N.P. The Burden of Schizophrenia on Caregivers: A Review, *PharmacoEconomics*, 26(2), (2008), 149-162.

Balfour D.J.K. and Fagerström K.O. Pharmacology of nicotine and its therapeutic use in smoking cessation and neurodegenerative disorders, *Pharmacology and Therapeutics*, 72(1), (1996), 51-81.

Banks W.A. Characteristics of compounds that cross the blood-brain barrier. *BMC Neurology*, 9(Suppl 1), (2009), S3.

Begley D.J. Delivery of therapeutic agents to the central nervous system: the problems and the possibilities. *Pharmacol. Ther,* **104,** (2004), 29–45.

Benoita J. Faisanta N. and Venier-Juliennea M. Development of microspheres for neurological disorders: From basics to clinical applications. *Journal of Controlled Release,* 65, (2000), 285–296.

Birnbaum M. and Sharif Z. Medication adherence in schizophrenia: patient perspectives and the clinical utility of paliperidone ER. *Patient Prefer Adherence*, 2, (2008), 233–240.

Boison D., Scheurer L., Tseng J.L., Aebischer P., Mohler H. Seizure suppression in kindled rats by intraventricular grafting of an adenosine releasing synthetic polymer. *Exp Neurol.,* 160, (1999), 164–174.

Boison D. Cell and gene therapies for refractory epilepsy. *Current Neuropharmacology,* 5, (2007), 115–125.

Boison D. Adenosine augmentation therapies (AATs) for epilepsy: prospect of cell and gene therapies. *Epilepsy Res.,* 85(2-3), (2009), 131–141.

Breen R. and Thornhill J.T. Noncompliance with Medication for Psychiatric Disorders: Reasons and Remedies, *CNS Drugs*, 9(6), (1998), 457-471.

Brem H., Piantadosi S., Burger P.C., Walker M., Selker R., Vick N.A., Black K., Sisti M., Brem S., Mohr G., Muller P., Morawetz R. and Schold S.C. Placebo-controlled trial of safety and efficacy of intraoperative controlled delivery by biodegradable polymers of chemotherapy for recurrent gliomas, *Lancet*, 345, (1995), 1008–1012.

Bryan J. Crossing the blood-brain barrier: drug delivery to the brain is still elusive, *The pharmaceutical journal*, 273, (2004), 475-476.

Bulloch A.G.M. and Pattern S.B. Non-adherence with psychotropic medications in the general population. *Soc Psychiat Epidemiol.*, 45(1), (2009), 47-56.

Centorrino F., Hernán M.A., Drago-Ferrante G, Rendall M., Apicella A., Längar G., and Baldessarini R.J. Factors Associated With Noncompliance With Psychiatric Outpatient Visits, *Psychiatr Serv*, 52, (2001), 378-380.

Chan Young Park C.Y., Park J.W and Lee J.J. Development of totally implantable pulsatile biventricular assist device. *Artificial Organs* 27(1), (2003),119–123.

Chen W., He J., Olson J.J. and Lu D.R. Carboplatin-loaded PLGA microspheres for intracerebral implantation: in vivo characterization, *Drug delivery*, 4(4), (1997), 301-311.

Cheng H., Fraidakis M., Blomback B., Lapchak P., Hoffer B. and Olson L. Characterization of a fibrin glue-GDNF slow-release preparation, *Cell Transplant*, 7(1), (1998), 53–61.

Choi-Lundberg D.L., Lin Q., Chang Y.N., Chiang Y.L., Hay C., Mohajeri H., Davidson B.L. and Bohn M.C. Dopaminergic neurons protected from degeneration by GDNF gene therapy, *Science*, 275(5301), (1997), 838–41.

Chopda G. Transdermal drug delivery systems: A review, *Latest reviews*, 4(1), (2006).

Cole S.L. and Vassar R. The Basic Biology of BACE1: A Key Therapeutic Target for Alzheimer's disease. *Current Genomics,* 8, (2007), 509-529.

Conners C.K., Levin E.D., Sparrow E., Hinton S.C., Erhardt D., Meck W.H. *et al.* Nicotine and attention in adult ADHD. *Psychopharmacol Bull,* **32,** (1996), 67–73.

Cudmore S. Elan drug delivery: delivery across the blood-brain barrier, The Irish Scientist, 2002.

Cunnane S., Nugent S., Roy M., Courchesne-Loyer A et al. Brain fuel metabolism, aging, and Alzheimer's disease, *Nutrition,* (2010). doi:10.1016/j.nut.2010.07.021.

Decina P., Caracci G., Sandik R., Berman W., Mukherjee S. and Scapicchio P. Cigarette smoking and neuroleptic-induced Parkinsonism. *Biol Psychiatry,* **28**, (1990), 502–508.

Domb A.J., Rock M., Schwartz J., Perkint C., Yipchuck G., Broxup B. and Villemure J.G. Metabolic disposition and elimination studies of a radiolabel led biodegradable polymeric implant in the rat brain, *Biomaterials*,15(9), (1994), 681-688.

Domb A.J. Polymeric carriers for regional drug therapy. *Mol. Med. Today,* **1**, (1995), 134–139.

Domb A.J., Rock M., Perkint C., Yipchuck G., Broxup B. and Villemure J.G. Excretion of a radiolabel led anticancer biodegradable polymeric implant from the rabbit brain, *Biomaterials,* 16, (1995), 1069-1072.

During M.J., Freese A., Sabel B.A., Saltzman W.M., Deutch A., Roth R.H. and Langer R. Controlled release of dopamine from a polymeric brain implant: *in vivo* characterization, *Ann. Neurol.*, 25, (1989), 351-356.

Ewing G.W., What is regressive autism and why does it occur? Is it the consequence of multi-systemic dysfunction affecting the elimination of heavy metals and the ability to regulate neural temperature? *North American Journal of Medical Sciences,* (1)2, (2009), 28-47.

Fehm H.L., Perras B., Smolnik R., Kern W. and Born J. Manipulating neuropeptidergic pathways in humans: A novel approach to neuropharmacology. *Eur. J. Pharmacol*, **405,** (2000), 43–54.

Ferber D. Bridging the Blood-Brain Barrier: New Methods Improve the Odds of Getting Drugs to the Brain Cells That Need Them, *Plos Biology*, 5(6), (2007).

Ferrara M., Langiano E., Di Brango T., De Vito E., Di Cioccio L. and Bauco C. Prevalence of stress, anxiety and depression in with Alzheimer caregivers, *Health Qual Life Outcomes*, 6(93), 2008.

Freese A., Sabel B.A., Saltzman W.M., During M.J. and Langer R. Controlled release of dopamine from a polymeric brain implant: *in vitro* characterization, *Experimental Neurology*, 103, (1989), 234-238.

Frey W.H., Interview, Reuters Health, 16th February (2001).

Gavini E., Manunta L, Giua S., Achenza G., and Giunchedi P. Spray-Dried Poly(D,L-Lactide) Microspheres Containing Carboplatin for Veterinary Use: In Vitro and In Vivo Studies. *AAPS PharmSciTech.,* 6(1), (2005),108-114.

Gelperina S. Nanoparticulate Drug Delivery Systems for the Non-Invasive Chemotherapy of Brain Tumors, NSTI nanotechnology conference and trade show, 2006.

Goodwin G. and Vieta E., Effective maintenance treatment—breaking the cycle of bipolar disorder. *European Psychiatry,* 20 (2005), 365–371.

Gorell, J.M., Rybicki, B.A., Johnson, C.C., Peterson, E.L. Smoking and Parkinson's disease: a dose–response relationship. *Neurology*, 52, (1999), 115–119.

Guerin C., Olivi A., Weingart J.D., Lawson H.C. and Brem H. Recent advances in brain tumor therapy: local intracerebral drug delivery by polymers, *Investigational new drugs*, 22, (2004), 27-37.

Guo-yu P, Xiao-Dong L, Guo-Quing L. intracarotid infusion of hypertonic mannitol changes permeability of blood-brain barrier to methotrexate in rats, *Acta pharmacol sin*, 21(7), (2000), 613-616.

Henningfield J.E., Fant R.V., Buchhalter A.R. and Stitzer M.L. Pharmacotherapy for Nicotine Dependence. *A Cancer Journal for Clinicians,* 45(5), (2005), 281-299.

Hoane M.R., Gulwadi A.G., Morrison S., Hovanesian G., Lindner M.D. and Tao W. Differential in vivo effects of neurturin and glial cell-line-derived neurotrophic factor, *Experimental Neurology*, 160(1), (1999), 235–43.

Hoffman D., Wahlberg L., Aebischer P. NGF released from a polymer matrix prevents loss of ChAT expression in basal forebrain neurons following a fimbria-fornix lesion. *Exp Neurol.,* 110, (1990), 39–44.

Horan R.L, Antle K., Collette A.L., Wang Y., Huang J., Moreau J.E., Volloch V., Kaplan D.L., Altman G.H. *In vitro* degradation of silk fibroin. *Biomaterials,* 26, (2005), 3385–3393.

Howard M.A., Gross A., Grady M.S., Langer R.S., Mathiowitz E., Winn R. and Mayberg M.R. Intracerebral drug delivery in rats with lesion-induced memory detitits, *J. Neurosurg.*, 71, (1989), 105-112.

Hsu D, Chen C, Hsu M, and Beggs JM. An open hypothesis: is epilepsy learned, and can it be unlearned? *Epilepsy Behav.* 13(3), (2008), 511–522.

Illum l. Transport of drugs from the nasal cavity to the central nervous system. *Eur. J. Pharm. Sci*, **11,** (2000), 1–18.

Irani F., Dankert M., Brensinger C., Bilker W.B., Nair S.R., Kohler C.G., Kanes S.J., Turetsky B.I., Moberg P.J., Ragland J.D., Gur R.C., Gur R.E. and Siegel S.J. Patient Attitudes towards Surgically Implantable, Long-Term Delivery of Psychiatric Medicine, *Neuropsychopharmacology*, 29, (2004), 960–968.

Iyer S.S., Barr W.H., Dance M.E., Coleman P.R. and Karnes H.T. A 'biorelevant' system to investigate in vitro drug released from a naltrexone implant, *International Journal of Pharmaceutics*, 340, (2007), 104-118.

Jain J.P., Modi S., Domb A.J. and Kumar N. Role of polyanhydrides as localized drug carriers, *Journal of controlled release*, 103, (2005), 541-563.

Jayprakar M., Ramnar J. and Shah S. An overview on brain targeting drug delivery system, *Latest reviews*, 7(1), (2009).

Jeon Y., Kwak K., Kim S., Kim Y., Lim J and Baek W. Intrathecal Implants of Microencapsulated Xenogenic Chromaffin Cells Provide a Long-Term Source of Analgesic Substances, *Transplantation Proceedings,* 38(9), (2006), 3061-3065.

Jin H.J., Kaplan D.L. Mechanism of silk processing in insects and spiders. *Nature,* 424, (2003), 1057–1061.

Jollivet C., Aubert-Pouessel A., Clavreul A., Venier-Julienne M., Remy S., Montero-Menei C.N., Benoit J. and Menei P. Striatal implantation of GDNF releasing biodegradable microspheres promotes recovery of motor function in a partial model of Parkinson's disease, *Biomaterials*, 25, (2004), 933–942.

Jones G.M.M., Sahakian B.J., Levy, R., Warburton, D.M. and Gray, J.A. Effects of acute subcutaneous nicotine on attention, information processing and short-term memory in Alzheimer's disease. *Psychopharmacology* **108,** (1992), 485–494.

Kandanearatchi A, Williams B, Everall IP. Assessing the efficacy of highly active antiretroviral therapy in the brain, *Brain Pathol* **13,** (2003), 104–110.

Keith S.K., Pani L, D. Nick, B. and Emsley R., Practical Application of Pharmacotherapy With Long-Acting Risperidone for Patients With Schizophrenia. *Psychiatric Services,* 55(9), (2004), 997-1005.

Kokaia M., Aebischer P., Elmér E., Bengzon J., Kalèn P., Kokaia Z. and Lindvall O. Seizure suppression in kindling epilepsy by intracerebral implants of GABA-but not by noradrenaline-releasing polymer matrices. Experimental Brain Research, 100, (1994), 385–394.

Koo YL, Reddy GR, Bhojani M, Schneider R, Philbert MA, Rehemtulla A, Ross BD, Kopelman R. Brain cancer diagnosis and therapy with nanoplatforms, *Advanced Drug Delivery Reviews*, 58(14), (2006), 1556-1577.

Koziara JM, Lockman PR, Allen DD, Mumper RJ. The blood-brain barrier and brain drug delivery, *Journal of nanoscience and nanotechnology*, 6, (2006).

Krauze MT, Forsayeth J, Park JW, Bankiewicz KS. Real-time Imaging and Quantification of Brain Delivery of Liposomes, *Pharmaceutical Research*, 23(11), (2006), 2493-2504.

Kubek M.J., Liang D., Byrd K.E. and Domb A.J. Prolonged seizure suppression by a single implantable polymeric-TRH microdisk preparation, *Brain Research*, 809, (1998), 189 – 197.

Kumari A., Kumar S., Yadav S.K. and Yadav S.C. Biodegradable polymeric nanoparticles based drug delivery systems, *Colloids Surf. B: Biointerfaces,* (2009), doi:10.1016/j.colsurfb.2009.09.001.

http://www.scribd.com/doc/21746180/Biodegradable-Polymeric-Nano-Particles-Based-Drug-Delivery-Systems [Accessed November 21, 2010].

Kwan *P. and* Brodie *M.J.* Refractory epilepsy: mechanisms and solutions. *Expert Review of Neurotherapeutics*, 6(3), (2006), 397-406.

Lee J.S., An T.K., Chae G.S., Jeong J.K., Cho S.H., Lee H.B. and Khang G. Evaluation of in vitro and in vivo antitumor activity of BCNU-loaded PLGA wafer against 9L gliosarcoma, *European Journal of Pharmaceutics and Biopharmaceutics*, 59, (2005), 169-175.

Levin, E.D. Nicotine systems and cognitive function. *Psychopharmacology* **108,** (1992), 418–431.

Levin E.D. and Simon B.B. Nicotinic acetylcholine involvement in cognitive function in animals, *Psychopharmacology* **138,** (1998), 217–230.

Li Y., Owusu A., and Lehnert S. Treatment of intracranial rat glioma model with implant of radiosensitizer and biomodulator drug combined with external beam radiotherapy, *Int. J. Radiation Oncology Biol. Phys*., 58(2), (2004), 519-527.

Li C., Vepari C., Jin H.J., Kim H.J., Kaplan D.L. Electrospun silk-BMP-2 scaffolds for bone tissue engineering. *Biomaterials,* 27, (2006), 3115–3124.

Lindner M.D., Francis J.M. and Saydoff J.A. Intrathecal Polymer-Encapsulated Bovine Adrenal Chromaffin Cells Fail to Produce Analgeṣic Effects in the Hotplate and Formalin Test, *Experimental Neurology*, 165, (2000), 370–383.

Loscher W., Gernert M., Heinemann U. Cell and gene therapies in epilepsy - promising avenues or blind alleys? *Trends in Neurosciences,* 31, (2008), 62–73.

Lopez T., Ortiz E., Quintana P. and Gonzalez R.D. A nanostructured titania bioceramic implantable device capable of drug delivery to the temporal lobe of the brain. *Colloids Surf A Physicochem Eng Asp,* 300, (2007), 3–10.

Lopez T., Quintana P., Ascencio J. and Gonzalez R.D. The determination of dielectric constants of mixtures used in the treatment of epilepsy and the encapsulation of phenytoin in a titania matrix, *Colloids Surf A Physicochem Eng Asp,* 300, (2007), 99 – 105.

Mardor Y, Roth Y, Lidar Z, et al. Monitoring response to convection-enhanced taxol delivery in brain tumor patients using diffusionweighted magnetic resonance imaging. *Cancer Res.,* 61(13), (2001), 4971-3.

Mayberg M., Langer R., Zervas N. and Moskowitz M. Perivascular meningeal projections from cat trigeminal ganglia: possible pathway for vascular headaches in man, *Science*, 2(13), (1981), 228-230.

Menei P., Venier M.C., Gamelin E., Saint-Andr′e J.P., Hayek G., Jadaud E., Fournier D., Mercier P., Guy G. and Benoit J.P. Local and Sustained Delivery of 5-Fluorouracil from Biodegradable Microspheres for the Radiosensitization of Glioblastoma, *Cancer,* 86(2), (1999), 325-330.

Menei P., Jadaud E., Faisant N., Boisdron-Celle M., Michalak S., Fournier D., Delhaye M. and Benoit J.P. Stereotaxic Implantation of 5-Fluorouracil-Releasing Microspheres in Malignant Glioma, *Cancer,* 100(2), (2004), 405-410.

Menei P., Pean J.M., Nerrie`re-Daguin V., Jollivet C., Brachet P. and Benoit J.P. Intracerebral Implantation of NGF-Releasing Biodegradable Microspheres Protects Striatum against Excitotoxic Damage, *Experimental neurology,* 161, (2000), 259-272.

Montoya I.D. and Vocci F. Novel Medications to Treat Addictive Disorders, *Current Psychiatry Reports,* 10, (2008), 392– 398.

Naik A., Kalia Y.N. and Guy R.H. Transdermal drug delivery: overcoming the skin's barrier function, *Pharmaceutical Science and Technology Today*, 3(9), (2000), 318-326.

Nakao N., Yokote H., Nakai K. and Itakura T. Promotion of survival and regeneration of nigral dopamine neurons in a rat model of Parkinson's disease after implantation of embryonal carcinomaderived neurons genetically engineered to produce glial cell line-derived neurotrophic factor, *Journal of Neurosurgery*, 92(4), (2000), 659–70.

Nelson T.J., Alkon D.L. and Quattrone, A., Artificial low-density lipoprotein carriers for transport of substances across the blood-brain barrier U S Patent 7803400. *http://www.freepatentsonline.com/7803400.html (2010)* [Accessed November 19, 2010].

Pardridge WM. The Blood-Brain Barrier: Bottleneck in Brain Drug Development, *NeuroRX*, 2(1), (2005), 3-14.

Park C.Y., Park J.W. and Lee J.J. Development of Totally Implantable Pulsatile Biventricular Assist Device. *Artificial Organs,* 27(1), (2003), 119–123.

Park S. and Sinko P.J., P-Glycoprotein and Mutlidrug Resistance-Associated Proteins Limit the Brain Uptake of Saquinavir in Mice. *The Journal of Pharmacology and Experimental Therapeutics.* 312(3), (2005), 1249-1256.

Patel T.B. Recent Trends in Brain Targeted Drug Delivery Systems:An Overview, *Latest reviews*, 5(2), (2007).

Pèan J.M., Menei P., Morel O., Montero-Menei C.M. and Benoit J.P. Intraseptal implantation of NGF-releasing microspheres promote the survival of axotomized cholinergic neurons, *Biomaterials*, 21, (2000), 2097-2101.

Perras B., Marshall L., Kohler G., Born J.and Fehm H.L. Sleep and endocrine changes after intranasal administration of growth hormone-releasing hormone in young and aged humans. *Psychoneuroendocrinology*, **24,** (1999), 743–757.

Perras B., Pannenborg H., Marchall L., Pietrowsky R., Born J. and Fehm H.L. Beneficial treatment of age-related sleep disturbances with prolonged intranasal vasopressin. *J. Clin. Psychopharmacol*,**19,** (1999), 28–36.

Pillay S., Pillay V., Choonara Y.E., Naidoo D., Khan R.A., du Toit L.C., Ndesendo V.M.K., Modi G., Danckwerts M.P. and Iyuke S.E. Design, biometric simulation and optimization of a nano-enabled scaffold device for enhanced delivery of dopamine to the brain. *International Journal of Pharmaceutics*, 382, (2009), 277–290.

Pillay S., Design and Development of an Implantable drug Delivery polymeric Scaffold for the Treatment of Parkinson's Disease. A dissertation submitted to the Faculty of Health Sciences, University of the Witwatersrand, in fulfillment of the requirements for the degree of Master of Pharmacy, 2009. http://wiredspace.wits.ac.za/handle/10539/7436 [Accessed November 20, 2010].

Popperl G, Goldbrunner R, Gildehaus FJ, et al. O-(2-[18F]fluoroethyl)-L-tyrosine PET for monitoring the effects of convection-enhanced delivery of paclitaxel in patients with recurrent glioblastoma. *Eur. J. Nucl. Med. Mol. Imaging,* 32(9), 2005, 1018-25.

Powell E.M., Sobarzo M.R. and Saltzman W.M. Controlled release of nerve growth factor from a polymeric implant, *Brain Research,* 515, (1990), 309-311.

Quay S.C. Successful delivery of apomorphine to the brain following intranasal administration demonstrated in clinical study. *PRNewswire*, July 18th (2001).

Quik, M. Smoking, nicotine and Parkinson's disease. *Trends Neurosci,* 27 (9), (2004), 561–568.

Rabin C., Liang Y., Ehrlichman R.S., Budhian A., Metzger K.L., Majewski-Tiedeken C., Winey K.I. and Siegel S.J. In vitro and in vivo demonstration of risperidone implants in mice, *Schizophrenia Research*, 98(1-3), (2008), 67-78.

Raghavan R., Brady M.L., Rodríguez-Ponce M.I., Hartlep A., Pedain C., Sampson. Convection-Enhanced Delivery of Therapeutics for Brain Disease, and Its Optimization, *Neurosurg Focus*, 20(4), (2006), E12.

Ranganath S.H. and Wang C., Biodegradable microfiber implants delivering paclitaxel for post-surgical chemotherapy against malignant glioma, *Biomaterials*, 29, (2008), 2996–3003.

Ranganath S.H., Fu Y., Arifin D.Y. Kee I.,Zheng L., Lee H., Chow P.K.H., Wang C. The use of submicron/nanoscale PLGA implants to deliver paclitaxel with enhanced pharmacokinetics and therapeutic efficacy in intracranial glioblastoma in mice. *Biomaterials,* 31, (2010), 5199-5207.

Rao M.S, Hattiangady B., Shetty A.K. Fetal hippocampal CA3 cell grafts enriched with FGF-2 and BDNF exhibit robust long-term survival and integration and suppress aberrant mossy fiber sprouting in the injured middle-aged hippocampus. *Neurobiol. Dis.,* 21, (2006), 276–290.

Rapoport S.I. and Thompson H.K. Osmotic opening of the blood-brain barrier in the monkey without associated neurological deficits, *Science (New York, N.Y.)*, 180(89), (1973), 971.

Raedt R., Van Dycke A., Vonck K., Boon P. Cell therapy in models for temporal lobe epilepsy. *Seizure-European Journal of Epilepsy,* 16, (2007), 565–578.

Reece A.S. Comparative treatment and mortality correlates and adverse event profile of implant naltrexone and sublingual buprenorphine, *Journal of Substance Abuse Treatment*, (2009).

Reinhard C.S., Radomsky M.L., Saltzman W.M., Hilton J. and Brem H. Polymeric controlled release of dexamethasone in normal rat brain, *Journal of Controlled Release*, 16, (1991), 331-340.

Rezvani A.H. and Levin E.D. Nicotine mechanisms in Alzheimer's disease, Cognitive effects of nicotine, *Biological Psychiatry*, 49(3), (2001), 258-267.

Rosenblad C., Martinez-Serrano A. and Bjorklund A. Intrastriatal glial cell line-derived neurotrophic factor promotes sprouting of spared nigrostriatal dopaminergic afferents and induces recovery of function in a rat model of Parkinson's disease, *Neuroscience*, 82(1) , (1998), 129–37.

Roullin V.G., Mege M., Lemaire L., Cueyssac J.P., Venier-Julienne M.C., Menei P., Gamelin E. and Benoit J.P. Influence of 5-fluorouracil–loaded microsphere formulation on efficient rat glioma radiosensitization, *Pharmaceutical research*, 21(9), (2004), 1558-1563.

Rusted, J.M., Newhouse, P.A., Levin, E.D. Nicotinic treatment for degenerative neuropsychiatric disorders such as Alzheimer's disease and Parkinson's disease, *Behav. Brain Res*, 113 (1–2), (2000), 121–129.

Sahakian B.J. and Jones G.M.M.The effects of nicotine on attention, information processing, and working memory in patients with dementia of the Alzheimer type. In: Adlkofer F, Thruau K, editors. *Effects of Nicotine on Biological Systems.* Basel, Switzerland: Birkhauser Verlag, (1991), 623–230.

Sahoo S.K., Jain T.K., Reddy M.K. and Labhasetwar V. Nano-Sized Carriers for Drug Delivery, *NanoBioTechnology*, (2008), 329-348.

Scherrmann J.M. Drug delivery to brain via the blood–brain barrier, Vascular Pharmacology, 38, (2002), 349-354.

Saltzman W.M., Mak M.W., Mahoney M.J., Duenas E.T., Cleland J.L. Intracranial delivery of recombinant nerve growth factor: release kinetics and protein distribution for three delivery systems. *Pharm Res.,* 16, (1999), 232–240.

Sanberg P.R., Emerich D.F., Aebischer P., Amisetti S.M., Ouellette W., Koutouzis T.K., Cahill D.W., Norman A.B. Substance P containing polymer implants protect against striatal excitotoxicity. *Brain Res.,* 628, (1993), 327–329.

Saraon T.S., Gill-Saraon **N.,** Miller **A.C.** and Sinert, R.H., 2010. Pacemakers and Implantable Cardioverter Defibrillators. http://emedicine.medscape.com/article/780825-overview [Accessed November 21, 2010].

Sawyer A.J., Piepmeier J.M. and Saltzman W.M. New Methods for Direct Delivery of Chemotherapy for Treating Brain Tumors. *Yale Journal of Biology and Medicine*, 79 (2006), 141-152.

Schlegelmilch T. Henke K. and Peri F. Microglia inthedevelopingbrain: from immunity to behaviour *Current Opinionin Neurobiology*, 21, (2010), 1–6.

Shah S. Transdermal drug delivery technology revisited: Recent advances, *Latest reviews,* 6(5), (2008).

Shetty A.K. and Hattiangady B. Concise review: Prospects of stem cell therapy for temporal lobe epilepsy. *Stem Cells,* 25, (2007), 2396–2407.

Siegel S.J., Winey K.I., Gur R.E., Lenox R.H., Bilker W.B., Ikeda D., Gandhi N. and Zhang W. Surgically Implantable Long-term Antipsychotic Delivery Systems for the Treatment of Schizophrenia, *Neuropsychopharmacology,* 26, (2002), 817-823.

Silva G.A. Nanotechnology approaches for drug and small molecule delivery across the blood brain barrier, *Surgical Neurology*, 67, (2007) 113–116.

Singh N., Pillay V. and Choonara Y.E. Advances in the treatment of Parkinson's disease, *Progress in Neurobiology*, 81, (2007), 29–44.

Singh, A.C., Massey, A.J., Thompson, M.D., Rappa L.R. and Honeywell M.S., Addressing Nonadherence in the Schizophrenic Population. *Journal of Pharmacy Practice* 2006 19: 361-368.

Sittl R., Griessinger N. and Likar R. Analgesic Efficacy and Tolerability of Transdermal Buprenorphine in Patients with Inadequately Controlled Chronic Pain Related to Cancer and Other Disorders: A Multicenter, Randomized, Double-Blind, Placebo-Controlled Trial, *Clinical therapeutics*, (2002), 150-168.

SoppimathK.S.,Aminabhavi T.M.,Kulkarni A.R.,Rudzinski W.E., Biodegradable polymeric nanoparticles as drug delivery devices , *Journal of Controlled Release*, 70, 1-20, 2001.

Sugibayashi K. and Morimoto Y. Polymers for transdermal drug delivery systems, *Journal of Controlled Release*, 29(1-2), (1994), 177-185.

Suter D.M. and Krause K.H. Neural commitment of embryonic stem cells: molecules, pathways and potential for cell therapy. *Journal of Pathology,* 215, (2008), 355–368.

Tamargo R.J., Rossell L.A., Kossoff E.H., Tyler B.M., Ewend M.G. and Aryanpur J.J. The intracerebral administration of phenytoin using controlled-release polymers reduces experimental seizures in rats, *Epilepsy Research*, 48, (2002), 145–155.

Tamilvanan S., Venkateshan N and Ludwig A. The potential of lipid- and polymer-based drug delivery carriers for eradicating biofilm consortia on device-related nosocomial infections. *Journal of Controlled Release*, 128, (2008), 2–22.

The British Pain Society 2008. Intrathecal drug delivery for the management of pain and spasticity in adults; recommendations for best clinical practice.

Tiwari SB, Amiji MM. A review of nanocarrier-based CNS delivery systems, *Current Drug Delivery*, 3(2), (2006), 219-32.

Tomac A., Lindqvist E., Lin L.F., Ogren S.O., Young D., Hoffer B.J. and Olson L. Protection and repair of the nigrostriatal dopaminergic system by GDNF in vivo, *Nature*, 373(6512), (1995), 335–339.

Tornqvist N., Bjorklund L., Almqvist P., Wahlberg L. and Stromberg I. Implantation of bioactive growth factor-secreting rods enhances fetal dopaminergic graft survival, outgrowth density, and functional recovery in a rat model of Parkinson's disease, *Experimental Neurology*, 164(1), (2000), 130–8.

Unsicker K. Growth factors in Parkinson's disease, *Prog Growth Factor Res*, 5(1), (1994), 73–87.

Grossman, S.A., Leong, K.W., Lesser, G.J. and Lo, H. Subcutaneous implant. *US Patent 6126956. http://www.patentstorm.us/patents/6126956.html* [Accessed November, 2010].

Wahlberg L.U., Almqvist P.M., Glantz M.J. and Boethius J. Polymeric Controlled-Release Amsacrine Chemotherapy in an Experimental Glioma Model, *Acta Neurochir*, 138 (1996), 1323-1330.

Wang P.P., Frazier J. and Brem H. Local drug delivery to the brain, *Advanced Drug Delivery Reviews*, 54, (2002), 987–1013.

White, H.K. and Levin, E.D. Four week nicotine skin patch treatment effects on cognitive performance in Alzheimer's disease. *Psychopharmacology* **143,** (1999), 158–165.

White J., Bell J., Saunders J.B., Williamson P., Makowska M., Farquharson A. and Beebe K.L. Open-label dose-finding trial of buprenorphine implants (Probuphine)® for treatment of heroin dependence, *Drug and Alcohol Dependence*, 103, (2009), 37-43.

Wilson, A.L., Langley, L.K., Monley, J., Bauer, T., Rottunda, S., Mcfalls, E. *et al.* Nicotine patches in Alzheimer's disease: Pilot study on learning,memory, and safety, *Pharmacol. Biochem. Behav.,* **51,** (1995), 509–514.

Wilz A., Pritchard E.M., Li T., Lan J.Q., Kaplan D.L., Boison D. Silk polymer-based adenosine release: Therapeutic potential for epilepsy. *Biomaterials,* 29, (2008), 3609–3616.

Yamaguchi M, Nagashima T, Kinoshita I, Tsukamoto Y. Opening of blood-brain barrier by infusion of hypertonic mannitol solution did not disturb the active avoidance of learning rats, *Bulletin of allied medical sciences*, 6, (1990), 177-181.

Zhong Y. and Bellamkonda R.V. Biomaterials for the central nervous system, *J. R. Soc. Interface*, 5(26), (2008), 957–975.

In: Advances in Polymer Research
Editors: Losif Sidorov, Fedor Nikitin, pp. 109-155
ISBN: 978-1-61942-223-0

POLYMER PHASE BEHAVIOR IN NANOCOMPOSITES

G. V. Kozlov*

Kabardino-Balkarian State University,
Chernyshevskiy st., 173, Nal'chik – 360004,
KBR, Russian Federation.

ABSTRACT

Polymer phase (polymer matrix) behavior in nanocomposites in many respects defines the behavior of nanocomposite as a whole independently from the used nanofiller type (disperse particles, organoclay, nanotubes and so on). In connection with this it is necessary to account for polymer matrix structure changes at nanofiller introduction in initial matrix polymer. These changes can be realized with the aid of different processes, namely, crystallization, amorphous polymer phase structure change, interfacial regions formation. In its turn, such factors as polymer matrix chain flexibility, interfacial adhesion level, nanofiller particle shape and so on influence on characteristics and realization possibility of the mentioned processes. Hence, at polymer nanocomposites structure formation complex dynamics of polymer phase behavior in them is observed, that defines in the long run a nanocomposite properties. In this aspect particularly important is the role of interfacial regions, which are the same reinforcing element of structure in polymer nanocomposites as actually nanofiller. For semicrystalline polymer matrix nanofiller can play a nucleator role changing in reality the indicated matrices crystallinity degree. In the present review the quantitative relationships of the initial polymer characteristics and their modification at nanofiller introduction and their influence on nanocomposite final structure are considered.

Keywords: Nanocomposite, nanofiller, crystallization, polymer phase changes, interfacial adhesion, polymer chain flexibility.

INTRODUCTION

Polymer phase (polymer matrix) behavior has critical significance for the behavior of polymer nanocomposite as a whole. Widely applied earlier microcomposite models [1] showed their scanfiness that resulted to the development of percolation [2] and fractal [3] models for polymer composites reinforcement description in the last years. The most essential

* E-mail: i_dolbin@mail.ru

distinction of the mentioned models is that circumstance, that the latter do not take into consideration filler (nanofiller) elasticity modulus at polymer composites most important characteristic - reinforcement degree determination, i.e. composites (nanocomposites) elasticity modulus enhancement in comparison with corresponding parameter for matrix polymer. It is obvious, that in such treatment polymer phase role in nanocomposites properties complex increases repeatedly.

The other important aspect of polymer nanocomposites theoretical description is complex dynamical nature of polymer phase in these materials. As a rule, this circumstance at present is not taken into consideration (in any case, properly). A nanofiller introduction in the initial matrix polymer always results to the essential changes of its structure on all its levels: molecular, topological and supramolecular (or, more precisely, suprasegmental). This process can be expressed by a number of effects: interfacial regions formation, crystallinity degree change, variation of local order level in polymer matrix and so on. The additional aspect of the problem is interconnection of the changes on all indicated structural levels. It is natural, that it complicates the theoretical description of polymer phase in nanocomposites behaviour. Let us also note, that interfacial regions in polymer nanocomposites represent, actually, a separate phase, whose characteristics are differed essentially from the bulk polymer matrix properties. So, the authors [4] showed experimentally, that for nanocomposites butadiene-styrene rubber/shungite interfacial regions elasticity modulus exceeded in 8 times the corresponding parameter for the bulk polymer matrix.

The adduced above a brief list of problems, arising at the description of polymer phase in nanocomposites structure and properties, makes obvious their solution complexity. Nevertheless, the development lately of such new for this field theoretical approaches as solid body synergetics, fractal analysis, cluster model of polymers amorphous state structure, percolation theory and so on makes possible quantitative description of all structural changes of matrix polymer at nanofiller introduction. The mentioned approaches would be used in the present review at the description of three main effects, realized in polymer matrix (crystallization, amorphous polymer phase structure change and interfacial regions formation) on the example of three main classes of polymer nanocomposites, filled with disperse particles, organoclay and carbon nanotubes.

Features of Nanocomposites Semicrystalline Matrix Crystallization

At present there are well known strong enough and diverse changes of polymers crystalline structure, occurring at the introduction in to them fillers of various sorts [5]. As a rule, these changes are described within the frameworks of polymers crystalline morphology. However, lately the fractal model was developed, which takes into consideration the whole complexity of polymers structure. It is supposed, that changes occur not only on supramolecular level of structure, but on molecular and topological levels [6]. Simultaneously it is necessary to take into account, that crystalline phase morphology variation causes noncrystalline regions structure changes [7]. As a rule, the introduction in semicrystalline polymer of particulate nanofillers results to polymer matrix crystallinity degree increase, since nanofiller particles serve as nucleators. Such effect was observed in nanocomposites high density polyethylene/calcium carbonate (HDPE/$CaCO_3$) [8, 9]. Let us note an important feature of semicrystalline polymers filling: filler introduction can result to both reduction and enhancement of polymer matrix crystallinity degree *K*. So, the authors [10] found out *K* decrease from 0.72 up to 0.38 at the introduction of carbon fibers (CF) in HDPE at its volume content φ_f=0.303. Therefore the authors [11] performed quantitative description of HDPE crystalline and amorphous phases structural changes at the introduction in to it disperse nanofiller $CaCO_3$.

Polymers crystallinity degree *K* change is closely connected with their crystallization kinetics, which can be described quantitatively by well-known Kolmogorov-Avrami equation [12]:

$$K = 1 - e^{-zt^n}, \tag{1}$$

where *z* is crystallization rate constant, *t* is crystallization process duration, *n* is Kolmogorov-Avrami exponent, characterizing nucleation type and growing crystalline structures form for the given polymer.

As it has been shown in paper [13], the exponent *n* is connected with fractal dimension D_{ch} of a chain part between local order domains (clusters) as follows:

$$n = 3(D_{ch} - 1) + 1. \tag{2}$$

D_{ch} value, which characterizes polymer molecular mobility level, can be calculated with the aid of the following equation [14]:

$$\frac{2}{\varphi_{cl}} = C_\infty^{D_{ch}}, \tag{3}$$

where φ_{cl} is local order domains (clusters) relative fraction, C_∞ is characteristic ratio, which is equal to 5.7 for HDPE [15].

In its turn, the parameter φ_{cl} was determined according to the following percolation equation [14]:

$$\varphi_{cl} = 0.03(1 - K)(T_m - T)^{0.55}, \tag{4}$$

where T_m and *T* are melting and testing temperatures, which are equal to 396 and 293 K, respectively.

In Figure 1 the dependence *K*(*n*) is adduced for microcomposites HDPE/CF according to the data of paper [10] and nanocomposites HDPE/$CaCO_3$. As one can see, despite the opposite *K* changes at filler contents increase (CF contents increase results to *K* reduction [10] and $CaCO_3$ contents increase – to its enhancement [9]), the indicated dependence is described by a common linear correlation, which allows to determine the greatest possible *K* variation for the considered materials: *K*=0.25 at *n*=1 (or D_{ch}=1.0) and *K*=0.90 at *n*=4 (or D_{ch}=2.0).

Since the nanofiller $CaCO_3$ is simultaneously a nucleator in nanocomposite matrix crystallization process then one should expect the exponent n increase at its contents φ_n growth. In Figure 2 the dependence $n(\varphi_n^{2/3})$ is adduced, where the value $\varphi_n^{2/3}$ characterizes particles $CaCO_3$ surface total area, which has shown linear growth n at $\varphi_n^{2/3}$ increase (or φ_n enhancement). At φ_n=0 the dependence $n(\varphi_n^{2/3})$ extrapolates to $n\approx1.72$, that is Kolmogorov-Avrami exponent for the initial HDPE and at $\varphi_n\approx0.7$ the greatest value n=4 is reached [12]. It is significant that the indicated value $\varphi_n\approx0.7$ corresponds well to the magnitude φ_n=0.6-0.7, which gives top nucleation effect [8].

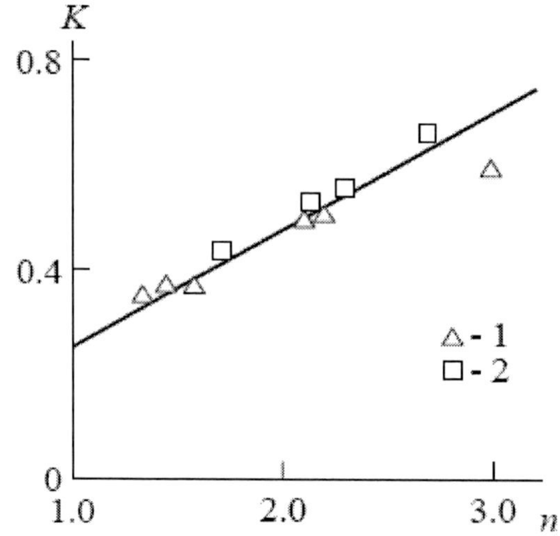

Figure 1. The dependence of crystallinity degree K on Kolmogorov-Avrami exponent n for carbon plastics HDPE/CF (1) and nanocomposites HDPE/$CaCO_3$ (2) [11].

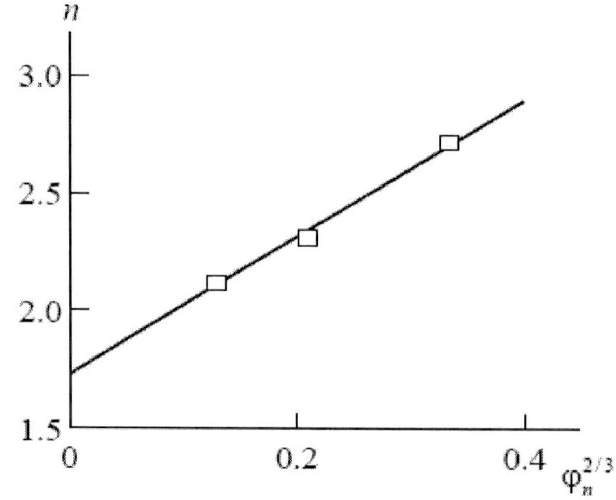

Figure 2. The dependence of Kolmogorov-Avrami exponent n on $CaCO_3$ contents φ_n for nanocomposites HDPE/$CaCO_3$ [11].

Let us note, that for CF in paper [10] the opposite effect was obtained, namely, n reduction at φ_f growth. Such discrepancy is explained by different surface structure of using fillers – CF and $CaCO_3$. If in the first case fibers have smooth surface with dimension $d_{surf}\approx 2.0$, then in the second case $CaCO_3$ disperse particles possess rough enough surface with dimension d_{surf}, close to 2.5 [16]. As Pfeifer [17] has shown, such difference of filler surface structure defines conformations difference of adjoining to surface macromolecular coils – they are stretched (are straightened) on smooth surface and maintain the initial conformation of statistical coil on a rough one. In its turn, this defines D_{ch} reduction in the first case and constant value or this dimension increase – in the second one. For $CaCO_3$ the dependence of n on its contents φ_n can be described by the following empirical equation [11]:

$$n = n_{\text{HDPE}} + 2.90\varphi_n, \tag{5}$$

where n_{HDPE} is n value for the initial HDPE, which is equal to 1.72.

In paper [10] for K calculation according to Kolmogorov-Avrami equation the constants following values were used: z=0.145 and t=2. In Figure 3 the comparison of experimental K and calculated according to the equation (1) K^T crystallinity degree is adduced for microcomposites HDPE/CF and nanocomposites HDPE/$CaCO_3$, moreover for both indicated materials the same constants z and t, adduced above, were used. As one can see, the theory and experiment good correspondence was obtained (the average discrepancy of K^T and K makes up ~ 10 %). From the equations (1) and (2) it follows, that K^T value is this case is defined by only fractal dimension D_{ch}, which, in its turn, depends on polymer matrix macromolecular coils conformation, which is formed at contact with filler surface [17]. Let us note that in this case filler type is of no importance. So, n reduction effect at φ_n increase (and, accordingly, K reduction) was observed for the system polyamide-6/quartz powder (particulate filler) [18].

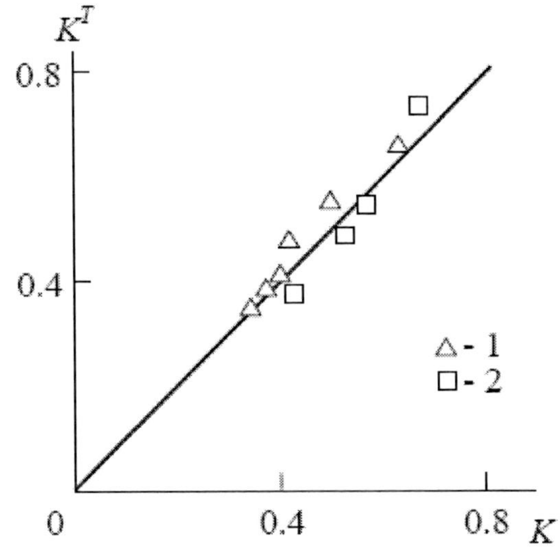

Figure 3. The comparison of experimental K and calculated according to the equation (1) K^T crystallinity degree for carbon plastics HDPE/CF (1) and nanocomposites HDPE/$CaCO_3$ (2) [11].

As it is known [12], the exponent n defines the forming crystalline phase morphology of polymers. In case of athermic nucleation at $n \le 2$ ribbons are formed by a two-dimensional growth mechanism, at $n \le 3$ – circles and at $n > 3$ – spheres. Fractional values n mean a combined mechanism of thermal/athermic nucleation, the fractional part decreasing indicates on athermic mechanism role enhancement, at that i.e. intensification of all crystallites growth simultaneous beginning [12]. As it follows from the data of Figure 2, precisely such effect is observed at φ_n growth, as was expected, since $CaCO_3$ is a nucleator [8, 9]. As electron microscopy data showed in case of nanocomposites HDPE/$CaCO_3$, the formation of large spherolites from flat fan-like lamellas was observed for the initial HDPE, whereas for nanocomposite HDPE/$CaCO_3$ with $CaCO_3$ content of 20 mass % a small pronounced bulk spherolites growth was realized [8].

The value φ_{cl} calculation according to the equation (4) proved found out its reduction from 0.23 up to 0.14 at $CaCO_3$ contents growth from zero up to 20 mass %, that is accompanied by crystallinity degree enhancement from 0.43 up to 0.67 [9]. In Figure 4 the dependence of the sum $(K+\varphi_{cl})$ on φ_n is adduced, which shows weak increase at φ_n growth. As it is known [19], the greatest relative fraction of densely-packed domains, i.e. in our case $(K+\varphi_{cl})$, can be calculated within the frameworks of thermal cluster concept:

$$(K+\varphi_{cl}) = \left(\frac{T_m - T}{T_m}\right)^{0.37}, \tag{6}$$

where the exponent 0.37 is equal to minimum value of percolation cluster order exponent [20].

The calculation of $(K+\varphi_{cl})$ value according to the equation (6) should take into consideration nanofiller availability in polymer nanocomposite and then this sum reduced value $(K+\varphi_{cl})^{red}$ was used:

$$(K+\varphi_{cl})^{red} = \frac{K+\varphi_{cl}}{1-\varphi_n}. \tag{7}$$

In Figure 4 the dependence of $(K+\varphi_{cl})^{red}$ on φ_n is shown by points and its comparison with the experimental dependence gives a good correspondence (the average discrepancy is within the limits of 7 %).

If for nanocomposites HDPE/$CaCO_3$ K increase at φ_n growth is observed (Figure 2), then for nanocomposites HDPE/Na^+-montmorillonite (HDPE/MMT) the opposite tendency was found, namely, crystallinity degree reduction from 0.50 up to 0.37 at organoclay mass contents enhancement up to 10 mass % [21]. Therefore the authors [22] studied the causes of nanocomposites HDPE/MMT crystallinity degree reduction at nanofiller contents growth within the frameworks of the considered above model [6, 11].

For analysis of value K variation the equations (1)-(3) were used, in the last of them the value φ_{cl} was replaced on densely-packed regions relative fraction φ_{dens}.

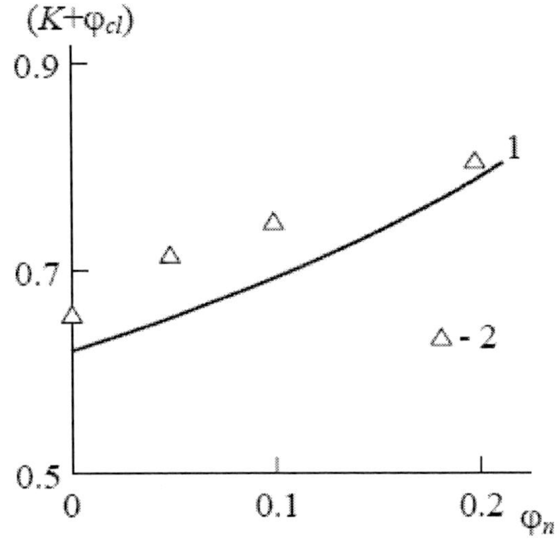

Figure 4. The dependences of densely-packed regions relative fraction $(K+\varphi_{cl})$ (1) and reduced sum $(K+\varphi_{cl})^{red}$ on $CaCO_3$ contents φ_n for nanocomposites HDPE/$CaCO_3$ [11].

As it is known [16], in nanocomposites polymer/organoclay local order domains (clusters) and interfacial regions, which are formed on organoclay platelets surface, with relative fractions φ_{cl} and φ_{if}, accordingly. The value φ_{cl} was calculated according to the percolation relationship (4) and interfacial regions relative fraction φ_{if} can be determined with the aid of the equation [16]:

$$\frac{E_n}{E_m} = 1 + 11\left(\varphi_n + \varphi_{if}\right)^{1.7}, \tag{8}$$

where E_n and E_m are elasticity moduli of nanocomposite and matrix polymer, accordingly, and the value φ_n can be estimated as follows [23]:

$$\varphi_n = \frac{W_n}{\rho_n}, \tag{9}$$

where W_n is organoclay mass contents, ρ_n is nanofiller density, which is equal to about 1800 kg/m^3 for Na^+-montmorillonite [16].

In Figure 5 the dependence of crystallinity degree K on the obtained experimentally Kolmogorov-Avrami exponent values n, accepted according to the data of paper [21], is adduced. As one can see, the linear increase K at n growth is observed, which allows to determine the limiting values K for the considered nanocomposites: K=0.33 at n=1 (or D_{ch}=1.0) and K=0.58 at n=4 (or D_{ch}=2.0).

In Figure 6 the comparison of the obtained experimentally n and calculated according to the equation (2) n^T Kolmogorov-Avrami exponent values for nanocomposites HDPE/MMT is adduced.

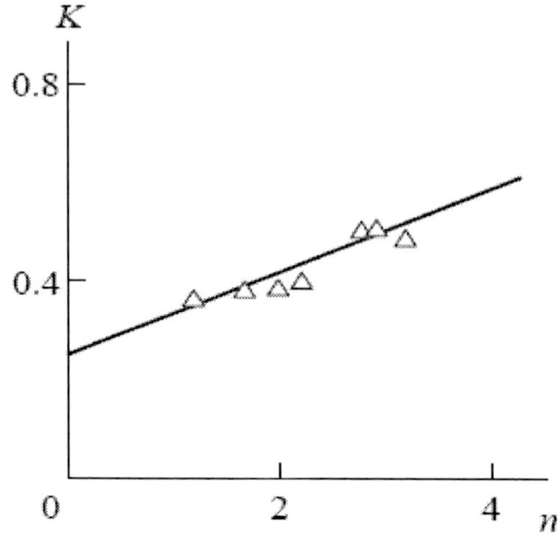

Figure 5. The dependence of crystallinity degree K on Kolmogorov-Avrami exponent n for nanocomposites HDPE/MMT and HDPE-M/MMT (HDPE-M is maleated HDPE) [22].

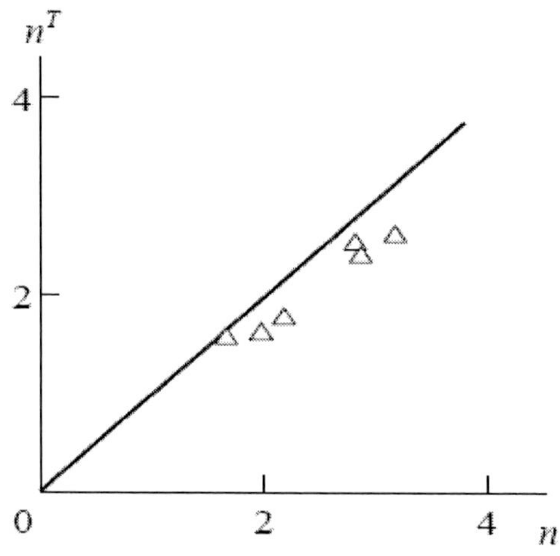

Figure 6. The comparison of experimental n and calculated according to the equation (2) n^T Kolmogorov-Avrami exponent values for nanocomposites HDPE/MMT and HDPE-M/MMT [22].

As one can see, the good enough correspondence between theory and experiment was obtained (the average discrepancy of n and n^T makes up 12 %). The comparison of Figs. 5 and 6 demonstrates clearly the cause of K reduction at W_n growth in case of the nanocomposites, filled with organoclay. Organoclay introduction in matrix polyethylene results to densely-packed interfacial regions formation, relative fraction of which increases essentially at φ_n growth. So, the authors [16] obtained the following relationships between φ_{if} and φ_n:

$$\varphi_{if} = 1.910\varphi_n \qquad (10)$$

for exfoliated organoclay and

$$\varphi_{if} = 0.955\varphi_n \qquad (11)$$

for intercalated one.

The increase φ_{if} results to densely-packed regions relative fraction φ_{dens} (φ_{dens}=φ_{cl}+φ_{if}) growth, reduction of polymer phase molecular mobility level, characterized by dimension D_{ch}, according to the equation (3) and the exponent n decrease according to the equation (2), that in its turn reduces K value (see the equation (1) and Figure 5). Simple empirical correlation can be obtained between parameter n and φ_n (Figure 3) [22]:

$$n = n_{\mathrm{HDPE}} - 40.5\varphi_n , \qquad (12)$$

where n_{HDPE} is Kolmogorov-Avrami exponent for matrix polyethylene, which is equal to 3.25 in the considered case. The equations (5) and (12) comparison shows, that unlike disperse nanofiller ($CaCO_3$), which is a nucleator [8, 9, 11], organoclay reduces the value n and, hence, K (Figure 5). Besides, organoclay action at n reduction is stronger by more than an order of the same $CaCO_3$ effect in the Kolmogorov-Avrami exponent enhancement. As it was noted above, the exponent n defined polymers forming crystalline phase morphology [12]. As it follows from the data of Figure 7, at Na^+-montmorillonite contents increase crystalline phase morphology transition from three-dimensional spherolites to one-dimensional ribbon structures is observed. This is explained by a flat structure of silicate platelets, on the surface of which crystallites are formed by one-dimensional growth mechanism, i.e. by epitaxial crystallization [24]. There is one more structural treatment, explaining nanocomposites crystalllinity degree reduction at organoclay contents growth – a thermal cluster model [25].

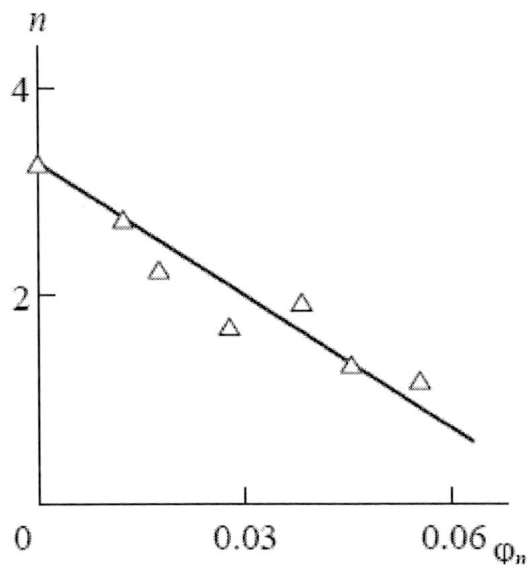

Figure 7. The dependence of Kolmogorov-Avrami exponent n on MMT volume contents φ_n for nanocomposites HDPE-M/MMT [22].

According to this model densely-packed regions contents, in the capacity of which the sum (K+φ_{cl}+φ_{if}) should be considered, cannot exceed definite value α, given by the modified equation (6) [25]:

$$\left(K+\varphi_{cl}+\varphi_{if}\right)\leq\alpha=\left(\frac{T_m-T}{T_m}\right)^{0.37}. \qquad (13)$$

In Figure 8 the comparison of parameters (K+φ_{cl}+φ_{if}) and α as a function W_n for the considered nanocomposites is adduced, from which their approximate equality follows. Hence, with the appreciation of the condition (K+φ_{cl}+φ_{if})≈α=const φ_n increase, resulting to φ_{if} growth (see the equations (10) and (11)) should give K reduction, that is observed experimentally. Let us note, that at W_n=10 mass % calculation according to the equation (10) shows φ_{if} increase on the value 0.11, that corresponds to K decrease on the value 0.13.

Hence, the stated above results demonstrate decisive influence of molecular mobility level on crystallinity degree value for nanocomposites with semicrystalline polymer matrix. In its turn, the fractal dimension of a chain part between clusters, characterizing the mentioned mobility, is defined by nanofiller particles surface structure and interfacial regions formation mechanism. The offered fractal model of crystallization process is universal for polymer composites (nanocomposites) with semicrystalline matrix irrespective of nanofiller type [11, 22].

As it was noted above, in the filling process changes occur and in the molecular level, that cannot influence on polymer nanocomposites crystallization process. The authors [26] studied polymer phase chains statistical flexibility on $CaCO_3$ particles efficiency as nucleator on the example of nanocomposites HDPE/$CaCO_3$.

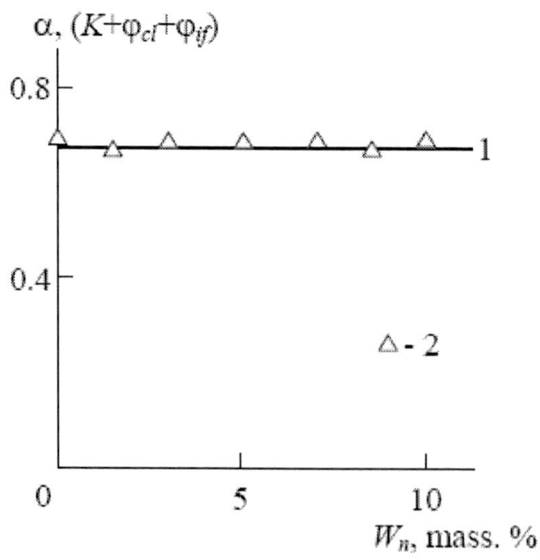

Figure 8. The dependence of parameters α (1) and (K+φ_{cl}+φ_{if}) (2) on MMT mass contents W_n for nanocomposites HDPE-M/MMT [22].

For nucleation efficiency by $CaCO_3$ particles N determination the method, based on the application of usual cycles of DSC cooling was used. The value N was determined according to the equation [8]:

$$N = 100 \times \frac{T_{\mathrm{CNA}} - T_c}{T_{c\max} - T_c}, \tag{14}$$

where T_{CNA} is the peak crystallization temperature of the polymer with the nucleating agent, T_c is the crystallization temperature, and $T_{c\max}$ is the optimum self-nucleation temperature.

As the results of papers [8, 9] showed, the introduction in HDPE of both micro- and nanoparticles $CaCO_3$ results to simultaneous increase of both crystallinity degree K (K=0.43-0.67) and nucleation efficiency N (N=35-52 %). Let us note, that $CaCO_3$ microparticles with diameter of 1200 nm give higher values N (N=44-52 %) and nanoparticles $CaCO_3$ with diameter of 50-60 nm give higher crystallinity degree. The common physical principles of these changes of parameters K and N can be determined within the frameworks of the model [6], according to which the value K depends on polymer chain flexibility and is determined as follows:

$$K = 0.32 C_\infty^{1/3}, \tag{15}$$

where C_∞ is characteristic ratio, which serves as polymer chain statistical flexibility indicator [27].

The change of the molecular characteristic C_∞ at $CaCO_3$ particles introduction in HDPE can be estimated as follows. For this purpose it is necessary to calculate fractal (Hausdorff) dimension d_f of nanocomposites structure, that can be made according to the equation [28]:

$$d_f = (d-1)(1+\nu), \tag{16}$$

where d is dimension of Euclidean space, in which a fractal is considered (it is obvious, that in our case d=3), ν is Poisson's ratio, which is estimated by mechanical tests results with the aid of the relationship [29]:

$$\frac{\sigma_Y}{E_n} = \frac{1-2\nu}{6(1+\nu)}, \tag{17}$$

where σ_Y is yield stress, E_n is nanocomposite elasticity modulus.

The value C_∞ can be calculated according to the following equation [14]:

$$C_\infty = \frac{2d_f}{d(d-1)(d-d_f)} + \frac{4}{3}. \tag{18}$$

The comparison of experimental K and calculated according to the equation (15) crystallinity degree values is adduced in Figure 9. As one can see, the good enough correspondence of theory and experiment is obtained (the average discrepancy of K and K^T makes up 10 %). Thus, the adduced above results demonstrate, that crystallinity degree change is due to molecular characteristic (polymer phase chain flexibility) variation, which is due to the disperse filler (nanofiller) introduction.

In Figure 10 the dependence of nucleation efficiency N on characteristic ratio C_∞ value for nanocomposites HDPE/$CaCO_3$ is adduced. As one can see, the linear correlation $N(C_\infty)$ is obtained, which can be approximated by the following empirical equation [26]:

$$N = 21 + 3.5C_\infty, \%. \tag{19}$$

Let us consider two limiting cases of the dependence $N(C_\infty)$, which can be obtained by the indicated dependence extrapolation to C_∞=0 and N=100 %. The value C_∞ according to the equation (18) cannot be equal to zero (for this the dimension d_f should be a negative one, that has no physical significance) and at d_f=0 its minimum value makes up 4/3. Thus, from the plot of Figure 10 it follows that minimum value of nucleation efficiency for $CaCO_3$, modified by stearic acid, in HDPE matrix makes up ~ 26 %. According to the definition [30] the value C_∞ can be written as follows:

$$C_\infty = \frac{\langle R_g^2 \rangle}{n\langle l_0^2 \rangle}, \tag{20}$$

where $\langle R_g \rangle$ is molecule gyration radius, n and $\langle l_0 \rangle$ are a number and mean length of molecule skeletal bond, accordingly.

The condition d_f=0 at d=3 is true for point object [31]. This, means, that in the equation (20) n=1 and then $C_\infty = \langle R_g^2 \rangle / \langle l_0^2 \rangle$, where l_0=1.54 Å for polyethylenes [32]. At minimum value C_∞=4/3 and l_0=1.54 Å the value R_g=1.778 Å or such low-molecular particle diameter will be equal to ~ 3.56 Å, that is a typical size for particles of such kind.

According to the equation (19), at N=100 % the value C_∞=22.6 and then the calculation according to the equation (18) gives corresponding to the indicated magnitude C_∞ value d_f=2.954. This value d_f practically precisely corresponds to maximum for real solids, i.e. for ideal rubbers, fractal dimension d_f=2.95 [28]. Hence, the equation (19) includes the entire structural range for polymer materials, beginning from low-molecular particles (for example, monomers) up to the ideal rubbers (d_f=0-2.95).

The authors [33] carried out the study of the efficiency of nucleation by nanoparticles $CaCO_3$ on the example of nanocomposites LDPE/$CaCO_3$ [34]. Since the values N and K are defined by polymer phase chain flexibility, characterized by parameter C_∞ (the equations (19) and (15), accordingly), then correlation between them should be expected, the indicated parameters characterize one and the same crystallization process at that. The estimations showed that $CaCO_3$ contents enhacement in nanocomposites LDPE/$CaCO_3$ up to 50 mass % resulted to C_∞ growth from 2.18 (for the initial LDPE) up to 6.0.

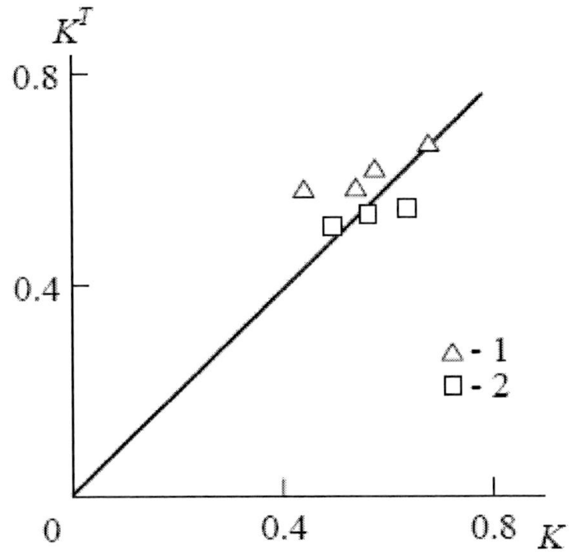

Figure 9. The comparison of experimental K and calculated according to the equation (15) K^T crystallinity degree values for nanocomposites HDPE/$CaCO_3$ with particles size of 1200 nm (1) and 50-60 nm (2) [26].

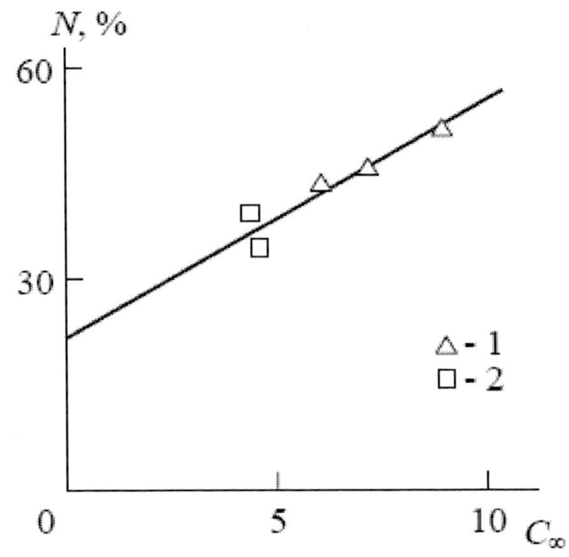

Figure 10. The dependence of nucleation efficiency N on characteristic ratio C_∞ for nanocomposites HDPE/$CaCO_3$ with particles size of 1200 nm (1) and 50-60 nm (2) [26].

According to the equations (19) and (15) this means N increase from 28.6 up to 42 % and K – from 0.389 up to 0.545. This assumes, that nanoparticles $CaCO_3$ are nucleation centers for crystalline regions in nanocomposites LDPE/$CaCO_3$ and their efficiency in this quality grows at $CaCO_3$ cotents increase, resulting in the end to crystallinity degree K enhancement.

In Figure 11 the correlation $K(N)$ for the considered nanocomposites is adduced, which has proved to be linear and pass through coordinates origin. The indicated correlation can be expressed analytically as follows [33]:

$$K = 0.013N, \tag{21}$$

where N is given in per cent.

The equation (21) allows to estimate the limiting characteristics of crystallization process for LDPE. At K=1.0 let us obtain the greatest value N=76.9 %, that according to the equation (19) corresponds to the greatest magnitude C_∞=16. According to the equation (15) the greatest value K can be determined as equal to 0.77. The minimum value C_∞ is accepted equal to 2 [27]. This allows to estimate the minimum magnitudes N and K as equal to 28 % and 0.40, respectively. The limiting values K=0.40-0.77 correspond well to the literary data, where K=0.50-0.65 for LDPE [35]. As it was expected, theoretical range of K variation is somewhat wider than a realized one in practice.

Let us estimate further the conditions, at which the greatest crystallinity degree, i.e. K=0.77, can be received for LDPE. Polymers glass transition temperature T_g can be estimated according to the empirical equation [14]:

$$T_g = 129\left(\frac{S}{C_\infty}\right)^{1/2}, \text{K}, \tag{22}$$

where S is macromolecule cross-sectional area, which is equal to 14.35 Å^2 for polyethylenes [36].

At present T_g estimations for polyethylenes are contradictory enough: T_g=170-250 K. The calculation C_∞ according to the equation (22) in this case gives limiting values 8.25-3.82 and K limiting magnitudes according to the equation (15) 0.50-0.65, that corresponds excellently to the literary data [35]. For the greatest value K=0.77 reaching the condition C_∞=16 (see above) and value T_c=122 K realization is required [33].

Let us note that N enhancement is not due to $CaCO_3$ contents increase, but to its particles aggregates diameter D_{ag} growth, which can be calculated according to the technique [16]. In Figure 11 the experimental data for composites HDPE/$CaCO_3$ [8] are adduced, which correspond well to the present paper results. Although for the latter the values N are higher, than for nanocomposites LDPE/$CaCO_3$, but $CaCO_3$ contents in them is smaller (5-20 mass %) [8]. At the same time $CaCO_3$ particles size in composites HDPE/$CaCO_3$ makes up 1200 nm. In Figure 12 the dependence $N(D_{ag})$ for both types of the indicated composites is adduced, which is approximated well by one linear correlation, described by the equation [33]:

$$N = 28 + 18D_{ag}, \%, \tag{23}$$

where D_{ag} is given in mcm.

The reasons of the shown in Figure 12 correlation are obvious enough. Firstly, the larger D_{ag}, the is smoother nanofiller particles (aggregates of particles) surface. Secondly, the higher

D_{ag} is, the smaller nanofiller particles surface fractal dimension d_{surf} is (for D_{ag}=80 nm d_{surf}=2.6 and for D_{ag}=1.20 mcm d_{surf}=2.2), i.e. the surface is less rough. Both indicated factors create conditions for polymer phase chains "stretching" on filler particles surface, that increases C_∞ and, hence, N. Thus, the adduced above data have shown, that particles $CaCO_3$ are crystallization effective nucleator for polyethylenes. The contents $CaCO_3$ increase results to nucleation efficiency N growth and crystallinity degree enhancement for nanocomposites polyethylene/$CaCO_3$. The polymer chain flexibility is a decisive factor in this process. The value N grows at nanofiller particles aggregation intensification, which accompanies $CaCO_3$ contents enhancement, and corresponds to the increase of $CaCO_3$ particles aggregates diameter [33].

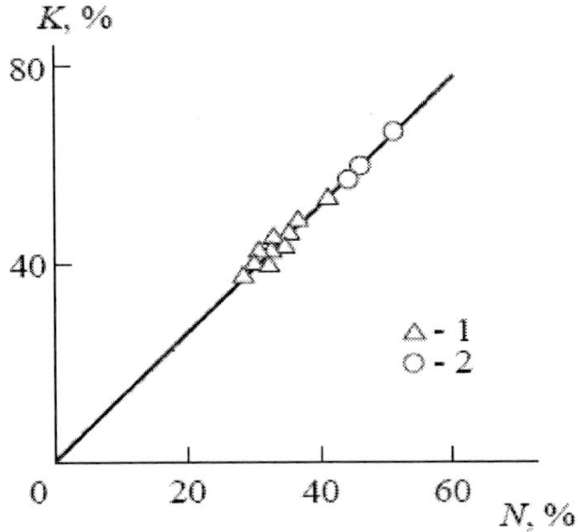

Figure 11. The dependence of crystallinity degree K on nucleation efficiency N for nanocomposites HDPE/$CaCO_3$ (1) and composites HDPE/$CaCO_3$ (2) [33].

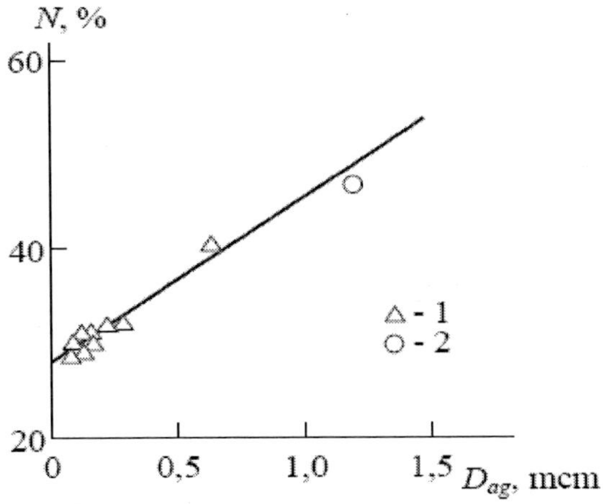

Figure 12. The dependence of nucleation efficiency N on particles (aggregates of particles) size D_{ag} for nanocomposites HDPE/$CaCO_3$ (1) and composites HDPE/$CaCO_3$ (2) [33].

The Influence of Polymer Phase Molecular Characteristics Change on Nanocomposites Mechanical Properties

At present there are many enough experimental observations, supposing influence of polymer phase molecular characteristics on reinforcement degree E_n/E_m of nanocomposites. It was noticed, that the best results in nanocomposites with layered nanofiller were obtained for matrices, consisting of rigid-chain polymers. So, in Figure 13 the dependences of reinforcement degree E_n/E_m on montmorillonite volume contents φ_n for nanocomposites on the basis of poly(butylene terephtalate) [37] and thermotropic liquid crystalline polyester [38] are adduced. As one can see, the value E_n/E_m increases at φ_n growth much faster for the second of the indicated nanocomposites. The authors [39] received for nanocomposites on the basis of polyimide record reinforcement degree - $E_n/E_m \approx 3.7$ at montmorillonite content 3 mass % only. These results did not obtain proper explanation and even did not attract any attention, since the authors confined themselves to the fact statement and general assumptions.

Another interesting experimental observation also remains without proper attention. As it is known, the reinforcement degree of the same polymer is essentially higher in a rubbery state than in a glassy one. For example, for nanocomposites on the basis of epoxy polymer the value E_n/E_m in glassy state does not exceed ~ 1.65, whereas at the temperature, exceeding on 40 K the glass transition temperature, the value E_n/E_m for the same nanocomposite achieves ~ 3.8 [40]. The indicated effect is also observed without nanocomposites polymer matrix phase state change, but simply at testing temperature raising.

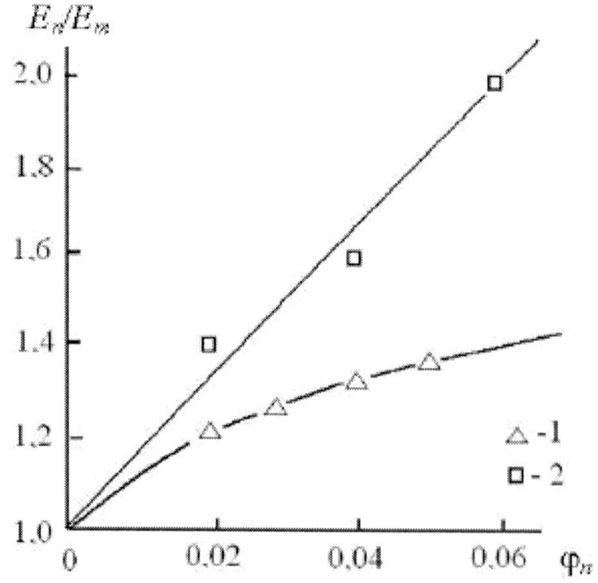

Figure 13. The dependences of reinforcement degree E_n/E_m on montmorillonite contents φ_n for nanocomposites on the basis of poly(butylene terephthalate) (1) [37] and thermotropic liquid crystalline polyester (2) [38].

As the example in Figure 14 the dependences of elasticity modulus E_n on amine groups ratio for nanocomposites on the basis of polyimide for two testing temperatures (293 and 373 K) are adduced. As one can see, the value E_n/E_m is essentially higher at T=373 [16]. The stated results assumed, that polymer phase molecular characteristics is a key parameter at nanocomposites reinforcement description.

In virtue of strong anisotropy of layered silicates particles shape Halpin-Tsai and Mori-Tanaka models are often used for theoretical estimation of filled with them nanocomposites reinforcement degree [23]. Halpin-Tsai equation and the closed-form solution based on Mori-Tanaka model have the following form [23]:

$$\frac{E_n}{E_m}=\frac{1+2(L/d_{pl})\varphi_n\eta}{1-\varphi_n\eta}, \tag{24}$$

$$\eta=\frac{(E_n/E_m)-1}{(E_n/E_m)+2(L/d_{pl})},$$

and

$$\frac{E_n}{E_m}=\frac{1}{1-\varphi_n\xi}, \tag{25}$$

$$\xi=\frac{-2\nu_m A_3+(1-\nu_m)A_4+(1+\nu_m)A_5 A}{-2A},$$

where L/d_{pl} is an aspect ratio of anisotropic particles (L is length, d_{pl} is thickness of silicate platelet), ξ is a positive coefficient depending on the matrix Poisson's ratio ν_m and constants A and A_i, which can be calculated from the matrix/particle properties and components of Eshelby tensor, which depends on the particle aspect ratio (L/d_{pl}) and dimensionless elastic constants of the polymer matrix [16].

For the case of isotropic (spherical) filler particles E_n/E_m estimation can be carried out according to the equation [42]:

$$\frac{E_n}{E_m}=1+11.6\varphi_n-44.4\varphi_n^2+96.3\varphi_n^3. \tag{26}$$

The theoretical dependences (lines), corresponding to the equations (24) and (25) at L/d_{pl}=50 and 100, and also to the equation (26) are adduced in Figure 15. As abscissa axis Na^+-montmorillonite contents W_n in mass % is chosen as most often used in practice. Besides, in the same Figure the experimental values E_n/E_m (points) for seven nanocomposites are plotted. Let us note first of all, that the transition from theoretical dependences with larger values L/d_{pl} to smaller ones at W_n increase (in Figure it is indicated by arrows) is observed.

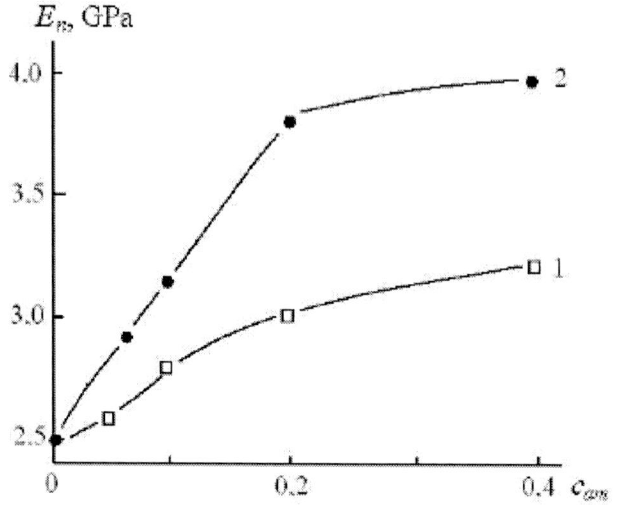

Figure 14. The dependence of elasticity modulus E_n on amine groups ratio c_{am} for nanocomposites on the basis of polyimide at testing temperature 293 (1) and 373 K (2) [41].

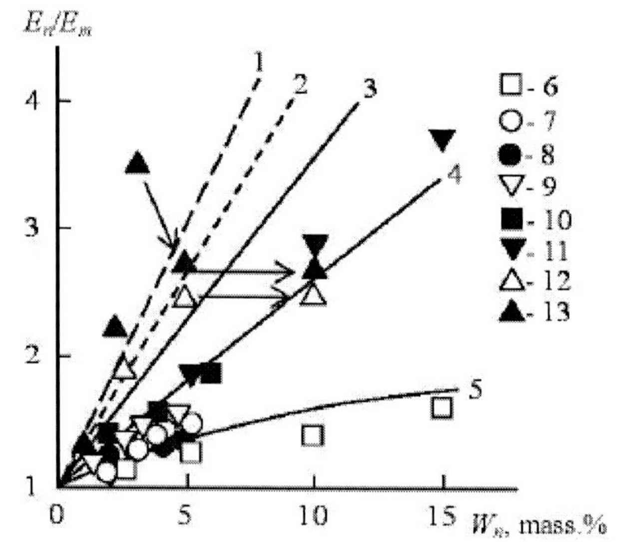

Figure 15. The dependences of reinforcement degree E_n/E_m on filling degree W_n for nanocomposites, filled with Na^+-montmorillonite. 1-5 – the theoretical dependences, corresponding to the equations (24) (1, 2) and (25) (3, 4) at L/d_{pl}=100 (1, 3) and 50 (2, 4) and to the equation (26) (5). 6-13 – the experimental data for nanocomposites on the basis of epoxy polymer at $T<T_g$ (6), polyamide-6 (7), poly(butylene terephthalate) (8), polycarbonate (9), thermotropic liquid crystalline polyester (10), epoxy polymer at $T>T_g$ (11), polypropylene (12) and polyimide (13) [16].

Such transition was expected in virtue of the layered silicate platelets aggregation, which is Na^+-montmorillonite, at W_n increase. The aggregation means platelets number rise in "stack" (tactoid) d_{pl} growth at L=const and, as consequence, the ratio L/d_{pl} decrease. But the most interesting aspect of the dependences of Figure 15 is the fact, that different polymer matrices correspond to different theoretical curves, showing in addition widely variating E_n/E_m values [16].

It is significant that reinforcement degree corresponds to a class of polymer, forming nanocomposites polymer phase. The greatest ratios E_n/E_m are obtained for polymers, whose chains are able to stretch on silicate platelet surface (rigid-chain polyimide, crystallizing polypropylene and thermotropic liquid crystalline polyester), intermediate – for polymers, whose chains are able to stretch only partly (polycarbonate, poly(butylene terephtalate) and amorphous polyamide-6), and the smallest ones – for nanocomposites on the basis of epoxy polymer, the capability of which to chains stretching decreases sharply by the availability of transverse covalent bonds network.

Hence, the data of Figure 15 indicate that the nanocomposite ability to reinforcement is defined not actually by filler particles shape anisotropy, but by polymer phase ability to reflect (reproduce) this anisotropy. In other words, the filler role comes to polymer phase structure modification in comparison with the initial matrix polymer. The similar concept was used for polymer microcomposites reinforcement description [43]. However, this general treatment distinction consists in the fact that in the microcomposites case bulk polymer matrix structure changes (its fractal dimension d_f is increased) [43] and in the nanocomposites case – only interfacial regions structure at common condition d_f=const [44].

The indicated ability of polymer matrix or, more precisely, interfacial regions to reflect filler particles form anisotropy one can express quantitatively with the polymer chain statistical segment length l_{st}: the larger l_{st}, the higher E_n/E_m. Let us remind, that the value l_{st} is connected with characteristic ratio C_∞ by the following equation [30]:

$$l_{st} = C_\infty l_0 , \tag{27}$$

where l_0 is the main chain skeletal bond length and since C_∞ is a polymer chain statistical flexibility indicator [27], then l_{st} plays the same role. In Figure 16 the dependences of E_n/E_m on l_{st} for the seven adduced in Figure 15 nanocomposites at two contents of Na^+-montmorillonite (W_n=2 and 5 mass %) are shown. As one can see, these dependences are linear and at l_{st}=0 are extrapolated to E_n/E_m=1.0, that means the reinforcement absence for low-molecular matrices (at any rate, by interfacial regions formation mechanism). These dependences can be described analytically as follows [16]:

$$\frac{E_n}{E_m} = 1 + A l_{st} , \tag{28}$$

where A is constant, which is equal to 1.08 and 1.36 for nanocomposites with W_n=2 and 5 mass %, if l_{st} is given in nm.

The common dependence of E_n/E_m on W_n and l_{st} (with appreciation of A increase with W_n growth) for nanocomposites with exfoliated (nonaggregated) organoclay has the following form [16]:

$$\frac{E_n}{E_m} = 1 + 0.32W_n^{1/2}l_{st}, \tag{29}$$

where W_n is given in mass %, l_{st} – in nm.

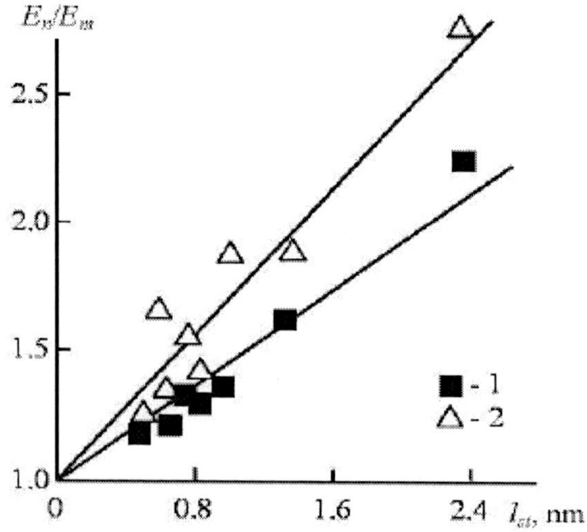

Figure 16. The dependences of reinforcement degree E_n/E_m on statistical segment length l_{st} for nanocomposites with Na^+-montmorillonite contents 2 (1) and 5 (2) mass % [16].

In Figure 17 calculated according to the equation (29) and obtained experimentally reinforcement degree values for the considered nanocomposites at $W_n \leq 5$ mass % are adduced. As one can see, the estimation according to the equation (29) gives a good correspondence to experiment.

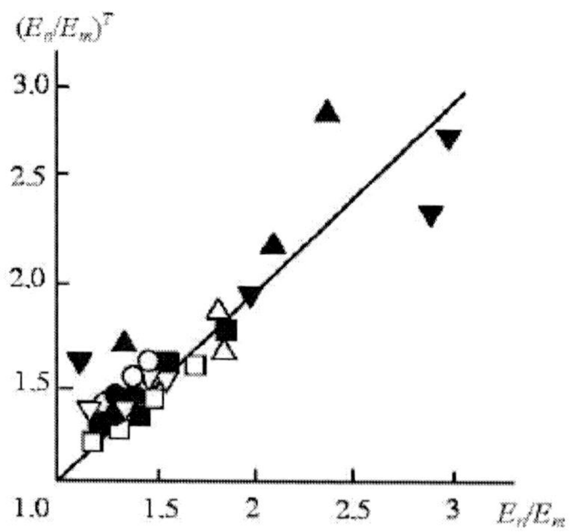

Figure 17. The comparison of experimental E_n/E_m and calculated according to the equation (29) $(E_n/E_m)^T$ reinforcement degree for nanocomposites with $W_n \leq 5$ mass %. Designations are the same as in Figure 15 [16].

In Figure 18 the dependence $E_n/E_m(W_n)$, calculated according to the equation (29) (the solid line) and experimental data (points) for nanocomposites on the basis of phenylone, filled with carbon nanotubes (CNT) is adduced. A good enough correspondence of theory and experiment is observed [45]. Accounting for the fact, that the equation (29) was obtained for nanocomposites, filled with organoclay, the data of Figure 18 allow to confirm, that CNT possess the same ability to rise polymer elasticity modulus as well as exfoliated layered silicates.

From the data of Figure 18 it follows, that at W_n=10 mass % the experimental value E_n/E_m is essentially lower than its theoretical magnitude. The cause of this effect can be elucidated by the equation (8) usage. For nanocomposites phenylone/CNT the values φ_{if}=0.029; 0.096; 0.112 and 0 for CNT contents W_n=3, 5, 7 and 10 mass %, accordingly, were obtained. The condition φ_{if}=0 is due to nanocomposites structure synergetic behaviour: for W_n=10 mass % its behaviour becomes chaotic one, CNT orientation factor η, which is a governing parameter of the mentioned structure, turns into zero and according to the equation [47]:

$$\varphi_{if} = 1.09\eta \tag{30}$$

the value φ_{if} is also equal to zero.

In Figure 18 the experimental dependences $E_n/E_m(W_n)$ are also adduced for two nanocomposites, filled with organoclay: polycarbonate/Na^+-montmorillonite (PC/MMT) [48] and polyamide-6/Na^+-montmorillonite (PA-6/MMT) [23]. The choice of these nanocomposites is due to closeness of PC, PA-6 and phenylone molecular characteristics [32].

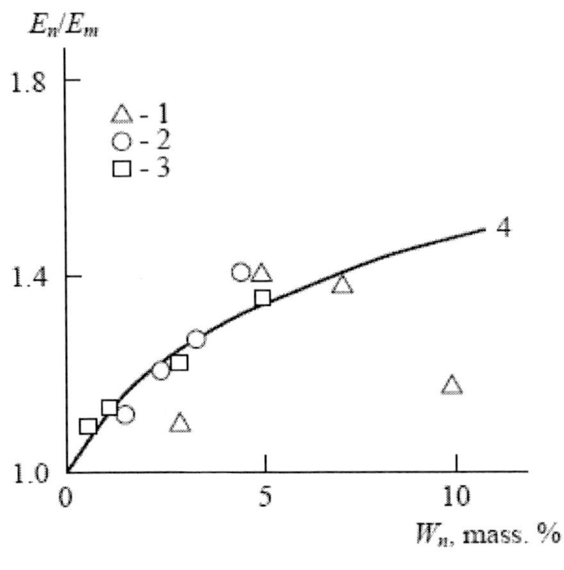

Figure 18. The dependence of reinforcement degree E_n/E_m on nanofiller contents W_n for nanocomposites phenylone/CNT (1), PC/MMT (2) and PA-6/MMT (3). 4 – calculation according to the equation (29) [45].

As it follows from the data of Figure 18, the dependences $(E_n/E_m)(W_n)$ for nanocomposites, filled with Na^+-montmorillonite, correspond to both results for nanocomposites phenylone/CNT and theoretical calculation according to the equation (29).

The close results, obtained for the three mentioned above nanocomposites, are due to the identical values of the received values φ_{if} owing to comparable contact area of nanofiller-polymer matrix. The equation (10) gives the relation between φ_{if} and φ_n for exfoliated layered silicates.

For CNT the same relation can be obtained from the following considerations. As it is known [49], nanocomposites polymer matrix and nanofiller particles surface, interacting at interfacial regions formation, are fractal objects. At their interaction there is a sole linear scale l, defining such objects interpenetration distance [50]. Since in polymer composites filler elasticity modulus, as a rule, is higher than the corresponding parameter for polymer matrix, it is assumed [49], that in this case filler indentation in polymer matrix occurs and then l is equal to interfacial layer thickness l_{if}. It can be written [50]:

$$l_{if} \sim a\left(\frac{r_n}{a}\right)^{2(d-d_{surf})/d}, \tag{31}$$

where a is a lower linear scale of polymer matrix fractal behaviour, r_n is a nanofiller particles radius, d_{surf} is fractal dimension of nanofiller particles surface.

For CNT the mean value r_n=17.5 nm, the value a is accepted equal to l_{st} and d_{surf} magnitude can be accepted in the first approximation equal to 2.0 [16]. As it is known, d_{surf}=2.0 means nanofiller smooth surface, that results to stretching of macromolecular coils on it [17] and, as consequence, to C_∞ growth in comparison with the similar parameter for bulk polymer matrix. Therefore for the estimations according to the equation (31) C_∞=9 was accepted [51] and then $a=l_{st}$=1.26 nm. At the indicated parameters the equation (31) gives l_{if}=7.34 nm. The equation can be used further [16]:

$$\varphi_{if} = \varphi_n\left[\left(\frac{r_n + l_{if}}{r_n}\right)^3 - 1\right], \tag{32}$$

that at r_n=17.5 nm and l_{if}=7.34 nm gives [45]:

$$\varphi_{if} = 1.86\varphi_n. \tag{33}$$

The comparison of the equations (10) and (33) shows approximate equality of the coefficients in them, that defines close values of reinforcement degree for nanocomposites, filled with organoclay and carbon nanotubes.

It is necessary to indicate, that high values E_n/E_m for nanocomposites phenylone/CNT are due to nanotubes size nanometer scale. So, in case of short fibers (r_n=4000 nm [47]) the calculation according to the equations (31) and (32) at other equal conditions gives [45]:

$$\varphi_{if} = 0.18\varphi_n, \quad (34)$$

that defines essentially lower microcomposites reinforcement degree in comparison with nanocomposites at other equal conditions.

And let us note in conclusion matrix polymer molecular characteristics strong influence on nanocomposites reinforcement degree, which follows from the equation (29). The indicated factor acts stronger, than nanofiller content, since the value l_{st} is included in the equation (29) to the first power and W_n – to the half power.

Thus, the stated above results allow to make three main conclusions. Firstly, polymers reinforcement degree by CNT (nonaggregated) is comparable with corresponding characteristic for exfoliated layered silicates (organoclays). Secondly, polymer molecular characteristics, namely, chain statistical flexibility, influence on reinforcement degree is much stronger, than it assumes earlier. And, thirdly, interfacial phenomena, i.e. interfacial regions formation, is the main factor, defining nanocomposites reinforcement [45].

The polyurethanes, widely used in industry, represent two-phase polymer materials, consisting of rigid and flexible blocks. At room temperature the first are in a glassy state, the second – in rubber-like. An organoclay small amounts introduction (within the limits of 1-5 mass %) improves essentially polyurethanes properties [52, 53]. So, these rubber-like polymers modulus at strain of 100 % and organoclay content 5 mass % increases approximately in 3.6 times, and strength – in about 1.5 times [53]. The questions about structure of the received from them nanocomposites arise in connection with polyurethanes two-phase state. The first from them is linked to reinforcement high degree causes. The second question is related with nanofiller (organoclay) concentration in either phase of polyurethane. And at last, the third question is related to reinforcement degree sharp increase at testing temperatures higher than rigid blocks glass transition temperature [53]. The authors [54] gave the answer to the stated above questions within the frameworks of the offered above reinforcement model (see the equation (29)).

The polyurethanes (PU) and nanocomposites on its basis glass transition temperature was determined by the dynamic mechanical spectroscopy method. It has been found out [53], that PU have two glass transition temperatures, which are equal to 233 and 324 K, i.e. the studied polyurethane is a two-phase material, where T_g=233 K is related to flexible blocks and T_g=324 K – to rigid ones.

It has been reported, that for PU on the basis of poly(tetramethylene glycol) (PTMG) T_g value is equal to 185 K. Higher values T_g for the considered PU and wide damping peaks, received in dynamical mechanical tests, assume partial mixing of flexible and rigid segments phases. The nanocomposites on the basis of PU glass transition temperatures are approximately equal to these temperatures for the initial matrix polymer [53].

Using the stated above data, the value C_∞ and, accordingly, l_{st} can be calculated according to the equations (22) and (27) for polyurethane each phase separately. For rigid blocks C_∞=9.64, l_{st}=1.47 nm and for flexible ones - C_∞=18.6, l_{st}=2.83 nm. Further the reinforcement degree $(E_n/E_m)^T$ calculation according to the equation (29) gives this parameter values at using l_{st}=1.47 nm, if organoclay is concentrated in rigid blocks phase, and at using l_{st}=2.83 nm, if organoclay concentration is realized in flexible blocks phase.

In Figure 19 the comparison of the experimental E_n/E_m and calculated by the indicated mode $(E_n/E_m)^T$ reinforcement degree values for nanocomposites PU/MMT.

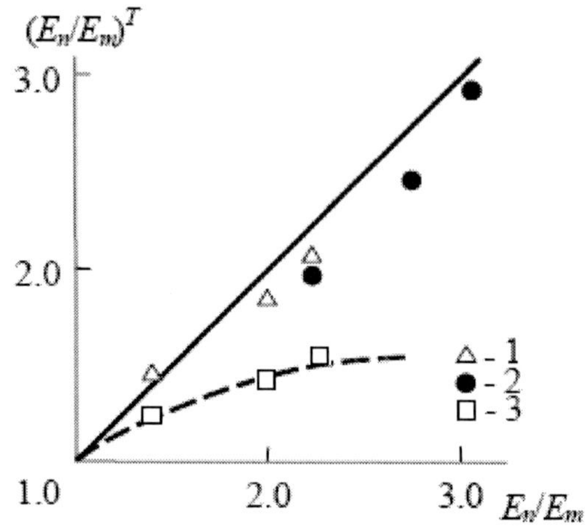

Figure 19. The comparison of experimental En/Em and calculated according to the equation (29) (En/Em)T at lst=2.83 (1), 4.51 (2) and 1.47 (3) nm reinforcement degree values for nanocomposites PU/MMT [54].

As it follows from the data of this Figure, the good correspondence of theory and experiment was obtained in that case, if in the equation (29) the value l_{st}=2.83 nm for flexible blocks was used. In case of value l_{st}=1.47 nm for rigid blocks application the equation (29) gives understated values of reinforcement degree. This comparison demonstrates, that for the considered polyurethane organoclay is concentrated in devitrificated at testing temperature (293 K) flexible blocks phase (T_g=233 K) and high reinforcement degree E_n/E_m at organoclay small contents (W_n=1-5 mass %) is defined by high values l_{st} or C_∞ (by polymer chain high flexibility), typical for rubber-like polymer (see above).

Let us consider the effect of reinforcement degree sharp increase at testing temperatures higher than rigid blocks glass transition temperature (T_g=324 K). Such effect was observed earlier as well, for the example, for nanocomposites epoxy polymer/organoclay. The authors [16] explained the observed effect by sharp increasing of chain flexibility in rubber-like state, that is reflected by C_∞ and, respectively, l_{st} enhancement. The flexible blocks phase T_g increase in comparison with the value obtained earlier (from 185 up to 233 K) the authors [53] explained by flexible and rigid blocks mixing definite degree and, accordingly, by the influence of the latter on the first. The rigid blocks phase devitrification removes the indicated limitations and therefore for the value C_∞ calculation in this case (at temperatures higher than 324 K) according to the equation (22) T_g=185 K is accepted. Then the estimations with the indicated equation usage give C_∞=29.7, l_{st}=4.51 nm. Further value $(E_n/E_m)^T$ can be calculated according to the equation (29). The comparison of experimental E_n/E_m and calculated by the indicated mode $(E_n/E_m)^T$ reinforcement degree values for nanocomposites PU/MMT at temperatures higher than rigid block glass transition temperature (T_g=324 K) is also adduced in Figure 19, from which a good correspondence of theory and experiment follows.

Hence, the adduced above data allow to give an answer to the three stated questions. Firstly, organoclay particles are concentrated in devitrificated flexible blocks phase. Secondly, polymer chains high flexibility in the indicated phase defines a large enough reinforcement degree of nanocomposites polyurethane/organoclay. And, thirdly, still larger polymer chain flexibility, realized after rigid blocks devitrification, results to corresponding enhancement of reinforcement degree for this state of the considered nanocomposites [54].

It is obvious, that for particulate nanofiller the equation (29) should be modified by including into it the nanoparticles diameter D_p. Such modification allows to obtain the following equation for particulate-filled polymer nanocomposites reinforcement degree estimation [55]:

$$\frac{E_n}{E_m} = 1 + \frac{0.19 W_n l_{st}}{D_p^{1/2}}, \tag{35}$$

where W_n is given in mass %, l_{st} and D_p – in nm.

The comparison of experimental E_n/E_m and calculated according to the equation (35) $(E_n/E_m)^T$ reinforcement degree for particulate-filled nanocomposites (Figure 20) shows the good correspondence of theory and experiment.

As it is known [56-58], a particulate-filled polymer nanocomposites reinforcement degree is a function of nanofiller particles diameter. In Figure 21 the calculated according to the equation (35) dependence of E_n/E_m on D_p at the following values of the included into this equation parameters: W_n=30 mass %, l_{st}=1 nm is adduced. As one can see, the sharp growth of reinforcement degree at D_p reduction is observed, that was expected [56-58]. This correspondence has not only qualitative, but also quantitative character: in Figure 21 the boundaries of "reinforcing" and "super-reinforcing" nanofillers according to the classification, offered in paper [56], are indicated by vertical shaded lines. As it follows from the plot of Figure 21, these values correspond well to gradient change of reinforcement degree, calculated according to the equation (35).

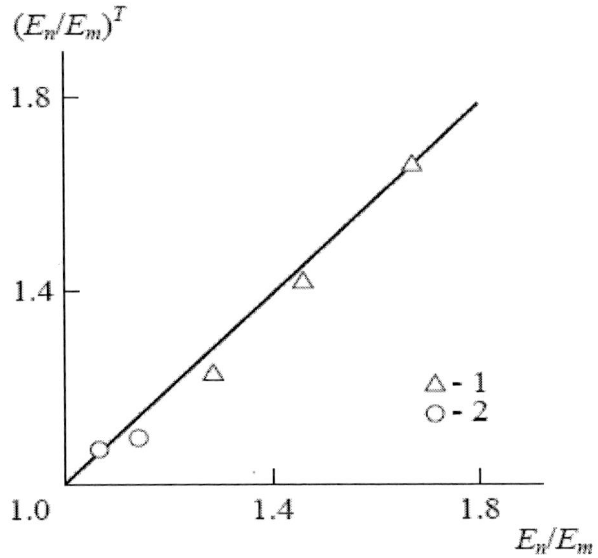

Figure 20. The comparison of experimental E_n/E_m and calculated according to the equation (29) $(E_n/E_m)^T$ reinforcement degree for nanocomposites on the basis of PP (1) and phenylone (2) [55].

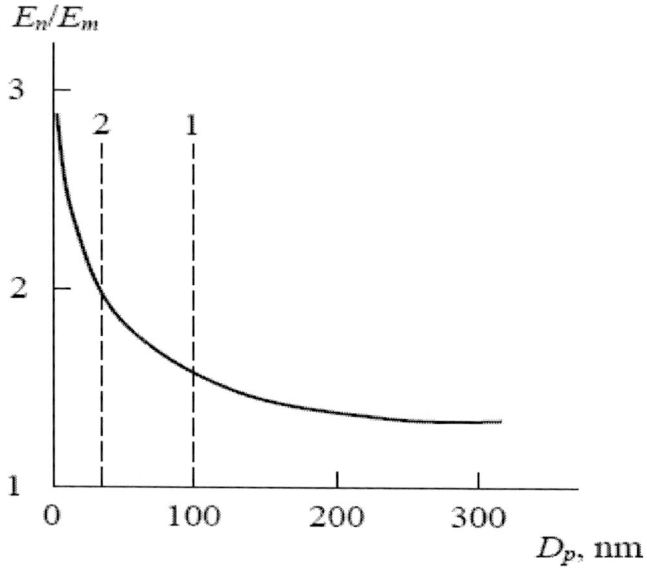

Figure 21. The dependence of reinforcement degree $(E_n/E_m)^T$ on nanofiller particles diameter D_p calculated according to the equation (35) for nanocomposites, filled with disperse particles. The vertical shaded lines indicate the boundary for reinforcing (1) and superreinforcing (2) nanofillers [55].

The dependences of E_n/E_m on W_n, plotted according to the equations (29) and (35) (l_{st}=1 nm, D_p=7 nm) for polymer nanocomposites, filled with layered (or CNT) and particulate nanofiller, accordingly, are presented in Figure 22.

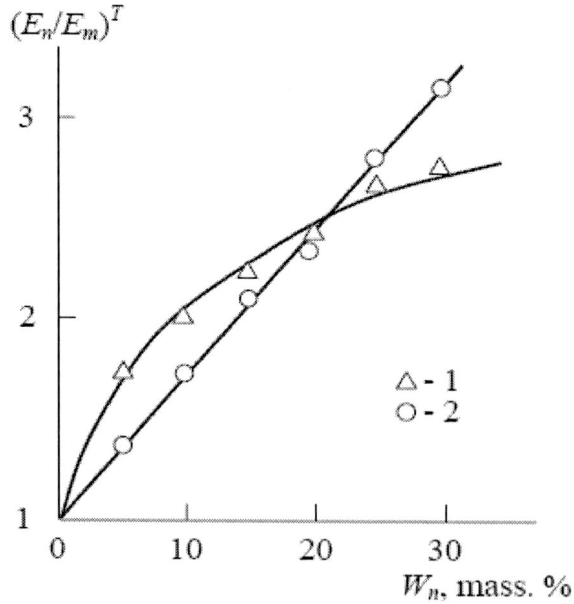

Figure 22. The dependences of reinforcement degree $(E_n/E_m)^T$ on nanofiller contents W_n for nanocomposites with layered (1) and disperse (2) nanofiller, calculated according to the equations (29) and (35), accordingly [55].

As one can see, at relatively small W_n (< 20 mass %) the reinforcement degree at the same W_n values is higher for nanocomposites, filled with layered silicate or CNT, but at W_n>20 mass % particulate nanofiller gives a higher reinforcement degree. This transition is explained as follows. It is known [16], that the layered silicates and CNT at contents W_n>10 mass % display strong tendency to aggregation. Since the equations (29) and (35) are empirical ones, which were received on the basis of the experimental data analysis, then they take into consideration this effect by the dependence of E_n/E_m on W_n to the half power for layered nanofillers and CNT (see the equation (29)) and to first power for particulate particles (the equation (35)). In the general case filled with organoclay polymer nanocomposites plasticity, characterized by strain up to failure, decreases sharply at small (of order of 5-10 mass %) nanofiller contents [24, 48]. For this nanocomposites brittleness effect explanation the authors [59] used the fractal analysis methods, within the frameworks of which the value of strain up to failure ε_f was described according to the following equation [60]:

$$\varepsilon_f = C_\infty^{D_{ch}} - 1, \tag{36}$$

where the characteristic ratio C_∞ was determined according to the equation (18) and dimension D_{ch} was calculated according to the equation (3) with φ_{cl} replacement by densely-packed regions φ_{dens} of nanocomposite structure relative fraction.

As it is known [16], in nanocomposite structure local order domains (clusters), interfacial regions and actually nanofiller with relative fractions φ_{cl}, φ_{if} and φ_n, accordingly, are densely-packed regions. The value φ_{cl} was calculated with the aid of the equation [14]:

$$d_f = 3 - 6\left(\frac{\varphi_{cl}}{C_\infty S}\right)^{1/2}, \tag{37}$$

where S is macromolecule cross-sectional area, which is equal to 25 Å^2 for PC [36].

As the data of paper [48] showed, interlayer spacing between MMT platelets d_{001} for the considered nanocomposites PC/MMT makes up 17.1-24.3 Å, that allows to attribute them to the exfoliated ones class, for which the value φ_{if} is determined according to the equation (10). In Figure 23 the dependence of ε_f on MMT contents W_n, received experimentally is adduced. As one can see, for nanocomposites PC/MMT within the range of W_n=2.5-3.4 mass % sharp change of the dependence $\varepsilon_f(W_n)$ character and absolute values ε_f essential reduction is observed. Therefore the assumption was made [59], that at MMT small contents (W_n<3.2 mass %) only the clusters should be accepted as densely-packed regions and at W_n>3.2 mass % as the indicated regions the sum is accepted [59]:

$$\varphi_{dens} = \varphi_{cl} + \varphi_{if} + \varphi_n = \varphi_{cl} + 2.91\varphi_n. \tag{38}$$

In Figure 24 the comparison of ε_f obtained experimentally and calculated according to the considered above fractal model ε_f^T values of limiting strain up to failure for PC and seven nanocomposites PC/MMT with MMT contents 2.4 and 4.5 mass %.

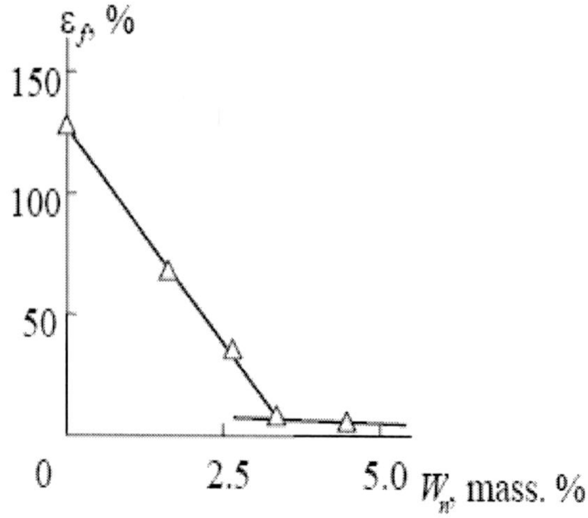

Figure 23. The dependence of strain up to failure ε_f on MMT contents W_n for nanocomposites PC/MMT [59].

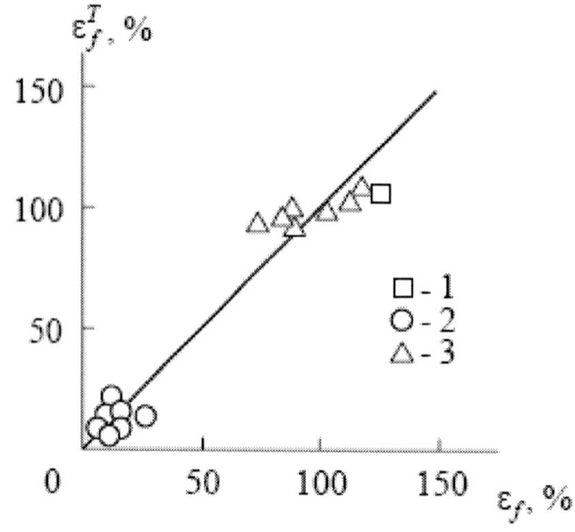

Figure 24. The comparison of experimental ε_f and calculated according to the equation (36) ε_f^T strain up to fracture values for PC (1) and nanocomposites PC/MMT with MMT contents 2.4 (2) and 4.5 (3) mass % [59].

As one can see, the well enough correspondence of theory and experiment is obtained (the average discrepancy between ε_f and ε_f^T makes up ~ 20 %). This correspondence supposes correctness of the made above assumption: at small W_n (≤3.2 mass %) availability of MMT and interfacial regions does not impose restrictions on molecular mobility (D_{ch}≈1.39) and, hence, does not decrease essentially the value D_{ch} in comparison with PC. This results to high enough ε_f values comparable with the corresponding parameter for PC. And on the contrary, MMT contents increase larger than 3.2 mass % reduces sharply molecular mobility level (D_{ch}≈1.02-1.05) in virtue of φ_{dens} growth (see the equation (38)), that defines corresponding reduction of nanocomposites plasticity, characterized by strain ε_f. Let us note, that similar sharp ε_f reduction at $W_n \geq 2.5$ mass % was observed for other classes of nanocomposites, as well for example, polypropylene/MMT [24]. The described above technique was applied successfully for plasticity treatment of particulate-filled nanocomposites poly(vinyl chloride)/$CaCO_3$ [61] and nanocomposites polypropylene/MMT [62].

For polymer nanocomposites, filled with CNT, strain up to failure ε_f increase in comparison with the initial matrix polymer or nanocomposites plasticity enhancement was noted repeatedly. The indicated effect was found out for nanocomposites epoxy polymer/CNT [63], phenylone/CNT [64] and so on. In other words, plasticity increasing effect for nanocomposites indicated class has common enough character. This effect is very important from the practical point of view, since the main deficiency of polymer composites in general is their brittleness, i.e. ε_f reduction at filler contents increasing [62, 65]. Therefore it is important to give theoretical estimation of plasticity enhancement effect for nanocomposites polymer/CNT since in future this gives possibility of production of polymer nanocomposites with operating characteristics unique set: with these materials simultaneous increasing of stiffness, strength and plasticity. The authors [66, 67] carried out the structural analysis of nanocomposites phenylone/CNT plasticity enhancement within the frameworks of the considered above model. Calculated according to the equation (36) ε_f^T values for the indicated nanocomposites are adduced in table 1 together with necessary for their calculation characteristics C_∞, φ_{cl} and D_{ch}. As it follows from the data of this table, the comparison of experimental ε_f and theoretical ε_f^T values of limiting strain at failure, characterizing nanocomposites phenylone/CNT plasticity, showed their good correspondence – the average discrepancy makes up 8.5 %, that corresponds to the experimental error of this parameter determination.

Let us consider physical principles of the studied plasticity increasing effect for nanocomposites, filled with CNT. CNT distinguishing feature from other nanofillers (disperse particles, organoclay) is their smooth in atomic scale surface [63], that results to matrix polymer macromolecules stretching on this surface and densely-packed interfacial regions polymer matrix – CNT formation [17]. The values of interfacial regions relative fraction φ_{if} for the considered nanocomposites according to the data, of paper [64] are also adduced in table 1.

Table 1. The molecular, structural and mechanical characteristics of nanocomposite phenylone/CNT [67]

CNT contents, mass %	d_f	C_∞	φ_{cl}	D_{ch}	ε_f, %	ε_f^T, %	φ_{if} [64]	d_f^T
0	2.247	2.33	0.646	1.336	21.8	21.9	-	2.247
3	2.230	2.30	0.667	1.318	22.0	20.2	0.022	2.265
5	2.354	2.55	0.520	1.439	26.3	33.8	0.098	2.316
10	2.280	2.38	0.605	1.376	25.0	25.8	-	2.247

These regions dense packing supposes, that fractal dimension d_f for them (d_f^{if}) achieves the greatest possible value for real solid bodies, namely, d_f^{if} =2.95 [28]. Further the calculation of theoretical dimension d_f (d_f^T) can be fulfilled according to the mixtures rule [66]:

$$d_f^T = d_f^p\left(1-\varphi_{if}\right)+d_f^{if}\varphi_{if}, \tag{39}$$

where d_f^p is fractal dimension of phenylone structure, which is equal to 2.247 (table 1).

The calculated according to the equation (39) d_f^T values are also adduced in table 1 and their comparison with d_f values was shown the good correspondence (the average discrepancy of d_f and d_f^T makes up 1.5 %). Taking into account, that the offered above technique of ε_f^T theoretical calculation is founded as a matter of fact on dimension d_f knowledge, one should make the conclusion, that the observed for nanocomposites polymer/CNT plasticity increase effect is due to densely-packed interfacial regions formation on nanotubes surface and, as consequence, to polymer matrix molecular and structural characteristics change (C_∞, d_f, φ_{cl} and D_{ch}, see table 1). Let us also note, that ε_f increase is accompanied by dimension D_{ch} growth (table 1), i.e. molecular mobility intensification. As it has been shown by Kausch [68], such interconnection is common for polymer materials.

And in conclusion let us consider the prediction of properties for nanocomposite phenylone/CNT, contained 30 mass % nonaggregated CNT. In this case the value φ_{if} can be estimated according to the equation (33) and elasticity modulus E_n – according to the equation (8) at the condition E_m=1.25 GPa in tension testing for phenylone. Then the value E_n for the indicated above hypothetical nanocomposite phenylone/CNT contained 30 mass % CNT, makes up 11.85 GPa. The calculation d_f^T according to the equation (39) gives value 2.64 or ν=0.32 according to the formula (16). Hence, yield stress σ_Y value for such nanocomposite according to the relationship (17) makes up 820 MPa. And at last, the calculation according to the equations (3), (18), (36) and (37) gives the value ε_f^T =120 %. To obtain such unique by their mechanical properties totality nanocomposite two technological difficulties hinder: very high nanotubes cost, making up several tens and even hundreds of dollars per gram, and the latter production technology, when nanotubes can be received only in the shape of tangled bundles and available at present dispersion methods work only up to CNT contents of order of several per cents [63, 64, 69].

Thus, the stated above results confirm correctness of structural analysis at limiting strain (plasticity) estimation for nanocomposites phenylone/CNT. The offered model gives the good correspondence of theory and experiment. The physical ground of plasticity increase for nanocomposites, filled with CNT, is densely-packed interfacial regions formation on smooth in atomic scale nanotubes surface. The offered treatment allows to predict nanocomposites properties and to plan their practical realization ways.

The authors of papers [8, 9] found out, that particulate nanofiller $CaCO_3$ introduction in HDPE resulted to nanocomposite HDPE/$CaCO_3$ impact toughness A_p increasing in comparison with the initial polymer approximately by 20 %. The authors [8, 9] carried out a detailed fractographic analysis of this effect and explained A_p observed increase by plastic deformation mechanism change for nanocomposite HDPE/$CaCO_3$ in comparison with the initial HDPE. Without going into details of the indicated analysis, some grounds for doubts in its correctness it is necessary to note. In Figure 25 the schematic diagrams load-time (*P-t*) for polymer materials samples fracture two cases are adduced: by instable (a) and stable (b) cracks. As it is known [70], the value A_p is characterized by area under *P-t* diagram, which gives mechanical energy, spent on samples fracture. Polymer materials macroscopic fracture process, defined by the main crack, begins at the greatest load *P*. From the schematic diagrams *P-t* it follows, that actually fracture process does not practically influences on the value A_p in case of crack instable propagation and influences only partly – in case of stable crack. Although the authors [8, 9] carried out impact tests on instrumented apparatus, allowing to obtain *P-t* diagrams, these diagrams were not adduced. Besides, fracture process structural aspect in papers [8, 9] is considered with secondary structures (crazes, shear zones and so on) using. Their interconnection with the initial undeformed material structure is purely speculative. It is obvious, that to obtain quantitative relationships structure-properties (that is the main goal of polymers physics [71]) at such analysis method is not possible. Therefore the authors [72] carried out quantitative structural analysis of HDPE and nanocomposites HDPE/$CaCO_3$ impact tests results within the frameworks of fractal models.

As it is known [73], the fractal dimension d_f is the most general object (in our case – polymer material) structure informant and true structural characteristic, describing structure elements distribution in space. The value d_f can be determined according to the equations (16) and (37). In the second from the indicated equations the value φ_{cl} was determined according to the relationship (4). Estimations according to the equations (16) and (37) gave the following d_f values at testing temperature 293 K: for HDPE 2.68 and 2.73, for nanocomposite HDPE/$CaCO_3$ 2.73 and 2.75, accordingly. As one can see, the good enough correspondence was obtained – the discrepancy by fractional part d_f, which performs the main information amount about structure, does not exceed 7 %.

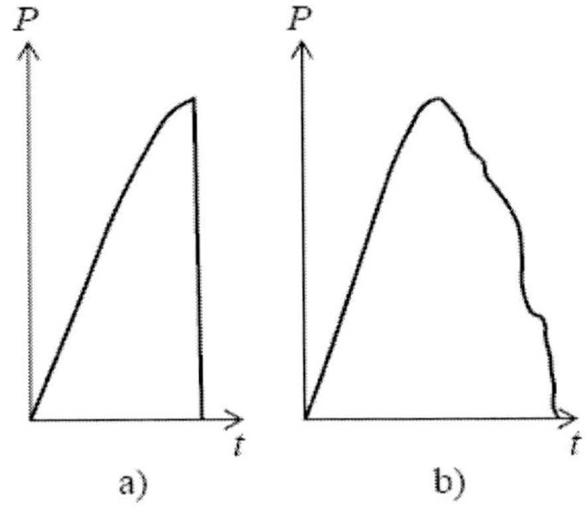

Figure 25. Schematic diagrams load-time (*P-t*) in instrumented impact tests. The fracture by instable crack (a) and stable one (b) [72].

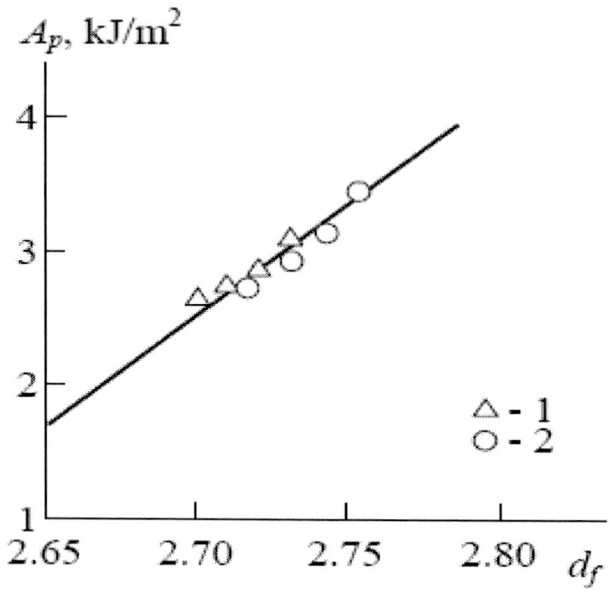

Figure 26. The dependence of impact toughness A_p on structure fractal dimension d_f for HDPE (1) and nanocomposite HDPE/$CaCO_3$ (2) [72].

In Figure 26 the dependence $A_p(d_f)$ is adduced for the considered polymer materials, which turns out to be linear, common for initial HDPE and nanocomposite HDPE/$CaCO_3$ and is described by the following empirical correlation [72]:

$$A_p = 13.5(d_f - 2.5),\ \text{kj/m}^2. \tag{40}$$

From the equation (40) it follows, that at d_f=2.5 the value A_p=0. The indicated fractal dimension corresponds to the ideal brittle fracture condition [28], that defines the condition A_p=0. For real solid bodies the greatest fractal dimension of their structure is equal to 2.95 [28], that allows to determine the greatest value A_p according to the equation (40), which is equal to ~ 6.1 kj/m^2 [72].

As Kausch showed [68], the energy dissipation at impact grew at polymer material molecular mobility level enhancement. As it was noted above, within the frameworks of fractal analysis this level can be characterized with the aid of the fractal dimension D_{ch} of polymer chain part between its fixation points (chemical cross-linkings, physical entanglement nodes, clusters and so on) [14]. Such analysis method was applied successfully for impact toughness description in case of particulate-filled nanocomposites phenylone/β-sialone [74] and amorphous glassy polymers, treated as natural nanocomposites [75]. The value D_{ch} can be determined with the aid of the following equation [14]:

$$D_{ch} = \frac{\ln N_{cl}}{\ln(4 - d_f) - \ln(3 - d_f)}, \tag{41}$$

where N_{cl} is statistical segments number per chain part between clusters, which is determined as follows. First physical entanglements cluster network density ν_{cl} was determined [14]:

$$\nu_{cl} = \frac{\varphi_{cl}}{C_\infty l_0 S}, \tag{42}$$

where l_0 is the main chain skeletal bond length, which is equal to 1.54 Å for polyethylenes [32].

Then the estimation of polymer chains total length per polymer volume unit L was fulfilled as follows [14]:

$$L = S^{-1}. \tag{43}$$

The chain part between clusters length L_{cl} is determined according to the equation [14]:

$$L_{cl} = \frac{2L}{\nu_{cl}}. \tag{44}$$

And at last, the value N_{cl} can be determined as the ratio [14]:

$$N_{cl} = \frac{L_{cl}}{l_{st}} = \frac{2}{S\nu_{cl} l_{st}}. \tag{45}$$

In Figure 27 the dependence of impact toughness A_p on chain fractal dimension D_{ch} is adduced for the considered materials. As it was expected, A_p growth at D_{ch} increase was observed, described analytically by the following relationship [72]:

$$A_p = 6.75(D_{ch} - 1), \text{ kj/m}^2. \tag{46}$$

The equation (46) allows to determine the greatest value A_p for the considered materials at the condition D_{ch}=2.0: this value is equal to 6.75 kj/m^2, that is close to the adduced above estimation according to the equation (40) – the average discrepancy makes up less than 10 %.

Let us consider the condition of zero impact toughness reaching at d_f=2.50, but not at d_f=2.0 ($2.0 \le d_f < 3$ [31]).

As it is known [76], at φ_{cl} growth irrespective of its causes structure of polymers in general and HDPE in particular, reaches its quasiequilibrium state, when φ_{cl} growth is balanced by polymer chains entropic stretching and ceases. As it was shown in paper [76], for HDPE structure quasiequilibrium state dimension was equal to 2.50. The calculation according to the adduced above technique (the equations (4) and (41)-(45)) showed, that in this case $D_{ch}\approx$1.0 and, hence, polymer material became ideally brittle (see the equation (46)).

The equations (40) and (46) usage allows to estimate theoretical values A_p (A_p^T) and to compare them with experimental magnitudes of impact toughness.

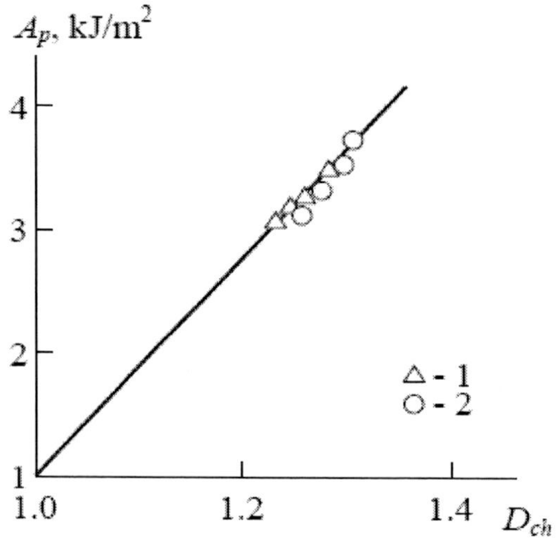

Figure 27. The dependence of impact toughness A_p on fractal dimension D_{ch} of chain part between clusters for HDPE (1) and nanocomposite HDPE/$CaCO_3$ (2) [72].

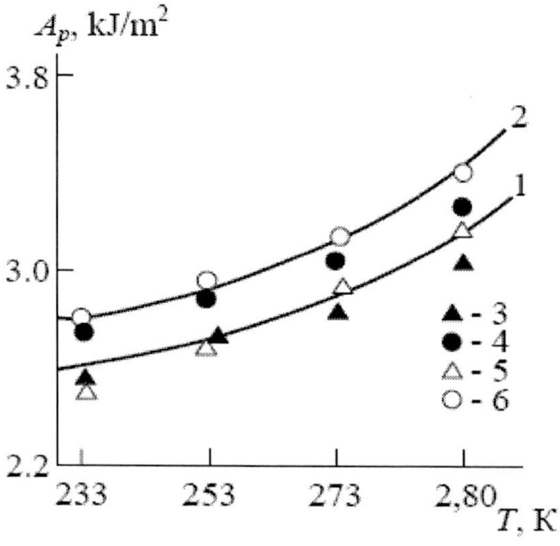

Figure 28. The comparison of experimental (1, 2) and calculated according to the equations (40) (3, 4) and (46) (5, 6) temperature dependences of impact toughness A_p for HDPE (1, 3, 5) and nanocomposite HDPE/$CaCO_3$ (2, 4, 6) [72].

Such comparison is presented in Figure 28, from which a good correspondence of theory and experiment follows (for calculation according to the equation (40) the average discrepancy A_p^T and A_p makes up 3.5 %, according to the equation (46) – 3.9 %, that is essentially less than impact toughness experimental determination usual error, which makes up ~ 10 %).

Let us consider physical grounds of impact toughness increase effect for nanocomposites HDPE/$CaCO_3$ in comparison with matrix polymer HDPE. As it follows from the plots of Figures 26 and 27, A_p increase is defined by dimensions d_f and D_{ch} enhancement, accordingly. In its turn, as it follows from the equations (37) and (41), the indicated dimensions enhancement is defined by φ_{cl} reduction, since the molecular characteristics C_∞ and S were adduced as constants. As it follows from the equation (4), φ_{cl} reduction depends on crystallinity degree K increase only, since the value T_m for HDPE and nanocomposite HDPE/$CaCO_3$ are practically equal [8, 9]. This means, that A_p increasing effect for nanocomposite HDPE/$CaCO_3$ in comparison with matrix polymer is due to nanofiller $CaCO_3$ action as a nucleator, promoting to crystallinity degree enhancement (see above).

Thus, within the frameworks of fractal analysis the quantitative correlation between impact toughness and structural characteristics is obtained for high density polyethylene and particulate-filled nanocomposites on its basis. An impact toughness increase is due to calcium carbonate action as a nucleator and corresponding change of amorphous phase. The theoretical calculation has shown a good correspondence to the experimental data [72].

THE INFLUENCE OF POLYMER PHASE MOLECULAR CHARACTERISTICS ON INTERFACIAL REGIONS FORMATION IN POLYMER NANOCOMPOSITES

As it is well known [16], the interaction degree or adhesion level for polyphase systems, including nanocomposites polymer/organoclay, influences essentially on their properties. The high level of interfacial adhesion polymer-organoclay is an obligatory condition of necessary for these materials characteristics reaching. The authors [77] carried out computer simulation of interaction energy for nanocomposites PA-6/MMT at 7 quaternary ammonium salts usage as coupling agent. This simulation demonstrated binding energy E_b between PA-6 and MMT growth at the indicated salts molecular volume V_m reduction. So, V_m decrease from 690.8 Å^3 up to 201.2 Å^3, i.e. in 3.4 times, results to E_b growth from 8.22 up to 215.1 kcal/mole, i.e. more than in 26 times. Such essential E_b change should be due to the change of quaternary salts molecules interaction with MMT platelet surface, which is a fractal object with dimension d_{surf}=2.78 [78]. Let us note, that the indicated d_{surf} value was obtained at using as a probe of nitrogen N_2 molecules, whose sizes are essentially smaller than quaternary salts molecules (N_2 molecule volume makes up ~ 17 Å^3), in case of method BET application. Therefore V_m increase from 17 up to 201.2-690.8 Å^3 will result to the appearance on MMT platelet surface a large number of "screened" (inaccessible for quaternary salts) contact sites, that will decrease sharply both effective value of its specific surface S_u and effective dimension of MMT platelet surface d_{surf}^{ef}. This effect results to the reduction of contacts binding agent – MMT number and, as consequence, to their interaction energy E_b decrease. The authors [79] carried out quantitative estimation of the indicated effect within the frameworks of fractal analysis.

As it is known [80], the effective specific surface of fractal object S_u depends on the adsorbed molecule cross-section σ_m as follows:

$$S_u \sim \sigma_m^{(2-d_{surf})/2}. \qquad (47)$$

Value σ_m for the considered ammonium salts can be estimated from geometrical considerations, assuming in the first approximation their spherical shape. The proportionality coefficient in the relationship (47) can be determined proceeding from BET results, where N_2 was used as an adsorbent [79]. In this case S_u=73.7 m^2/g, σ_m=8.04 Å^2, d_{surf}=2.78 and then the indicated coefficient is equal to 167.7 m^2/g. The estimations according to the equation (47) at the mentioned above proportionality coefficient showed that the effective value S_u for MMT and seven quaternary ammonium salts is changed within the limits of 28.5-39.4 m^2/g.

As it was expected E_b growth at S_u increasing was observed, moreover this growth was stronger than a linear one. Therefore in Figure 29 the dependence of E_b on S_u^2 is adduced (such correlation shape was chosen with the purpose of its linearization). At S_u≈29.6 m^2/g this dependence supposes zero E_b value and in general case the indicated correlation was described by the following empirical equation [79]:

$$E_b = 0.265\left(S_u^2 - 880\right), \text{ kcal/mole}, \qquad (48)$$

where value S_u, as earlier, is given in m^2/g. Thus, the equation (48) allows E_b estimation according to the known S_u values. The change of S_u results to the corresponding variation of MMT platelet surface fractal dimension d_{surf}^{ef}, which can be calculated with the aid of the following relationship [16]:

$$S_u = 410L^{(d_{surf}-d)}, \qquad (49)$$

where L is characteristic size of MMT platelet, which is accepted here equal to 100 nm [23].

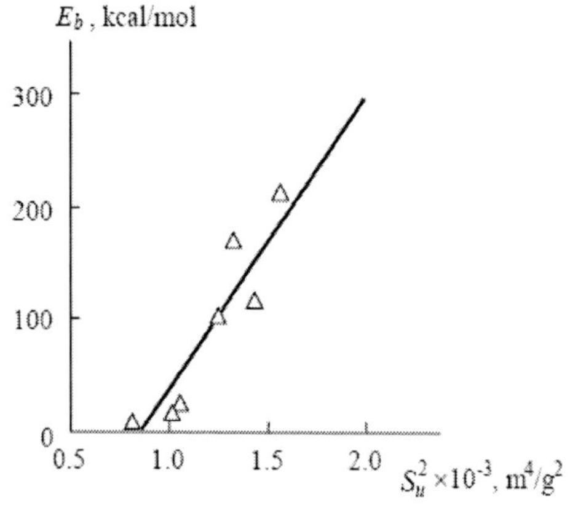

Figure 29. The dependence of PA-6-MMT binding energy E_b on MMT effective specific surface S_u [79].

The estimations according to the equation (49) for initial MMT (S_u=73.7 m^2/g) give the value d_{surf}^{ef}=2.61, that is close to this dimension magnitudes, measured by other methods [78]. Reduction S_u up to 28.5-39.4 m^2/g in case of quaternary ammonium salts usage causes d_{surf}^{ef} decrease up to 2.25-2.34, that defines MMT platelets "unscreened" (accessible for contact) surface dimension d_u reduction. This dimension is estimated as follows [81]:

$$d_u = \left(d_{surf}^{ef} - 1\right) + \frac{d - d_{surf}^{ef}}{d_w}, \quad (50)$$

where d_w is salts molecule random walk dimension, which is accepted equal to 2.

The number of ammonium salts molecules with MMT platelets contacts N_u can be estimated according to the relationship [81]:

$$N_u \sim L^{d_u}, \quad (51)$$

where L is accepted equal to half-sum of MMT platelet length and width, i.e. 67.5 nm.

The estimations according to the relationships (50) and (51) show N_u reduction for the smallest S_u value in comparison with the initial MMT in 3 times, that corresponds to N_u reduction for the same conditions approximately in 2.6 times [79].

Although at present theoretical models for elasticity modulus change description for nanocomposites, filled with organoclay, based on the notion of the so-called "effective particle" [23], exist, but these models do not take into consideration two factors, which are the most important for theoretical description of two-phase polymer systems: interfacial adhesion level and polymer matrix molecular characteristics. The first from the indicated factors gains a particular meaning after discovering of the nanoadhesion effect, which supposes availability of interfacial adhesion much higher level in polymer nanocomposites in comparison with traditional composites, filled with particles (fibers) of micron sizes [82]. Therefore the authors [83] carried out the theoretical research of elasticity modulus enhancement at the introduction of montmorillonite in polyimide, accounting for the indicated above factors.

The first series of organically modified MMT (MMT-16C) was prepared via ion exchange reaction in water using concentrated hydrochloric acid, the second MMT series (MMT-OM-m) was obtained also via reaction in water using N-[4-(4'-aminophenoxy)] phenyl phthalimide. Nanocomposites on the basis of polyimide (PI), filled with MMT (PI/MMT-16C and PI/MMT-OM-m) were prepared by polymerization in situ, the details of which are adduced in papers [39, 84].

The reinforcement degree E_n/E_m calculation for the considered nanocomposites on the basis of PI according to the equation (29) gave this parameter understated values: E_n/E_m=1.15-1.59, whereas reinforcement degree experimental magnitudes are within the range of 1.11-3.65 [39, 84]. The cause of this discrepancy is obvious enough: the equation (29) accounts for nanocomposite polymer chain flexibility with the aid of the parameter l_{st}, but does not account for the interfacial adhesion real level. The last factor accounting can be fulfilled with the aid of the parameter b, which is used at composites thermal expansion description and is determined according to the formula [85]:

$$\alpha_n = \alpha_n^{mix} - b\left(\alpha_n^{mix} - \alpha_n^T\right), \tag{52}$$

where α_n, α_n^{mix} and α_n^T are nanocomposite thermal expansion coefficient values, obtained experimentally and calculated according to the mixtures rule and Terner equation, accordingly.

Since the calculation according to Terner equation gives lower limiting magnitude of thermal expansion coefficient, then from the equation (52) it follows b=1 at $\alpha_n=\alpha_n^T$, i.e. at perfect adhesion reaching (see the equation (26)). The values b within the range of ~ -0.19-1.39 were obtained for a large number of polymer composites with various polymer matrices and fillers [85]. The nanoadhesion effect supposes much higher values b in nanocomposites case. So, for particulate-filled nanocomposite phenylone/aerosil the value b can reach ~ 12, that corresponds to the condition $\alpha_n<\alpha_n^T$ [86]. The value of parameter b or interfacial adhesion level can be estimated by mechanical tests results as follows. A polymer nanocomposites reinforcement degree can be determined according to the percolation relationship (8). The following relationship between φ_{if} and φ_n was obtained for exfoliated layerend silicates [16]:

$$\varphi_{if} = 1.91\varphi_n b. \tag{53}$$

Thus, the equation (53) demonstrates, that in the nanoadhesion case the value φ_{if} increases in b times in comparison with the perfect adhesion, for which b=1. The equation (53) substitution in the relationship (8) gives the following formula [83]:

$$\frac{E_n}{E_m} = 1 + 11\left[\varphi_n(1 + 1.91b)\right]^{1.7}, \tag{54}$$

which accounts for interfacial adhesion (nanoadhesion) level at reinforcement degree determination and it is true for exfoliated organoclays.

Let us note, returning to the equation (29), that the sole parameter, on which interfacial adhesion level can influence, in this equation is statistical segment length. One can suppose, that the enhancement of interfacial adhesion level, characterized by parameter b growth, or polymer matrix interaction with montmorillonite platelets surface intensification restricts polymer chains mobility in interfacial layer and results to larger rigidity of their chains in comparison with bulk polymer matrix or l_{st} growth, that can be expressed with the aid of the statistical segment effective length l_{st}^{ef} [83]:

$$l_{st}^{ef} = l_0 C_\infty b \tag{55}$$

Accounting for the equation (55) the relationship (29) can be written as follows [83]:

$$\frac{E_n}{E_m} = 1 + 0.19 W_n^{1/2} l_{st} b \tag{56}$$

The value b can be calculated with the aid of the equation (54). These estimations showed, that for the considered nanocomposites the parameter b varied within the limits of 4.09-5.31, i.e. for these materials the nanoadhesion was observed (b>1).

In Figure 30 the comparison of experimental E_n/E_m and calculated according to the equation (56) $(E_n/E_m)^T$ reinforcement degree values for nanocomposites polyimide/montmorillonite is adduced. As one can see, in this case the good correspondence of theory and experiment is obtained: $(E_n/E_m)^T$ value varies within the limits of 1.56-3.71 and its average discrepancy with experimental values E_n/E_m makes up ~ 16 %.

Hence, the stated above results demonstrated strong influence of interfacial adhesion (nanoadhesion) level on polymer nanocomposites reinforcement degree, which are filled with organoclay. Intensification of interaction nanofiller-polymer matrix, characterized by parameter b, results to essential (in b times) polymer chains flexibility change in interfacial layer in comparison with bulk polymer matrix. The offered modified equation for nanocomposites reinforcement degree determination, accounting for both polymer matrix molecular characteristics and interfacial adhesion level, gives the good correspondence to experiment. At present it is supposed, that interfacial interactions intensification (increase b) facilitates polymer chains arrangement on organoclay platelets surface, increases interfacial regions fraction and rises nanocomposites reinforcement degree [16]. The authors [87] considered more general variant of polymer matrix molecular characteristics (C_∞) and interfacial adhesion level (b) interconnection on the example of nanocomposites LDPE/organoclay [88].

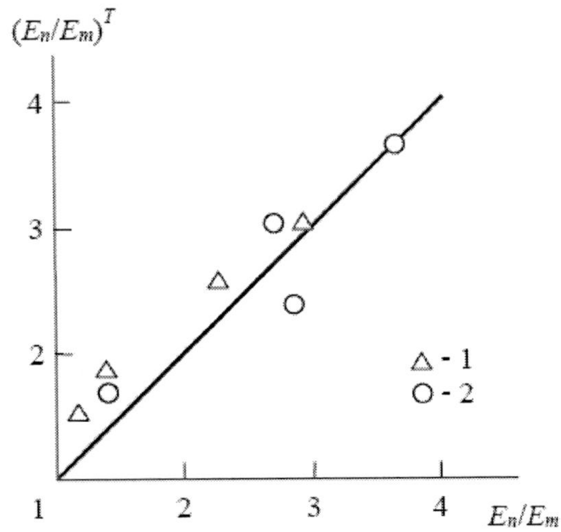

Figure 30. The comparison of experimental E_n/E_m and calculated according to the equation (56) $(E_n/E_m)^T$ reinforcement degree values of nanocomposites PI/MMT-16C (1) and PI/MMT-OM-m (2) [83].

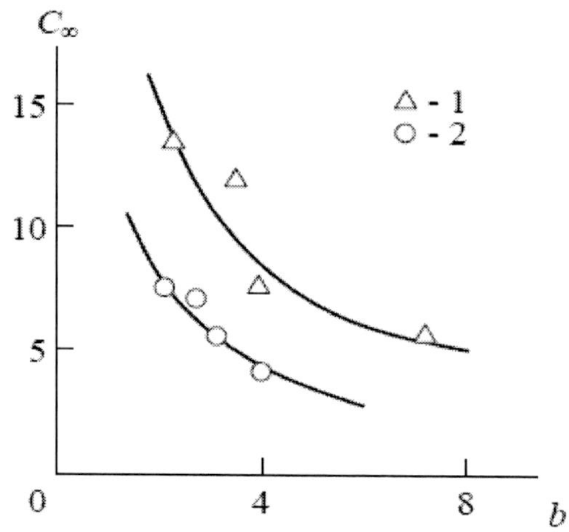

Figure 31. The dependence of characteristic ratio C_∞ on parameter b for nanocomposites LDPE/MMT-1 (1) and LDPE/MMT-2 (2) [87].

Two kinds of Na^+-montmorillonite were used as nanofiller: the first of them was industrial sample of mark Cloisite 20A (MMT-1) and the second was an experimental sample with application as compatibilized agent of trimethyl hydrogenated-tallow ammonium (MMT-2). In Figure 31 the dependence of characteristic ratio C_∞ on parameter b for two considered series of nanocomposites LDPE/MMT is adduced. As one can see, interfacial adhesion level, characterized by parameter b, reduction results to C_∞ growth, i.e. to l_{st} increase (the equation (27)), since l_0 value is constant and equal to 0.154 nm for polyethylenes [32]. At the first sight such result seems unexpected. Reduction b is due to organoclay contents increase, this effect is common one for nanocomposites polymer/organoclay and is defined by silicate platelets aggregation [16]. In practice this effect is reflected in E_n/E_m growth decelaration or its cessation at W_n increasing within the range of 5-10 mass % [24]. According to the stated above results, interfacial adhesion intensification should be promoted to polymer macromolecular coil "stretching" on montmorillonite platelets surface and, hence, to C_∞ increase. The observed discrepancy can be explained by the obvious fact, that macromolecular coil "stretching" is defined not only by interfacial interaction degree or parameter b, but by nanofiller particles surface area, on which this "stretching" is realized. Since the indicated area is proportional to nanofiller contents W_n, then in Figure 32 the dependence of C_∞ on complex parameter bW_n is adduced, which, as it has been expected, shows C_∞ increase (and, hence, l_{st} enhancement) at bW_n growth. In other words, the increase of both factors (b and W_n, where W_n characterizes contacts nanofiller-polymer number enhancement) promotes to C_∞ growth and nanocomposites reinforcement degree rising (see the equations (27) and (56)). The relation between C_∞ and bW_n, shown in Figure 30, can be expressed analytically as follows [87]:

$$C_\infty = 2 + 0.6bW_n \tag{57}$$

The characteristic feature of the dependence $C_\infty(bW_n)$ is its extrapolation to C_∞=2 at bW_n=0. Since the condition W_n=0 for nanocomposites has no physical significance, then the indicated extrapolation means, that at interfacial adhesion absence or b=0 macromolecular coil has only tetrahedral valent angles, i.e. limitingly compact macromolecular coil is formed, incapable to "stretching" on silicate platelets surface.

Since bW_n increase results to C_∞ growth, then in this case, interfacial regions relative fraction φ_{if} enhancement, which can be determined with the aid of the equation (8), should be expected. In Figure 33 the dependence of φ_{if} on complex parameter bW_n for the considered nanocomposites is adduced, which is approximated well enough by linear correlation, passing through coordinates origin and described analytically as follows [87]:

$$\varphi_{if} = 1.75\times10^{-2} bW_n \tag{58}$$

The equations (8) and (58) combination demonstrates clearly an interfacial adhesion level role in nanocomposites polymer/organoclay reinforcement process. So, for nanocomposites LDPE/MMT-1 W_n increase from 1 up to 7 mass % results to b reduction from 7.2 up to 2.1. If the value b were not reduced at W_n growth and at W_n=7 mass % were equal to 7.2, then nanocomposite elasticity modulus E_n would reach the value of 2167 MPa, whereas experimental E_n magnitude in this case is equal only to 569 MPa. As it follows from the equation (8), the greatest possible magnitude E_n/E_m=12. At value b=7.2 conservation this magnitude might reach at W_n only 7.6 mass % [87].

Thus, the adduced above results demonstrate that characteristic ratio C_∞, which is polymer phase important molecular parameter, depends equally on both interfacial adhesion level and organoclay contents. The minimum value C_∞=2 is reached at interfacial interactions absence irrespective of organoclay contents. The interfacial adhesion level crucial role in nanocomposites polymer/organoclay reinforcement process is shown. It is well known [23], that interlayer spacing d_{001} is one of the most important characteristics of nanofiller in nanocomposites polymer/organoclay. This parameter characterizes dispersion degree of organoclay in polymer matrix.

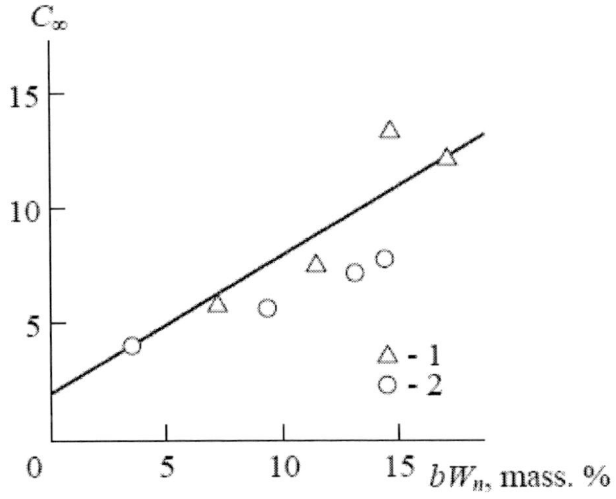

Figure 32. The dependence of characteristic ratio C_∞ on complex parameter bW_n for nanocomposites LDPE/MMT-1 (1) and LDPE/MMT-2 (2) [87].

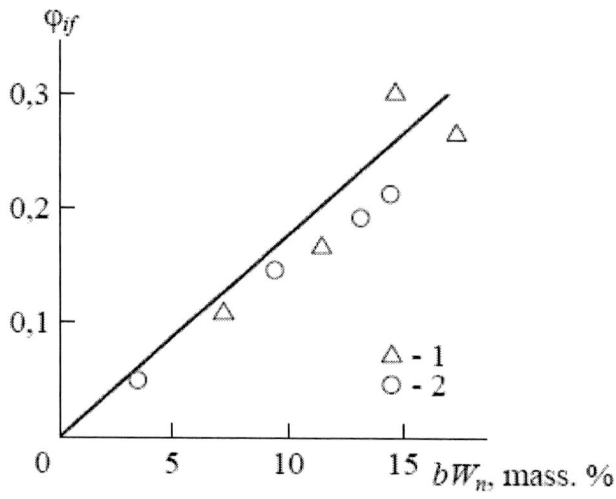

Figure 33. The dependence of interfacial regions relative fraction φ_{if} on complex parameter bW_n for nanocomposites LDPE/MMT-1 (1) and LDPE/MMT-2 (2) [87].

The value $d_{001}\approx$2-3 nm corresponds to intercalated organoclay and $d_{001}\approx$8-10 nm – to exfoliated one. Layered nanofillers dispersion often (if not always) has the cruiral significance in reinforcement process of nanocomposites polymer/organoclay [16, 23]. It is obvious, that interfacial adhesion level will be influenced on organoclay dispersion degree. An organoclay by no chance is introduced in polymer matrix only with a modifier, that allows to rise interfacial adhesion nanofiller-polymer matrix level [89].

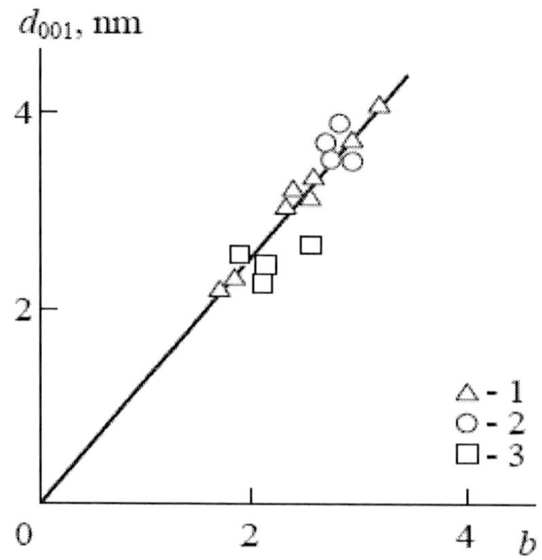

Figure 34. The dependence of interlayer spacing d_{001} on parameter b for nanocomposites HDPE/organoclay (1), PP/organoclay (2) and HDPE/montmorillonite [91] (3) [90].

In Figure 34 the dependence $d_{001}(b)$ is adduced, from which interlayer spacing linear increase follows at interfacial adhesion level growth. This dependence is approximated by the following empirical formula [90]:

$$d_{001} = 1.27b \text{ ,nm.} \quad (59)$$

At b=0 (interfacial adhesion absence) d_{001}=0, i.e. packet (tactoid) of silicate platelets is not split up (not dispersed). Exfoliation, i.e. the condition d_{001}=8-10 nm reaching, is realized at b=6.30-7.90. The adduced results confirm, that nanoadhesion effect realization is necessary for exfoliated organoclay receiving. In Figure 34 the dependence of d_{001} values, obtained experimentally (with the aid of X-raying), on b for nanocomposites HDPE/MMT according to the data of paper [91]. As one can see, these data correspond well to the correlation $d_{001}(b)$, calculated theoretically [90].

Conclusions

Thus, the obtained in the present review results have demonstrated, that polymer phase in polymer nanocomposites of any class is a complex system, moreover dividing into two structurally distinguishing parts: bulk polymer matrix and interfacial regions. As it has been expected, the characteristic ratio, which characterizes polymer chain statistical flexibility, is the main parameter, defining polymer phase structure changes (crystallinity degree, interfacial regions relative fraction and so on) at nanofiller introduction. The interfacial adhesion nanofiller-polymer matrix level influences strongly on the indicated characteristic. The relationships, obtained with accounting for these factors real values, allow to predict polymer nanocomposites limiting characteristics.

References

[1] Ahmed S., Jones F.R. *J. Mater. Sci.*, 1990, v. 25, № 12, p. 4933-4942.
[2] Bobryshev A.N., Kozomazov V.N., Babin L.O., Solomatov V.I. *Synergetics of Composite Materials.* Lipetsk, NPO ORIUS, 1994, 154 p.
[3] Novikov V.U., Kozlov G.V. *Mekhanika Kompozitnykh Materialov*, 1999, v. 35, № 3, p. 269-290.
[4] Yanovskii Yu.G. *Nanomechanics and Composite Materials Strength*. Moscow, Publishers IPRIM RAS, 2008, 180 p.
[5] Solomko V.P. A Filled Crystallizing Polymers. Kiev, *Naukova Dumka*, 1980, 264 p.
[6] Aloev V.Z., Kozlov G.V. The Physics of Orientation Phenomena in Polymer Materials. Nal'chik, *Polygraphservice and T,* 2002, 288 p.
[7] Kozlov G.V., Zaikov G.E. *Izvestiya* KBSC RAS, 2003, № 1, p. 54-57.
[8] Tanniru M., Misra R.D.K. *Mater. Sci. Engng*., 2005, v. 405A, № 1, p. 178-193.
[9] Deshmane C., Yuan Q., Misra R.D.K. *Mater. Sci. Engng*., 2007, v. 452-453A, № 3, p. 592-601.
[10] Dolbin I.V., Burya A.I., Kozlov G.V. *Kompozitnye Materialy*, 2008, v. 2, № 1, p. 3-7.
[11] Aphashagova Z.Kh., Kozlov G.V., Malamatov A.Kh. *Nanotekhnika,* 2009, № 1, p. 40-44.

[12] Wunderlich B. Macromolecular Physics. Vol. 2. *Crystal Nucleation, Growth, Annealing.* New York, San Francisco, London, Academic Press, 1976, 556 p.
[13] Aloev V.Z., Kozlov G.V., Zaikov G.E. *Russian Polymer* News, 2001, v. 6, № 4, p. 63-66.
[14] Kozlov G.V., Zaikov G.E. *Structure of the Polymer Amorphous State*. Leiden-Boston, Brill Academic Publishers, 2004, 465 p.
[15] Wu S. *J. Appl. Polymer Sci.*, 1992, v. 46, № 4, p. 619-624.
[16] Mikitaev A.K., Kozlov G.V., Zaikov G.E. *Polymer Nanocomposites: the Variety of Structural Forms and Applications*. New York, Nova Science Publishers, Inc., 2009, 319 p.
[17] Pfeifer P. In book: *Fractals in Physics*. Ed. Pietronero L., Tosatti E. Amsterdam, Oxford, New York, Tokyo, North-Holland, 1986, p. 72-81.
[18] Privalko V.P., Kawai T., Lipatov Yu.S. *Polymer J.*, 1979, v. 11, № 9, p. 699-709.
[19] Kozlov G.V., Zaikov G.E. *Materialovedenie,* 2002, № 12, p. 13-17.
[20] Shklovskii B.I., Efros A.L. *Uspekhi Fizicheskikh Nauk*, 1975, v. 117, № 3, p. 401-436.
[21] Gopakumar T.G., Lee J.A., Kontopoulou M., *Parent J.S. Polymer*, 2002, v. 43, № 16, p. 5483-5491.
[22] Dzhangurazov B.Zh., Kozlov G.V., Ovcharenko E.N., Mikitaev A.K. *Prikladnaya Fizika,* 2011 (in press).
[23] Sheng N., Boyce M.C., Parks D.M., Rutledge G.C., Abes J.I., Cohen R.E. *Polymer,* 2004, v. 45, № 2, p. 487-506.
[24] Antipov E.M., Barannikov A.A., Gerasin V.A., Shklyaruk B.F., Tsamalashvili L.A., Fischer H.R., Razumovskaya I.V. *Vysokomolek. Soed.* A, 2003, v. 45, № 11, p. 1885-1899.
[25] Family F. *J. Stat. Phys.*, 1984, v. 36, № 5/6, p. 881-896.
[26] Kozlov G.V., Aphashagova Z.Kh., Malamatov A.Kh. *Nanotekhnika,* 2011 (in press).
[27] Budtov V.P. *Physical Chemistry of Polymer Solutions*. Sankt-Peterburg, Khimiya, 1992, 384 p.
[28] Balankin A.S. *Synergetics of Deformable Body*. Moscow, Publishers Ministry of Defence SSSR, 1991, 404 p.
[29] Kozlov G.V., Sanditov D.S. Anharmonic Effects and Physical-Mechanical Properties of Polymers. Novosibirsk, *Nauka,* 1994, 261 p.
[30] Wu S. *J. Polymer Sci.: Part B: Polymer Phys.*, 1989, v. 27, № 4, p. 723-741.
[31] Feder F. *Fractals.* New York, Plenum Press, 1989, 252 p.
[32] Aharoni S.M. *Macromolecules*, 1983, v. 16, № 9, p. 1722-1728.
[33] Kozlov G.V., Sultonov N.Zh., Mikitaev A.K. *Proceedings of VI Intern. Sci.-Techn. Conf. "Machines Study and Planning Problems*". Penza, PSU, 2010, p. 87-93.
[34] Sultonov N.Zh., Mikitaev A.K. Proceedings of VI Intern. Sci.-*Pract. Conf. "New Polymer Composite Materials*". Nal'chik, KBSU, 2010, p. 392-398.
[35] Kalinchev E.L., Sakovtseva M.B. *Properties and Processing of Thermoplastics*. Leningrad, Khimiya, 1983, 288 p.
[36] Aharoni S.M. *Macromolecules*, 1985, v. 18, № 12, p. 2624-2630.
[37] Chang J.-H., An Y.U., Kim S.J., *Im S. Polymer,* 2003, v. 44, № 15, p. 5655-5661.
[38] Chang J.-H., Seo B.-S., Wang D.-H. *Polymer,* 2002, v. 43, № 7, p. 2969-2974.
[39] Liang Z.-M., Yin J., Xu H.-*J. Polymer*, 2003, v. 44, № 4, p. 1391-1399.

[40] Chen J.-S., Poliks M.D., Ober C.K., Zhang Y., Wiesner U., Giannelis E. *Polymer*, 2002, v. 43, № 13, p. 4895-4904.
[41] Huang J., He C., Xiao Y., Mya K.Y., Dai J., Siow Y.P. *Polymer*, 2003, v. 44, № 12, p. 4491-4499.
[42] Tugov I.I., Shaulov A.Yu. *Vysokomolek. Soed.* B, 1990, v. 32, № 7, p. 527-529.
[43] Kozlov G.V., Yanovskii Yu.G., Zaikov G.E. *Structure and Properties of Particulate-Filled Composites: the Fractal Analysis*. New York, Nova Science Publishers, Inc., 2010, 349 p.
[44] Malamatov A.Kh., Burya A.I., Kozlov G.V. *Sovremennye Naukoemkie Tekhnologii*, 2005, № 11, p. 16-18.
[45] Kozlov G.V., Burya A.I., Lipatov Yu.S. Doklady NAN Ukraine, 2008, № 1, p. 132-136.
[46] Kozlov G.V., Burya A.I., Zaikov G.E. J. Balkan Tribologic. Assoc., 2007, v. 13, № 4, p. 475-479.
[47] Kozlov G.V., Yanovskii Yu.G., Zaikov G.E. *Synergetics and Fractal Analysis of the Polymer Composites Filled with Short Fibers.* New York, Nova Science Publishers, Inc., 2010, 248 p.
[48] Yoon P.J., Hunter D.L., Paul D.R. *Polymer*, 2003, v. 44, № 14, p. 5323-5339.
[49] Kozlov G.V., Yanovskii Yu.G., Lipatov Yu.S. *Mekhanika Kompozitsionnykh Materialov i Konstruktsii*, 2002, v. 8, № 1, p. 111-149.
[50] Hentschel H.G.E., *Deutch J.M. Phys. Rev.* A, 1984, v. 29, № 12, p. 1609-1611.
[51] Malamatov A.Kh., Kozlov G.V., Antipov E.M., Mikitaev A.K. *Perspektivnye Materialy*, 2006, № 5, p. 54-58.
[52] Chen T.K., Tien Y.I., Wei K.H. *Polymer,* 2000, v. 41, № 3, p. 1345-1353.
[53] Kim B.K., Seo J.W., Jeong H.M. *Eur. Polymer J.*, 2003, v. 39, № 1, p. 85-91.
[54] Dzhangurazov B.Zh., Kozlov G.V., Mikitaev A.K. P*roceedings of 29th Intern. Conf. "Composite Materials in Industry"*. Kiev, Nauka, Tekhnika, Tekhnologiya, 2009, p. 325-327.
[55] Aphashagova Z.Kh., Kozlov G.V., Burya A.I., Mikitaev A.K. *Materialovedenie,* 2007, № 9, p. 10-13.
[56] Edwards D.C. *J. Mater. Sci.,* 1990, v. 25, № 12, p. 4175-4185.
[57] Kozlov G.V., Burya A.I., Lipatov Yu.S. *Mekhanika Kompozitnykh Materialov*, 2006, v. 42, № 6, p. 797-802.
[58] Kozlov G.V., Yanovskii Yu.G., Karnet Yu.N. *Mekhanika Kompozitnykh Materialov i Konstruktsii,* 2005, v. 11, № 3, p. 446-450.
[59] Dzhangurazov B.Zh., Kozlov G.V., Mikitaev A.K. Proceedings of the Third Intern. Conf. "*Deformation and Fracture of Materials and Nanomaterials – DFMN-2009*", Moscow, 2009, p. 142-143.
[60] Kozlov G.V., Serdyuk V.D., Dolbin I.V. *Materialovedenie,* 2000, № 12, p. 2-5.
[61] Aphashagova Z.Kh., Kozlov G.V., Malamatov A.Kh. *International Journal of Applied and Fundamental Researches*, 2009, № 3, p. 38-41.
[62] Dzhangurazov B.Zh., Kozlov G.V., Zaikov G.E., Mikitaev A.K. *Nanomechanics Sci. and Techn.*, 2010, v. 1, № 1, p. 31-48.
[63] Khabashesku V.N., Barrera E.V., Lobo R.F.M. *Current Research on Nanotechnology*, 2007, v. 1, № 2, p. 165-182.

[64] Yanovskii Yu.G., Kozlov G.V., Burya A.I., Lipatov Yu.S. *Fizicheskaya Mezomekhanika*, 2007, v. 10, № 6, p. 63-67.
[65] Wu S.-H., Wang F.-Y., Ma C.-C.M., Chang W.-C., Kuo C.-T., Kuan H.-C., Chen W.-*J. Mater. Lett.*, 2001, v. 49, № 2, p. 327-333.
[66] Kozlov G.V., Burya A.I., Aphashagova Z.Kh., Mikitaev A.K. *Nanotekhnika,* 2008, № 2, p. 33-36.
[67] Zaikov G.E., Kozlov G.V., Aphashagova Z.Kh., Mikitaev A.K. *J. Appl. Polymer Sci.,* 2009, v. 111, № 5, p. 2621-2624.
[68] Kausch H.H. *Polymer Fracture*. Berlin, Heidelberg, New York, Springer-Verlag, 1978, 423 p.
[69] Ryabenko A.G. Proceedings of Intern. Conf. "*Deformation and Fracture of Materials and Nanomaterials*", DFMN-2007, Moscow, 2007, p. 411-412.
[70] Bucknall C.B. Toughened Plastics. London, *Applied Science*, 1977, 328 p.
[71] Bartenev G.M., Frenkel S.Ya. *Physics of Polymers*. Leningrad, Khimiya, 1990, 432 p.
[72] Kozlov G.V., Misra R.D.K., Aphashagova Z.Kh. *Nanotekhnika,* 2009, № 2, p. 71-74.
[73] Kuzeev I.R., Samigullin G.Kh., Kulikov D.V., *Zakirnichnaya* M.M. A Complex Systems in Nature and Engineering. Ufa, Publishers USNTU, 1997, 225 p.
[74] Aphashagova Z.Kh., Kozlov G.V., Burya A.I., Malamatov A.Kh. *Proceedings of Khar'kovskoi Nanotekhnologicheskoi Assamblei,* Khar'kov, KhSU, 2007, p. 204-205.
[75] Bashorov M.T., Kozlov G.V., Mikitaev A.K. *Deformatsiya i Razrushenie Materialov*, 2009, № 11, p. 37-39.
[76] Kozlov G.V., Zaikov G.E. In book*: Fractals and Local Order in Polymeric Materials*. Ed. Kozlov G., Zaikov G. New Yprk, Nova Science Publishers, Inc., 2001, p. 55-63.
[77] Fermeglia M., Ferrone M., Pricl S. *Fluid Phase Equilibrium*, 2003, v. 212, № 2, p. 315-329.
[78] Pernyeszi T., Dekany I. *Colloid Polymer Sci*., 2003, v. 281, № 1, p. 73-78.
[79] Dzhangurazov B.Zh., Kozlov G.V., Mikitaev A.K. Proceedings of V Intern. Sci.-Techn. Conf. "*Progressive Technologies in Modern Mechanical Engineering*". Penza, PSU, 2009, p. 35-38.
[80] Avnir D., Farin D., Pfeifer P. *Nature,* 1984, v. 308, № 5959, p. 261-263.
[81] Stanley H.E. In book: *Fractals in Physics*. Ed. Pietronero L., Tosatti E. Amsterdam, Oxford, New York, Tokyo, North-Holland, 1986, p. 463-477.
[82] Kozlov G.V., Burya A.I., Yanovskii Yu.G., Aphashagova Z.Kh. *Nanotekhnika,* 2008, № 1, p. 81-85.
[83] Dzhangurazov B.Zh., Kozlov G.V., Mikitaev A.K. *Proceedings of Inter. Meeting "Low Dimensional Systems*". Rostov-na-Donu, SKHTs VSh YuFU, 2010, p. 126-129.
[84] Liang Z.-M., Yin J., Wu J.-H., Qiu Z.-X., He F.F. *Eur. Polymer J.,* 2004, v. 40, № 2, p. 307-314.
[85] Holliday L., Robinson J.D. In book: Polymer Engineering Composites. Ed. Richardson M.O.W. London, *Applied Science Publishers* LTD, 1978, p. 241-283.
[86] Kozlov G.V., Burya A.I., Yanovskii Yu.G., Zaikov G.E. In book: *Progress in Monomers, Oligomers, Polymers, Composites and Nanocomposites*. Ed. Pethrick R., Zaikov G., Pielichowski J. New York, Nova Science Publishers, Inc., 2009, p. 343-359.
[87] Dzhangurazov B.Zh., Kozlov G.V., Mikitaev A.K. *Kondensirovannye Sredy i Mezhfaznye Granitsy*, 2010, v. 12, № 2, p. 119-122.
[88] Hotta S., Paul D.R. *Polymer*, 2004, v. 45, № 22, p. 7639-7654.

[89] Ranage A., Nayak K., Fairbrother D., D'Souza N.A. *Polymer,* 2005, v. 46, № 21, p. 7323-7333.
[90] Dzhangurazov B.Zh., Kozlov G.V., Mikitaev A.K. *Poverkhnost'*, 2011 (in press).
[91] Gianelli W., Ferrara G., Camino G., Pellegatti G., Rosenthal J., Trombini R.C. *Polymer*, 2005, v. 46, № 21, p. 7037-7046.

In: Advances in Polymer Research
Editors: Losif Sidorov, Fedor Nikitin, pp. 157-184
ISBN: 978-1-61942-223-0

TRIBOTESTING OF POLYMERS IN LINE CONTACT

***Pieter Samyn*[1]*, Gustaaf Schoukens*[1] *and Patrick De Baets*[2]**

[1] Ghent University, Department of Textiles, Technologiepark 907, B-9052 Zwijnaarde, Belgium.
[2] Ghent University, Department of Mechanical Construction and Production, Sint-Pietersnieuwstraat 41, B-9000 Gent, Belgium.

ABSTRACT

The friction and wear properties of polymers are difficult to predict theoretically, due to the many parameters that influence the tribological behaviour. Knowledge about the tribological properties of self-lubricating properties of polymers is therefore mainly gathered from experimental testing, which is often closely related to the final application conditions. In contrast to traditional pin-on-disc test set-ups, we present the analysis of a contraformal cylinder-on-plate contact with initial line contact providing high contact pressures and promoting polymer transfer to the counter-face at running-in. Especially, different types of polyimide cylinders are tested because of their good thermal and mechanical properties and resistance against high initial contact pressures. The statistical variation in test results is discussed in relation to variations in the sample geometry.

There is a statistical variation of about 7 to 9% on dynamic coefficients of friction, about 10 to 12% on static coefficients of friction, about 8 to 9% on wear rates from weight loss and about 12 to 15% from wear rates from dimensional measurements. The on-line and off-line wear measurements can be combined to distinguish between real material loss, visco-elastic deformation, plastic deforma-tion and thermal expansion of the polymer samples. The transition between running-in and steady-state sliding conditions are further detailed, in relation to the visco-elastic deformation and variations in macroscopic contact situation into a more conformal contact during sliding.

Finally, an analytical expression for calculating the maximum polymer surface temperature T* has been developed for the present sliding configuration.

Keywords: Friction, Wear, Sliding, Contact, Temperature.

---

* Corresponding author: Pieter.Samyn@UGent.be. Pieter Samyn is a Post-doctoral Research Fellow of the Research Fuondation Flanders (F.W.O.), Belgium

1. INTRODUCTION

Tribological properties of polymers are not intrinsic material properties, in contrast to strength, elasticity modulus or other mechanical properties. General models predicting the tribological response of a polymer/steel contact are scarcely available due to many influencing parameters. Some models only apply to a single polymer [1]. The friction and wear perfor-mance is strongly influenced by the entire test system (material pair, contact geometry, normal load, sliding velocity, mechanical stiffness, etc.). For analysis of polymer elements in relation to working conditions occurring in real constructions, test conditions should simulate the dominant sliding mode as close as possible, including the structures of the material sliding couple, contact geometries, contact pressures, type of sliding motion, environmental condition, mechanical stiffness, etc. at the macro- and microlevel. As a result, many tribotest configura-tions are not standardized and fit-to-purpose test equipment is often built. The following considerations should be made to select a tribological testing program:

- Material structure: The macro- and microstructure of two contact bodies (polymer wear sample and steel counterface) should agree with practical applications. For microstructures being small relatively to the contact surface, the material behaves homogeneously and test results do not strongly depend on the contact size. For large microstructures such as internally lubricated polymers and composites, the contact size influences test results importantly. Surface finishing by sand blasting, grinding or polishing should be adequately chosen with representative roughness. The surface conditions often are different at running-in compared to steady-state conditions due to wear and/or transfer.
- Contact conditions: Conformal contacts (bodies with similar radius of curvature, e.g. flat-on-flat in pin-on-disc) and counter formal contacts (bodies with different radius of curvature, e.g. line and point contact in cylinder-on-plate or ball-on-disc) influence the contact pressure distribution over the contact area and the mobility of wear debris in the sliding interface. Debris particles can be immediately removed or interact to form a polymer transfer film. When wear debris remains circulating into the contact, it is called a closed tribosystem such as fretting contact. When wear debris is immediately removed from the sliding contact, it is called an open tribosystem such as gear contact.
- Sliding motion and sliding velocity: Unidirectional, reciprocating or multi-axial sliding motions and the distance of the sliding motion relatively to the contact length (mutual overlap coefficient) influence the wear debris mobility and dissipation of frictional heat. When contact is intermediately disrupted, chemical reactions with the atmosphere, cooling and/or relaxation of polymer samples influence friction and wear properties. The sliding velocity influences the visco-elastic behaviour of polymer molecules.
- Energy dissipation: Friction and wear result from the dissipation of mechanical energy being transformed into deformation (hysteresis) and frictional heating. Sliding temperatures predominantly influence the sliding performance of polymers due to transitions from the glassy into rubbery and from the rubbery into melting phase. The sliding properties for polymers are generally characterized by a pv sliding condition

(contact pressure x sliding velocity). Additional heating or cooling of the contact surfaces is often required as dissipation of energy artially depends on the test-rig construction acting as a heat sink.

In conventional tribotesting, small-scale tests are preferred due to their cost- and time effectiveness with easy handling of test specimens (apparent contact area 1 to 10^2 mm^2). An appropriate test set-up can be selected according to ASTM and some DIN standards for each wear mode. The National Physics Laboratory (UK) identified worldwide more than 400 test methods with either scientific or industrial interest [2]. Under adhesive conditions either pin-on-disc (ASTM G99) or pin-on-plate (ASTM G133) is used in continuous or reciprocating motion. Hertz contact stresses are often desired during running-in, appearing in heavily loaded tribosystems such as gears, journal bearings or cams. Block-on-ring [3] (ASTM G88) tests are applied to simulate a line contact in unidirectional motion while cylinder-on-plate [4] (ASTM F732–98) tests are used in reciprocating motion. The effect of sliding geometry on tribological behaviour of polymers in line contact was recently compared [5]. Also small component tests are available to simulate strong conformal contacts (Figure 1).

Tribologists build fit-to-purpose equipments and many test configurations are not standardized. As a drawback, statistical variation in test results is often large and it is difficult to rely on a single experiment for design. At present, tribologists do not have a clear view on the statistical repeatability of test results and influencing factors. The purpose of this study is to investigate the reliability of sliding experiments of a polymer in line contact during reciprocating motion with large strokes. For oscillating sliding tests, the wear can be quantified more accurately compared to continuous sliding, because only a small material part is involved in the wear process. Polymer contacts are also influenced by visco-elastic deformation, being an additional parameter affecting the repeatability of tribological data. A discussion on the repeatability of test results and influencing parameters will contribute to the development of better tribotest procedures, being hardly available for polymers at present.

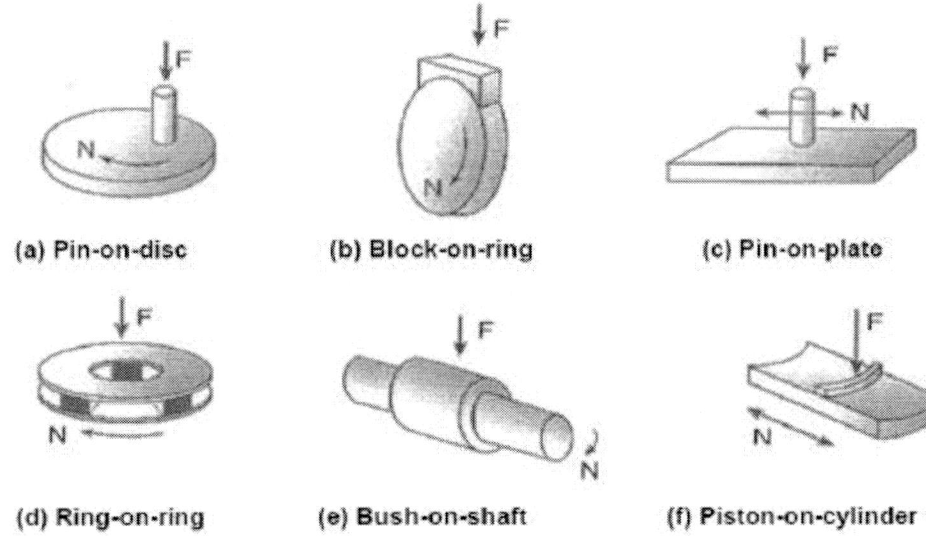

Figure 1. Conventional tribotesting configurations for adhesive wear testing.

2. Material Selection

Polyimides are candidates for friction and wear applications, presenting high thermal resistance, good mechanical strength and chemical inertness. Polyimide films were introduced under dry sliding for space applications [6] and tested under vacuum [7], in contact with cooling agents [8] or under fretting conditions [9]. In present research, sintered polyimides (SP) or thermoplastic polyimides (TP) are included, with properties listed in Table 1:

- Sintered polyimides under pure conditions (SP-1) consist of pure PMDA-ODA powders that are prepared from pyromellitic dianhydride (PMDA) and 4,40-diamino-diphenyl ether or oxydianyline (ODA). The sintered polyimides with graphite flake fillers (SP-2) contain 15 wt.-% powdery additives. The sintering process happens by pre-compacting the polyimide resins with initial grain size diameter of 10 to 20 μm into 100 μm powders for beneficial flow in the sintering mould. Sintering happens under pressures of 500 to 2000 bar at a temperature of about 300°C for 10 minutes.
- Thermoplastic polyimides under pure conditions (TP-1) are obtained by reacting pyromellitic dianhydride with 4,4'-bis(3-amino-phenoxy) biphenyl to give a polyamic acid precursor that is further cyclodehydrated. The ether links in the polymer backbone allow for higher molecular chain flexibility. The thermoplastic polyimides with PTFE internal lubricants (TP-2) contain 20 wt.-% of fillers. The samples are injection moulded parts with cylinder temperatures ranging between 380 to 410°C (typically 390°C), at 21 MPa moulding pressures and 100 to 200 rpm screw speeds.

Table 1. Mechanical and thermal properties of polyimide test materials

Property	Unit	SP-1	SP-2	TP-1	TP-2
Specific gravity	g/cm^3	1.34	1.42	1.33	1.44
Melting temperature	°C	None	None	388	388
Transition temperature	°C	None	None	250	250
Heat-deflection temperature	°C	360	360	238	230
Thermal conductivity	W/(mK)	0.30	0.46	0.17	0.17
Specific heat capacity	J/(gK)	1.13	1.13	1.01	1.01
Thermal diffusivity	10^{-7} m^2/s	1.98	1.87	1.27	1.16
Modulus of elasticity	MPa	2480	3170	2760	2400
Tensile stress at yield	MPa	No	No	94	72
Tensile stress at break	MPa	72	62	118	96
Tensile strain at break	%	7.5	5.5	90	60
Compressive stress at					
1% strain	MPa	24	23	25	25
2% strain	MPa	46	43	-	-
10% strain	MPa	112	104	122	122

3. Line-Contact Tribotesting Configuration

3.1. Small-Scale Testing Device

The small-scale cylinder-on-plate tester used for present studies is schematically shown in Figure 2 [10]. It is a universal PLINT TE 77 High Frequency tribotester for dry reciprocating sliding between a moving polymer cylinder and a fixed counterface plate. A line contact is chosen for easy alignment of the sliding pair and high contact pressures at running-in: it will promote transfer controlling steady-state sliding. The initial contact geometry is contra-formal with a contact area continuously growing as wear proceeds, called an 'increasing contact area' type of wear. The polymer cylinder (diameter 5 mm x width 15 mm) is mounted into a reciprocating moving arm by means of a clamp and aligning mechanism, preventing it from rolling during the test so that pure sliding is guaranteed. The oscillating motion of the arm is provided through an eccentric transmission, which converts the continuous rotational motion of a DC electromotor into a linear movement and is used for adjustment of the sliding stroke. The normal load is manually applied through a spring and lever system that pulls down a transverse bridge structure over the polymer specimen. Contact between the moving arm and the bridge structure is made by a roller bearing. The stationary steel counterface (58 mm x 38 mm x 4 mm) is fixed onto a plate holder that is connected to the base plate by means of four leaf springs with high stiffness in vertical direction and appropriate flexibility in horizontal direction. The counterface can be heated by four Vulstar heat cartridges (200 W, 400 Ω), continuously PID-controlled.

A piezo-electrical force transducer (range +/- 500 N, resolution 0.001 N) in contact with the stationary steel plate measures the horizontal friction force that the polymer specimen exerts on the counterface. The normal load is measured by a load cell connected to the spring and lever system.

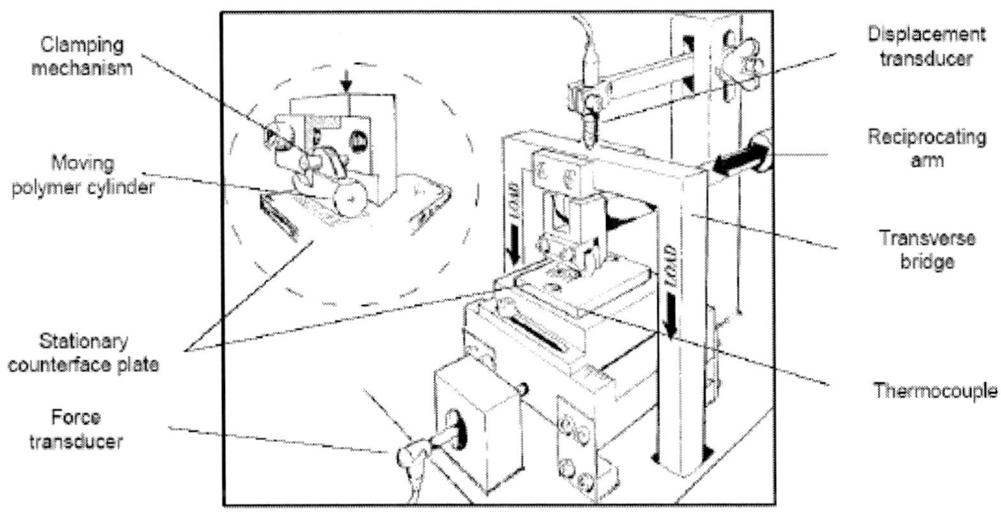

Figure 2. Reciprocating cylinder-on-plate tribotester.

A contactless inductive displacement transducer is positioned on top of the transverse bridge, measuring its displacement as the combined effect of wear and deformation of the polymer samples during sliding. Sliding temperatures are measured by a DIN 43710 K-type (nickel-chromium/nickel-aluminium) thermocouple positioned on top of the counterface at 10 mm from the end of the sliding stroke.

3.2. Small-Scale Test Conditions

Test conditions for small-scale tribotests are summarised in Table 2. The range for normal loads is 10 to 250 N, the sliding frequency range is 2,5 to 50 Hz and the sliding stroke can be varied between 0,1 to 15 mm. A selection of normal loads and sliding velocities includes both normal functioning and overload conditions for polymer elements. The sliding stroke is fixed at 15 mm as changing the eccentric mechanism could cause alignment problems and minimising the stroke would induce fretting. The test environment is surrounded by a climate box that is connected to an external climate conditioner. According to ASTM F732-98 “Reciprocating Pin-on-flat wear tests”, tests are done at controlled atmosphere of 23°C temperature and 60 % relative humidity. Counterfaces are rinsed with water and acetone and the polymer surface is cleaned with acetone after mounting in the test rig. Good alignment between the polymer sample and its counterface is ensured by applying a small preload (20 N) before tight fixation of the clamping mechanism. No external oil or grease lubricants are applied during sliding of polymers, relying on their self-lubricating abilities.

Table 2. Testing parameters for cylinder-on-plate sliding tests

Test parameter	Unit	Small-scale reciprocating cylinder-on-plate test
Normal load	N	50, 100, 150, 200
Sliding stroke	mm	15
Total sliding distance	m	15000
Sliding frequency	Hz	10, 20, 30, 40
Sliding velocity	m/s	0.3, 0.6, 0.9, 1.2
Counterface temperature	°C	no heating, 60, 80, 100, 120, 140, 180, 220, 260
Ambient temperature	°C	23 ± 2 (controlled standard atmosphere)
Relative humidity	%	60 ± 2 (controlled standard atmosphere)

3.3. Statistical Analysis and Influence of Test Geometry

The friction force is continuously measured as a function of the reciprocating sliding motion as shown in Figure 3 for five subsequent sliding cycles, with either a positive or negative value depending on the sliding direction. One sliding cycle (i.e. a double sliding stroke) contains 50 sample values, independent of the sliding velocity. This means that for a 10 Hz sliding test, where 100 sliding cycles take 10 seconds, the sample frequency is 500 Hz or the intermediate sampling time is 2 ms. For each 100 sliding cycles, however, only 5 are selected as shown in Figure 3 to calculate the friction force and the 95 following cycles are ignored. Within a single sliding stroke, the static friction force F_{stat} occurs at restart after reversal of the sliding direction and is calculated according to Formula (1). The dynamic friction force F_{dyn} occurs in the centre of the sliding stroke and is calculated as a root-mean-square value of the force transducer signal according to Formula (2), as the signal is often not centred around the zero value.

$$F_{stat} = \frac{\left|F_{stat,\min}\right| + \left|F_{stat,\max}\right|}{2} \tag{1}$$

$$F_{dyn} = \sqrt{\frac{1}{T}\int_0^T (F(t))^2\, dt} \tag{2}$$

The static and dynamic coefficient of friction are calculated as the ratio of friction force and applied normal load F_N, resulting in $\mu_s = F_{stat}/F_N$ and $\mu_d = F_{dyn}/F_N$. The evolution of the dynamic coefficient of friction μ_d as a function of the sliding distance is plotted in Figure 4a for sliding tests on sintered polyimide SP-1 at 50 N, 0.3 m/s. Test results forpolymer cylinders with fixed diameter 5 mm and width $2b$ either 12 or 15 mm are illustrated. Tests in this work will always be done on 15 mm cylinders, because of better alignment and larger available wear track to be analysed for transfer. For estimating the statistical repeatability of the small-scale test results, five tests are done on 15 mm cylinders with identical load and sliding velocity.

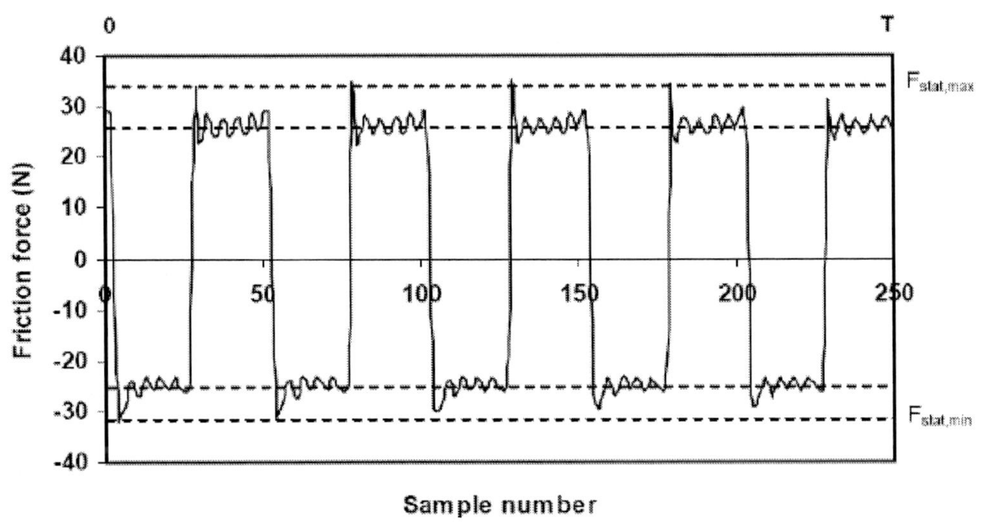

Figure 3. Sampling of the friction force over five reciprocating sliding cycles for SP-1, 50 N, 0.3 m/s on a small-scale cylinder-on-plate test against steel.

Coefficients of friction measured on 12 mm cylinders are systematically lower than friction measured on 15 mm cylinders under identical normal load, caused by the higher contact pressures on small cylinders. The second test on 12 mm samples provides lower friction, likely attributed to a lower counterface roughness. The 12 mm tests were done on the same counterface with intermediate polishing, reducing the initial counterface roughness Ra = 0.05 μm to Ra = 0.02 μm. For the statistical 15 mm tests, freshly machined counterfaces with identical roughness are used for each test and there is reasonable accordance between measurements: during running-in, there is some higher friction due to Hertz contact conditions, and during steady-state, friction attains a nearly constant value. The curves slightly deviate towards the end of the test possibly due to random interaction with wear debris particles in the interface. No coherent transfer film for present test conditions on the steel counterface was observed and the friction curve mainly indicates mechanical interactions.

The on-line measurements of vertical displacement are given in Figure 4b for 12 mm and 15 mm cylinders. The final vertical displacement for 12 mm samples is higher due to a higher contact pressure. The second test with lowest roughness Ra = 0.02 μm causes highest wear. The statistical 15 mm tests show good correspondence for the final vertical displacement value, however most scatter occurs during running-in due to deformation of the Hertz line contact. A statistically high or low friction does not necessarily correlate to a statistically high or low wear. The repeatability of test results for friction and wear on 15 mm cylindrical samples of SP-1 is analysed in Table 3. Wear is determined from weight measurements (analytical balance, 0.0001 g) and dimensional measurements (micrometer, 0.01 mm) both immediately and one week after testing, allowing for recovery and homogenisation at atmospheric temperature and humidity. As a measuring protocol for future data, weight and diameter reduction will be measured one week after testing to allow for equilibration.

Tribological test results typically show larger variation than intrinsic material properties such as mechanical strength, stiffness, thermal properties or electrical properties and are difficult to control.

The acceptable limit of statistical variation strongly depends on the application: more experiments generally improve the statistical repeatability of a test result, but it does not significantly improve the reliability of a real construction because laboratorial and on-the-field conditions are never completely similar. Therefore, a design factor of 1.25 to 1.50 is often applied on test results covering both statistical variation and safety [11]. The variation in experimental values does not only depend on the statistical variation in properties of both polymer and steel contact materials, but it is mainly dependent on the test rig design and contact situation. Most of them or not constant and change mutually during sliding. The contact and eventual material transfer between both sliding materials happens at some local roughness peaks changing in number and dimensions during sliding. The local contact conditions, however, strongly change with the surface finish of the test samples: close control on the composition, structure, shape and surface conditions is indispensable for good repeatability.

The polymer structure is strongly defined by its processing parameters, however as all samples presently used originate from industrial processes, this is considered as a minor influence. Particularly for polymer/steel mating pairs, the heat generation through the frictional process influences the sliding behaviour and depends on the local friction conditions, while any variation in friction coefficient causes a variation in temperature.

Table 3. Statistical analysis for small-scale tests on cylinders with 2*b* = 15 mm, SP-1 at 50 N, 0.3 m/s

Test number	Coefficient of friction	Vertical displacement (µm)	Weight loss (g)		Diameter reduction (mm)	
			immediately after testing	one week after testing	immediately after testing	one week after testing
1	0.50	472	0.0155	0.0152	0.42	0.42
2	0.51	510	0.0156	0.0156	0.46	0.45
3	0.48	550	0.0185	0.0179	0.51	0.51
4	0.52	518	0.0176	0.0173	0.49	0.48
5	0.55	493	0.0161	0.0160	0.45	0.45
Mean value	0.51	509	-	0.0164	-	0.46
Standard deviation	0.026	29.1	-	0.001	-	0.038
Max range	0.07	78	-	0.0027	-	0.10
Relative error around the average	± 7 %	± 8 %	-	± 8 %	-	± 12 %

A second reason for scatter in test results, and to our experience the most important, is the unreliable functioning of each part on the test rig. Most critical parts as the roller bearing in contact with the transverse bridge structure and the spring and lever system for loading the transverse bridge (Figure 1) should be regularly replaced. It is our experience that with good maintenance of the test rig, the steel surface conditions become the second most crucial influence.

Preparation according to ASTM G115-98 should be followed, including ultrasonic rinsing, polishing and degreasing. The final polishing step is done with grooves parallel to the sliding direction. As a variation of 10 to 15 % or even higher on experimental values is often observed [12], present data for SP-1 are "reasonably acceptable".

The relative error on friction coefficients seems somewhat lower than on wear values, as the latter are additionally influenced by creep and deformation. Thermoplastics showed somewhat higher statistical variations (± 10 % on friction to ± 16 % on wear). Klaffke [13] found for steel/steel pairings a scatter of ± 6.3 % on friction and a scatter of ± 30 % for wear. Friction and wear values should be interpreted relatively to each other and certain relationships with one test parameter should be established rather than relying on a single test result. E.g., the relation of dynamic friction with normal load agrees with power-law models from literature [14]: $\mu_{dyn} = 0.91\ F_N^{-0.15}$ with R^2=0.96.

For a normal distribution of the experimental values, the mean value can be statistically tested on its significance by means of the 't-test' [15]. Assuming that the standard deviation of the distribution is not a priori known, the confidence interval is given by Formula (3),

$$\mu_0 - t_{n-1,1-\alpha/2}\frac{S}{\sqrt{n}} \leq X_0 \leq \mu_0 + t_{n-1,1-\alpha/2}\frac{S}{\sqrt{n}} \tag{3}$$

with μ_0 the mean value calculated from experiments, $t_{n-1,\,1-\alpha/2}$ the *t*-value for n = 5 tests and a significance level α (tabulated) and *S* the standard deviation calculated from experiments. The 95 % and 90 % confidence intervals for on-line mean coefficient of friction μ_0 and off-line mean weight loss Δg_0 are given in Table 4. It means that 95 %, respectively 90 %, of the test results measured for SP-1 at 50 N, 0.3 m/s is within the given interval.

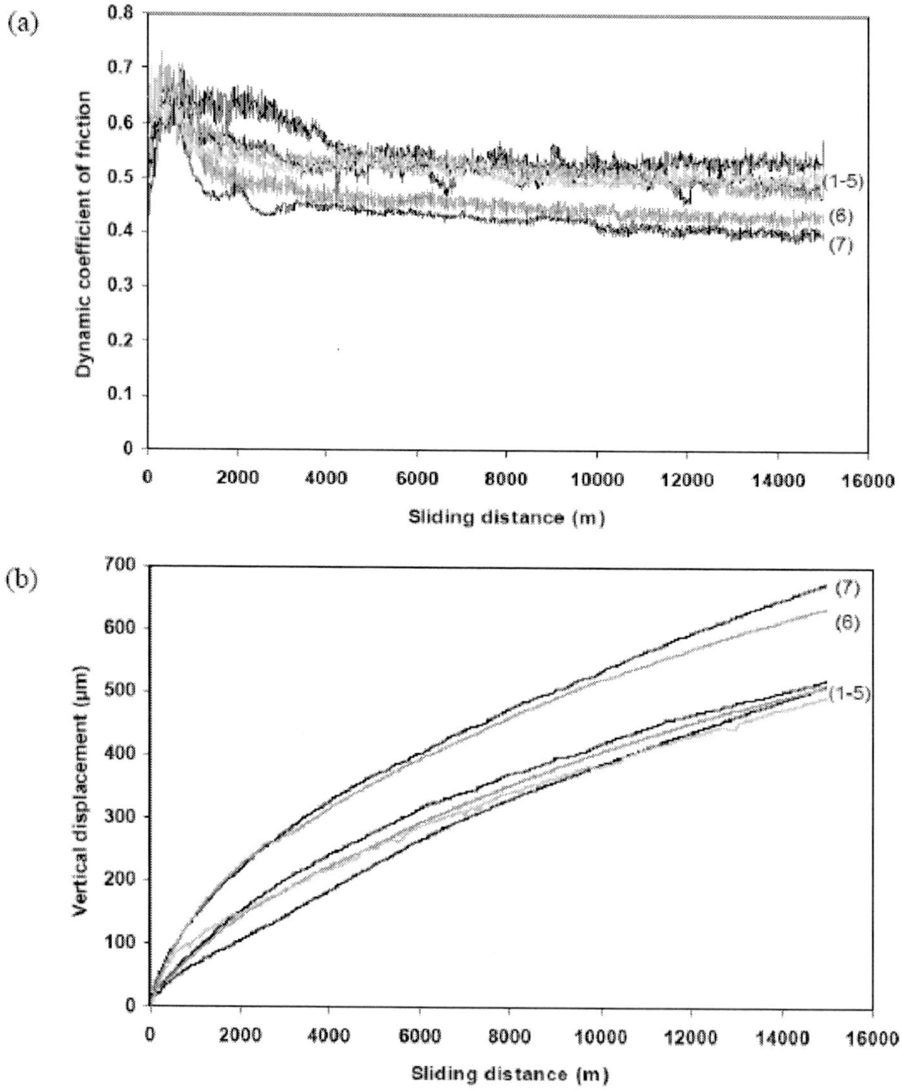

Figure 4. Preliminary statistical analysis of small-scale sliding measurements for SP-1, 50 N, 0.3 m/s on steel, cylinders with width $2b$ = 15 mm (tests 1-5) or $2b$ = 12 mm (tests 6 with Ra = 0.05 µm, test 7 with Ra = 0.02 µm) against steel, (a) friction, (b) wear.

Table 4. Confidence intervals for experimental friction and weight loss on small-scale tests

Parameter	95 % confidence interval	90 % confidence interval
Coefficient of friction	$\mu_0 \pm 0.032$	$\mu_0 \pm 0.025$
Weight loss	$\Delta g_0 \pm 0.0014$	$\Delta g_0 \pm 0.0011$

From the calculated mean value μ_0 and standard deviation S for the measurements of friction coefficients (Table 3), the number of tests required for a respective 10 %, 6 %, 5 % and 1 % relative error can be calculated according to Formula (4),

$$n = \left(\frac{t_{\alpha,v}\, S}{\mu_0\, E} \cdot 100 \right)^2 \qquad (4)$$

with n the number of measurements, $t_{\alpha,v}$ the t-value for confidence level $\alpha = 0.05$ and $v = 5\text{-}1 = 4$ degrees of freedom (tabulated), S the standard deviation, μ_0 the average value and E the percentage of relative error. For a relative error of 10 % the number of experiments n should be 1, for a relative error of 6 % the number of experiments n should be 3, for a relative error of 5 % the number of experiments n should be 5 and for relative error of 1 %, the number of experiments n should be 118. From the effectively available time, each test result in future work will be averaged from five repeated tests.

3.4. On-Line and Off-Line Wear Measurement

The wear measurements during a stepwise sliding test, with intermediate off-line examination of wear depths Δh_m after various sliding distances, are presented in Figure 5. The wear volumes are calculated from the weight measurements Δg and from the wear depths Δh according to Formula (5), for a cylinder with radius R and width $2b$.

$$2b\left(R^2 \sin^{-1}\frac{\sqrt{R^2-(R-\Delta h)^2}}{R} - (R-\Delta h)\sqrt{R^2-(R-\Delta h)^2} \right) \qquad (5)$$

The wear volumes from weight measurements increase linearly with sliding distance and can be extrapolated towards the origin, showing a constant volumetric wear rate and no transition during sliding of SP-1 against steel at 50 N, 0.3 m/s as a transfer film does not form for present sliding conditions. The wear volumes calculated from on-line vertical displacement Δh_v are the highest and do not follow a linear increase over the entire sliding distance: they indicate higher deformation at running-in while levelling off after 2000 m sliding.

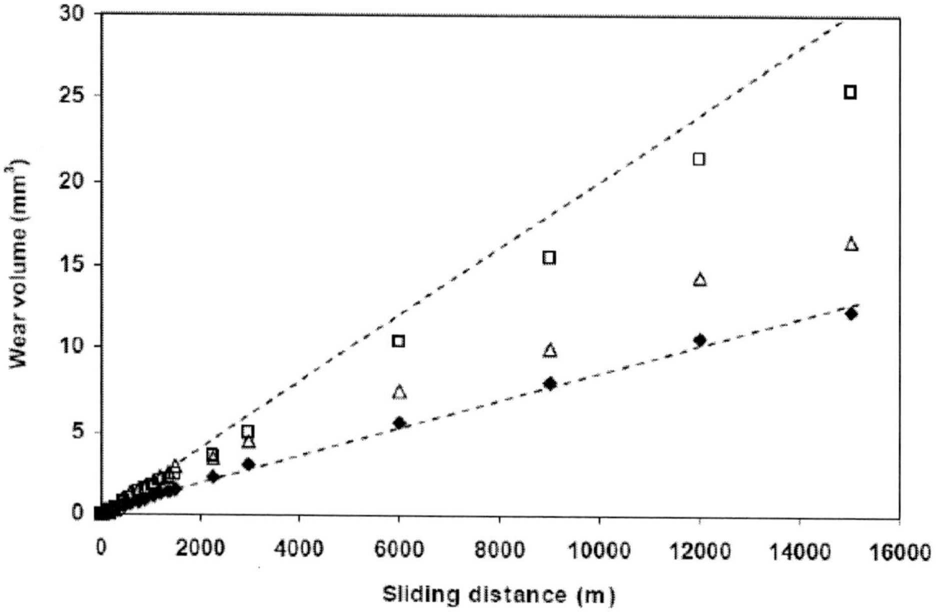

Figure 5. Evaluation of wear volumes from on-line vertical displacement Δh_v (□), off-line diameter reduction by micrometer measurement Δh_m (Δ), and off-line weight by balance measurement Δg (♦) for SP-1 at 50 N, 0.3 m/s against steel.

Due to the visco-elastic nature and creep of polyimides during sliding, the on-line vertical displacement Δh_v is strongly influenced by deformation of the Hertz line contact. A part of the deformation is recovered immediately after unloading as represented by the lower value of Δh_m. One week after testing, also most part of this deformation is recovered and there exists agreement between weight and dimensional measurements without significant permanent deformation.

The on-line vertical displacement Δh_v remains higher than off-line diameter reduction Δh_d and is further evaluated in Figure 6 and Table 5. Creep Δh_c (experimental) and thermal expansion Δh_t (calculated from maximum polymer temperature T* and linear thermal expansion α = 50 μm/mK) should be taken into account. Even the corrected on-line wear measurement $\Delta h_{v,corr} = \Delta h_v - \Delta h_c + \Delta h_t$ remains higher than diameter reduction Δh_d. The difference between on-line and off-line values indicates a part of visco-elastic deformation recovered during one week at zero load. The percentage of recovered deformation depends on the normal load, being 5 to 6% at 50 to 100 N, or 13 to 17% at 150 to 200 N. Statistical analysis shows a variation ±2% on the recovered visco-elastic deformation after sliding. It is important that the transition between low and high load noticed in visco-elastic deformation under static loading and creep also manifests in wear measurements. Small recovery during 1 week corresponds to small accumulation of visco-elastic deformation at low loads, while most deformation was recovered immediately after unloading. At high loads, less deformation was recovered imme-diately while a large percentage was recovered during one week. The average dimensional wear rates from on-line vertical displacement and off-line measurements show reasonable agreement (error < 6%) and confirm that most deformation is concentrated at running-in.

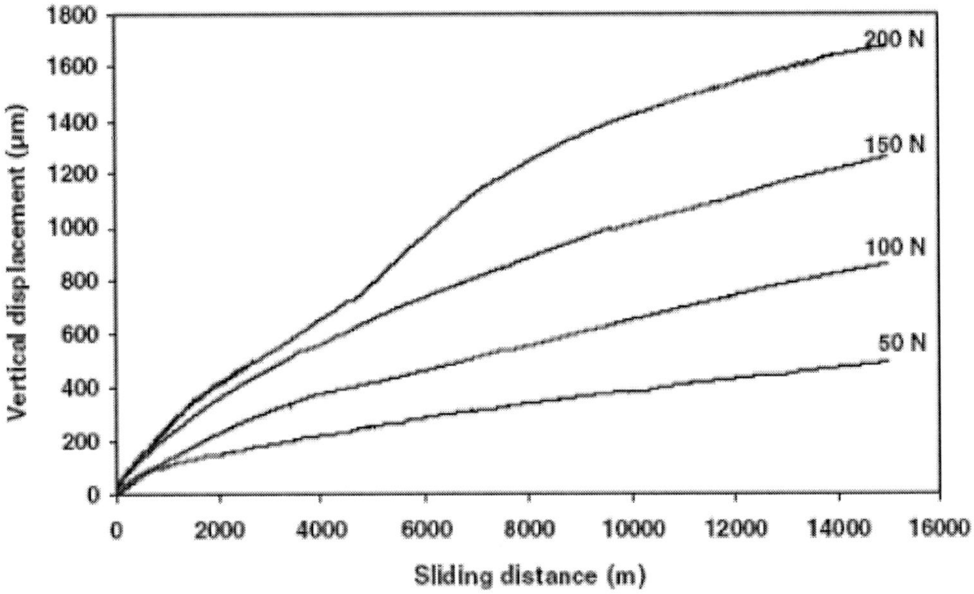

Figure 6. On-line vertical displacement as a function of sliding distance for SP-1 at 50–200 N.

Table 5. Evaluation of wear and deformation for SP-1 sliding in line contact under different normal loads and 0.3 m/s sliding velocity against steel

		50 N	100 N	150 N	200 N
Off-line diameter reduction Δh_d	µm	450	820	1050	1550
On-line wear depth Δh_v	µm	472	862	1264	1773
Creep Δh_c	µm	57	81	104	117
Maximum polymer surface temperature T*	°C	9	15	21	23
Thermal expansion Δh_t	µm	7	8	10	16
Corrected on-line wear depth	µm	474	869	1275	1780
Recovered visco-elastic deformation 1 week after test	µm	24	49	225	230
	%	5	6	17	13
On-line wear rates					
Running-in over first 1500 m	10^{-4} mm³/m	13	24	46	60
Steady-state	10^{-4} mm³/m	5	12	16	34
Average	10^{-4} mm³/m	9	18	31	47

4. CONTACT CONDITIONS

4.1. Running-in Conditions

The small-scale cylinder-on-plate macroscopic contact changes during sliding from a contra formal into a conformal situation. The initial line contact can be described by the analytical formulas developed by Hertz, based on the theory of elasticity [16]. A general model for line contact between two elastic cylinders is illustrated in Figure 7, where the radius of the second contact body (R_2) is taken infinite in case of a cylinder-on-plate configuration. The Hertz theory (1881) allows to calculate the contact geometry, contact pressures and elastic deformation when a static normal load F_N is applied between two elastic bodies apparently contacting at a single point. The contact area and contact pressure depend on whether the contact bodies are convex or concave and whether it are hard or soft bodies. In the calculations, the material properties of both elastic solids are represented by an equivalent modulus of elasticity E' according to Formula (6) with Poisson coefficients ν_1 = 0.4 for polymers and ν_2 = 0.3 for steel counterfaces, and Young's moduli E_1 for polymer and E_2 = 210 GPa for steel respectively. The reduced radius of curvature R' is given in Formula (7). A polymer cylinder pressed against a flat steel plate will produce a plane rectangular contact area A_H with width $2b$ and length 2ℓ, calculated from formula (8) to (10). The large difference in elasticity causes all deformation to be concentrated in the polymer sample.

Equivalent modulus of elasticity E':

$$\frac{1}{E'} = \frac{1}{2}\left[\frac{1-\nu_1^2}{E_1} + \frac{1-\nu_2^2}{E_2}\right] \quad (6)$$

Reduced radius R':

$$\frac{1}{R'} = \frac{1}{R_1} + \frac{1}{R_2} \quad (7)$$

Geometry:

$$2b = 15 \text{ mm} \quad (8)$$

$$\ell = \sqrt{\frac{4 F_N R'}{\pi b E'}} \quad (9)$$

Contact area $A_H = 4\, b\, \ell$ (10)

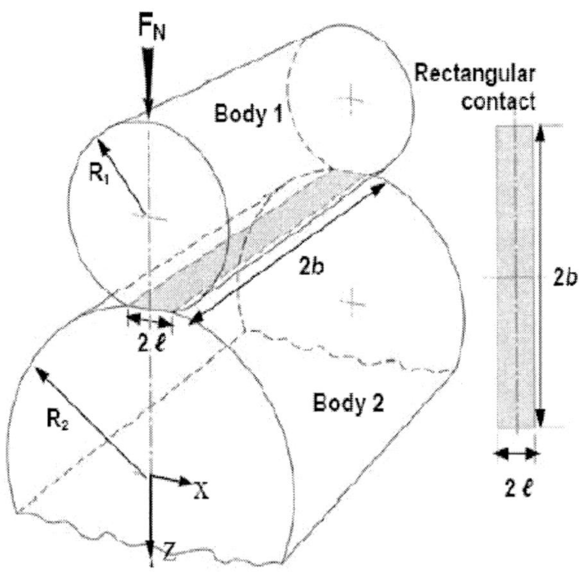

Figure 7. Hertz line contact situation ($R_2 = \infty$ for cylinder-on-plate contact) [17].

Under a normal load F_N, the contact pressure at the interface is parabolically distributed according to Formula (11) with a maximum value $p_{H,max}$ on the centre line where the load is applied and an average value $p_{H,avg}$ from Formulas (12) and (13).

$$p_H = p_{H,\max}\sqrt{1-\left(\frac{x}{\ell}\right)^2} \tag{11}$$

$$p_{H,\max} = \frac{F_N}{\pi b \ell} \tag{12}$$

$$p_{H,avg} = \frac{F_N}{4 b \ell} = \frac{\pi}{4} p_{H,\max} \tag{13}$$

The elastic indentation of the polymer cylinder can be estimated from the relative approach δ_H between polymer and steel according to Formula (14).

$$\delta_H = \frac{F_N}{\pi b}\left[\frac{1-\nu_1^2}{E_1}\left(\ln\frac{4R_1}{\ell} - \frac{1}{2}\right)\right] \tag{14}$$

As an illustration for sintered polyimide grades SP-1 and SP-2 and for thermoplastic polyimide grades TP-1 and TP-2 with different Young's moduli respectively (see Table 1), the Hertz deformation and contact pressures are calculated in Table 6 and Table 7.

The compressive deformation under static loading and unloading is given in Figure 8a. Each test was repeated three times with a statistical variation of ±2 μm on vertical displacement. The maximum Hertz contact pressures $p_{H,\ max}$ are calculated for 50, 100, 150 and 200 N and the behaviour is compared to a compression test according to ASTM D695 with flat contact (data not shown).

Table 6. Cylinder-on-plate contact conditions from Hertz theory for sintered polyimide

Normal load (N)	Pure sintered polyimide (SP-1)				Graphite-filled sintered polyimide (SP-2)			
	$p_{H,max}$ (MPa)	$p_{H,avg}$ (MPa)	A_H (mm^2)	δ_H (μm)	$p_{H,max}$ (MPa)	$p_{H,avg}$ (MPa)	A_H (mm^2)	δ_H (μm)
50	34	27	1.82	3.32	39	30	1.60	2.66
100	49	39	2.58	6.13	55	43	2.29	4.94
150	60	47	3.17	8.85	68	53	2.81	7.07
200	70	55	3.66	11.3	78	61	3.24	9.10

Table 7. Cylinder-on-plate contact conditions from Hertz theory for thermoplastic polyimide

Normal load (N)	Pure thermoplastic polyimide (TP-1)				PTFE-filled thermoplastic polyimide (TP-2)			
	$p_{H,max}$ (MPa)	$p_{H,avg}$ (MPa)	A_H (mm²)	δ_H (μm)	$p_{H,max}$ (MPa)	$p_{H,avg}$ (MPa)	A_H (mm²)	δ_H (μm)
50	36	29	1.73	3.02	34	27	1.86	3.42
100	52	41	2.45	5.59	48	38	2.63	6.32
150	64	50	3.00	8.00	59	46	3.22	9.04
200	73	58	3.47	10.3	68	53	3.72	11.6

(a)

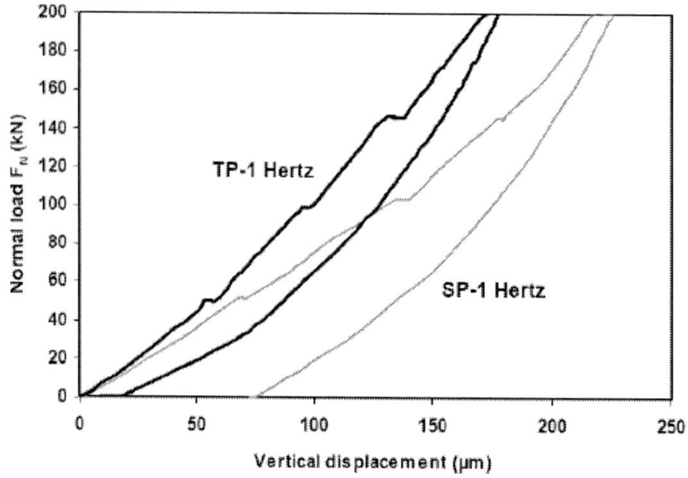

(b)

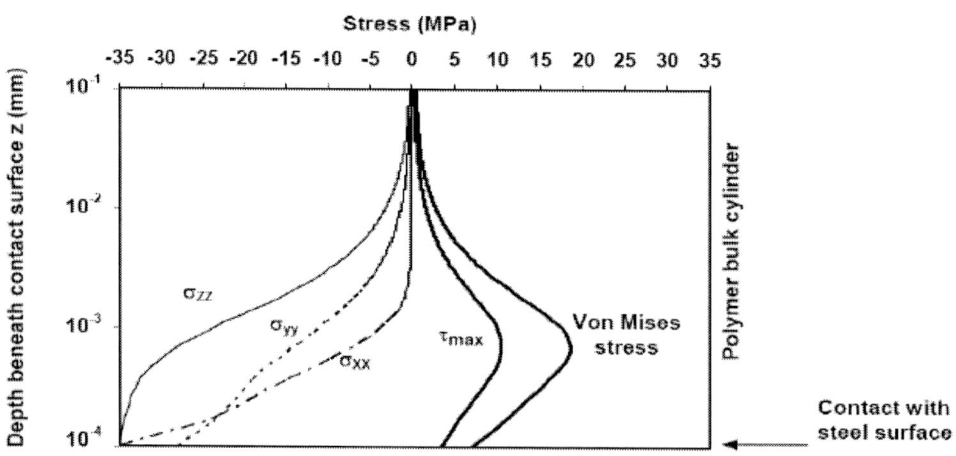

Figure 8. Static deformation of polymer cylinders in Hertz line contact, (a) experimental static compression tests, (b) calculated sub-surface stresses.

A transition between visco-elastic and plastic deformation occurs at 48 MPa, according to the ASTM compression test. A transition in slope of the normal load versus vertical displacement occurs at 100 N, corresponding to $p_{H,max} = 49$ MPa according to the maximum Hertz contact pressure. There is critical load at 100 to 150 N in line contact: creep measurements confirm that deformation is 9 to 10 μm at 50 to 100 N, while significantly higher deformation of 13 to 18 μm happens at 150 to 200 N. Recovery tests indicate nearly complete recuperation of the deformation at 50 to 100 N immediately after unloading, while it took about 48 h for complete recovery after 150 to 200 N loads due to visco-elasticity.

The experimental vertical displacement during loading of polyimide cylinders is up to a factor 10 larger than the estimated Hertz elastic deformation (Tables 6, and 7). The state of stress beneath the contact surface is illustrated in Figure 8b at 50 N ($p_{H,max}$ = 34 MPa), according to calculations from [18]. The principal stresses along the z-axis parallel to the line of loading symmetry are represented. The maximum shear stress attains $\tau_{max} = 0.33\ p_{H,max}$ at a depth of $z = 0.78\ \ell$. The maximum Von Mises stresses along the normal directions x, y, z in the subsurface attain 19 MPa at 50 N, 26 MPa at 100 N, 32 MPa at 150 N and 37 MPa at 200 N normal load.

4.2. Steady-State Contact Conditions

With ongoing small-scale sliding, the macroscopic contact area at steady-state is circumscribed by a narrow rectangle that progressively grows through wear. A geometrical relation between contact pressure p and wear depth Δh is given by Formula (15) for a cylindrical polymer specimen with radius R_1 (2.5 mm) tested under normal load F_N,

$$p = \frac{F_N}{A} = \frac{F_N}{2b\ 2\ell} = \frac{F_N}{4b\sqrt{R_1^2 - (R_1 - \Delta h)^2}} \qquad (15)$$

with A the macroscopic or apparent contact area, b the half-width of wear track (7.5 mm) and ℓ semi-length in the sliding direction, geometrically calculated from the wear depth Δh. Due to the increase of contact area during wear, the initial contact pressure decreases while running a cylinder-on-plate test. According to Formula (15) the contact pressure levels off with ongoing wear towards a steady-state situation with nearly constant contact pressure. The evolution of the contact pressure and the contact area as a function of the wear depth is plotted in Figure 9 for normal loads of 50, 100, 150 and 200 N. The average steady-state contact pressure on the macrocontact falls down to 1 to 3 MPa depending on the normal load. Its value at the end of the sliding test is calculated from measurements of diameter reduction (see Table 5). This contact pressure is further considered to determine the pv (contact pressure x sliding velocity) sliding conditions.

The influence of changing contact conditions on friction and wear of sintered polyimide is illustrated in Figure 10. Sliding against steel, there is hardly noted a transfer film, as such that the sliding conditions are minor affected by a third body in the interface and variation in friction and/or wear mainly depends on the variation in contact conditions.

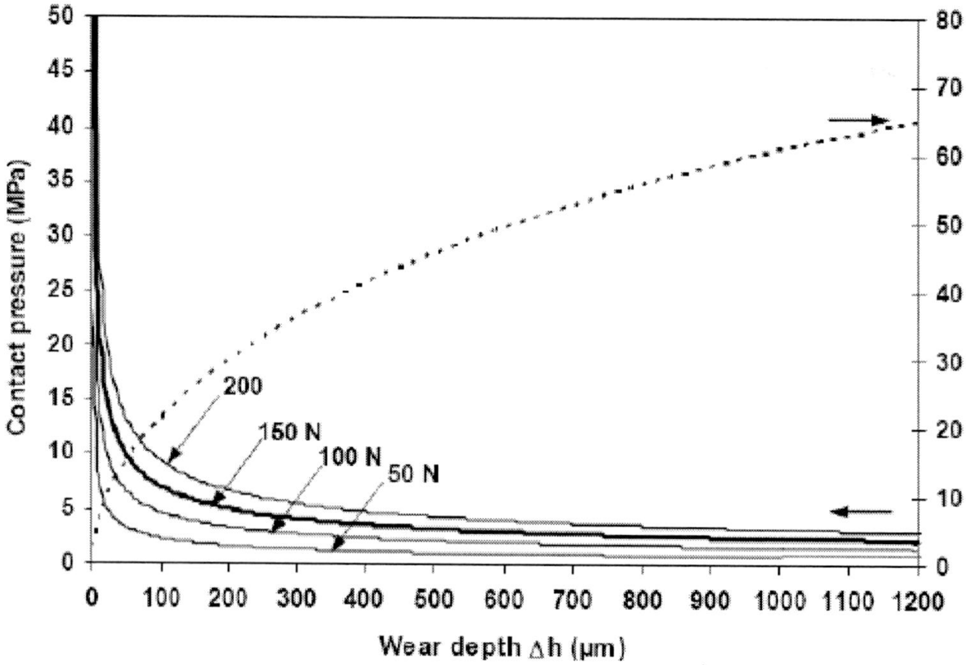

Figure 9. Evolution of contact pressure for different normal loads and contact area with wear depth for small-scale cylinder-on-plate tests at 50, 100, 150 and 200 N normal load.

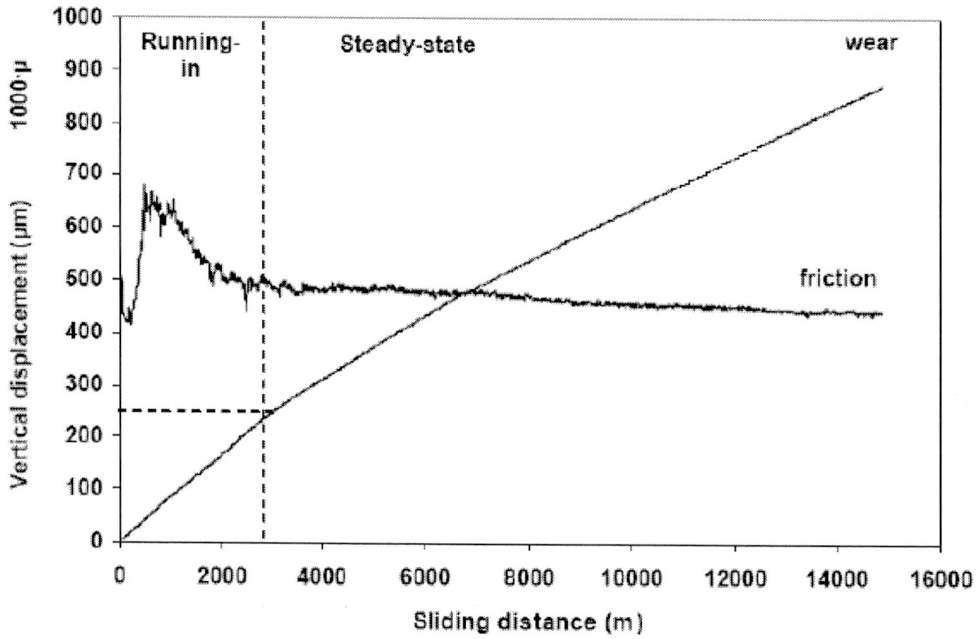

Figure 10. Determination of running-in and steady-state conditions for friction and wear parallel to the variation in contact conditions for tests on SP-1, 100 N 0.3 m/s against steel.

There is clearly distinguished a limit between running-in and steady-state wear, with a transition at about 2500 m sliding distance. Running-in is characterised by high and unstable friction with high wear rates (i.e. high slope of the wear curve), mainly attributed to the line contact situation with high contact pressures. After starting the sliding test, there is about 10 to 20 μm immediate vertical displacement due to deformation and wear of the line contact. Steady-state conditions with stable friction and lower wear rates establish after 250 μm wear depth. Referring to the variation of contact pressures with wear depth from Figure 9, this vertical displacement corresponds to a stabilisation in contact pressure at about 4 MPa. Present transition is thus mainly the effect of a change from contra formal line contact into conformal contact. In cases where polymer transfer interferes with the sliding contact, this transition was less clear and lower friction during running-in was noted as illustrated in future tests.

The characteristics of a line contact during the running-in process were studied by Horng et al. [19] for steel/steel pairings. For steels, it was experienced that the contact length 2ℓ rapidly increases at the initial stage of wear, no matter how large the normal load is, favouring low steady-state wear. The inaccuracy in contact area growth is, however, nearly 300 % compared to calculations from the Hertz theory. The effect of counterface roughness on the increase in contact length is inferior to the effect of normal load on contact length. During the line contact running-in, also the real contact area increases, but it reaches a stable value at the early middle stage of the running-in process. Also the influence of a temperature rise during running-in, approximately 40°C at $F_N = 200$ N and $\mu = 0.10$, was considered as inferior. The increase in contact during running-in seems to happen three times faster compared to steady-state. The running-in process and deformation of a line contact associated to a drop in contact pressure favourably controls the steady-state tribological properties. For polymer samples, the experimental vertical displacement during static loading is up to a factor 10 larger than the calculated Hertz elastic deformation. Polymers behave rather visco-elastic and deformation depends on the loading time and effective stress situation in the polymer bulk. The running-in features are also influenced by the morphology of the sliding and stress concentrations near the surface own to its production process.

4.3. Asperity Contact Conditions

The surface topography at micro level has important influence on the friction and wear properties, as it determines the contact conditions. Being softer than the steel counterface the polymer surface is less critical in this respect: roughness asperities deform or wear easily. The steel topography continuously changes during sliding as some roughness peaks are possibly flattened through abrasive wear or debris accumulates between the peaks. Possibly a directional roughness establishes relatively to the sliding direction under the action of, e.g., fillers. The detachment of steel particles and incorporation into the polymer surface reign the wear mechanism. The initial counterface roughness is however important for the running-in behaviour and influences further sliding, acting as a boundary condition for the wear debris motion and possible formation of a transfer film. Present counterfaces are characterised prior to testing and after testing by a two-dimensional roughness profile z(x) illustrated in Figure 11 for ground and polished steel. The height parameters determined from this profile are the centre-line-average or Ra value (Formula 16), and the maximum peak-to-valley height value Rt. The Ra value is the most commonly used parameter in engineering practice, representing the arithmetical roughness average over the total sample length L. The effect of a single spurious, non-typical peak or valley (e.g. scratch) is averaged out and has only a small effect on the final value. One of the main disadvantages of this parameter is that it can give identical values for surfaces with very different profiles. Since it is directly related to the area enclosed by the surface profile around the mean line, any redistribution of peaks has no effect on its value. A second parameter Rt is therefore also used, representing the extreme height over the entire profile. The highest asperities on the steel surface have importance in dry contacts because damage may be caused to the polymer body by penetration of few high asperities. Many other parameters can be calculated, although they are not further used in this work: the mean roughness depth Rz is the mean value of single roughness depths over n consecutive sampling lengths L_{min} (Formula 17), Rp is the maximum peak height relatively to the mean, and Rv is the maximum valley depth relatively to the mean. They are more frequently used for expressing the retention of oil.

$$Ra = \frac{1}{L}\int_0^L |z(x)|\, dx \tag{16}$$

$$Rz = \frac{1}{n}\sum_{i=1}^{n} R_{z,i} \tag{17}$$

The profilometry measuring protocol is taken form the DIN 4768 standard, performed on a 2D universal PERTHEN 5 SP equipment, or ISO 4287, done on a 3D-Surfascan, with a diamond tip scanning over the surface. Depending on the average roughness Ra, the minimum sample length L_{min} = 1.25 mm and the cut-off length λ_c = 0.25 mm for 0.02 µm < Ra < 0.1 µm , while L_{min} = 4.8 mm and λ_c = 0.8 µm for 0.1 µm < Ra < 2 µm. The total sample length L is divided in n sample lengths Lmin.

Roughness profiles are measured perpendicular to the sliding direction, unless stated otherwise. A normalised surface profile with Ra = 0.20 µm, Rt = 3.7 µm or Ra = 0.05 µm, Rt = 0.8 µm and is illustrated in Figure 11a and 11b. A detail of some surface asperities with an indication of their radius r_a is shown in Figure 11c for the smoothest counterface. Normalisation is done as such that the waviness is filtered and the mean profile line is parallel to the 0 µm reference height. The position of the mean line depends on equilibration of the 'material volumes' above and beneath this line. Theoretically the coefficient of friction and wear rate change with initial counterface roughness as illustrated in Figure 12a and 12b, according to Erhard et al. [20]. During sliding however, the contact conditions may change. He used the Rv value considering the local contact and deformation of the polymer over a single asperity peak. In relation to common information on bearing surfaces we prefer using the average roughness parameter Ra. As a rule of thumb and from experimental experience, it can be approximated that Rv ≈ 10 Ra for polished steel surfaces. A grey zone in Figure 12 indicates the present working range. The coefficient of friction drops at low roughness towards a minimum value and it further increases as a result from the transition between two sliding mechanisms: i.e. the adhesion component that is important at low roughness and decreases for high roughness and the deformation component that is low in contact with smooth counterfaces and becomes high for rough counterfaces.

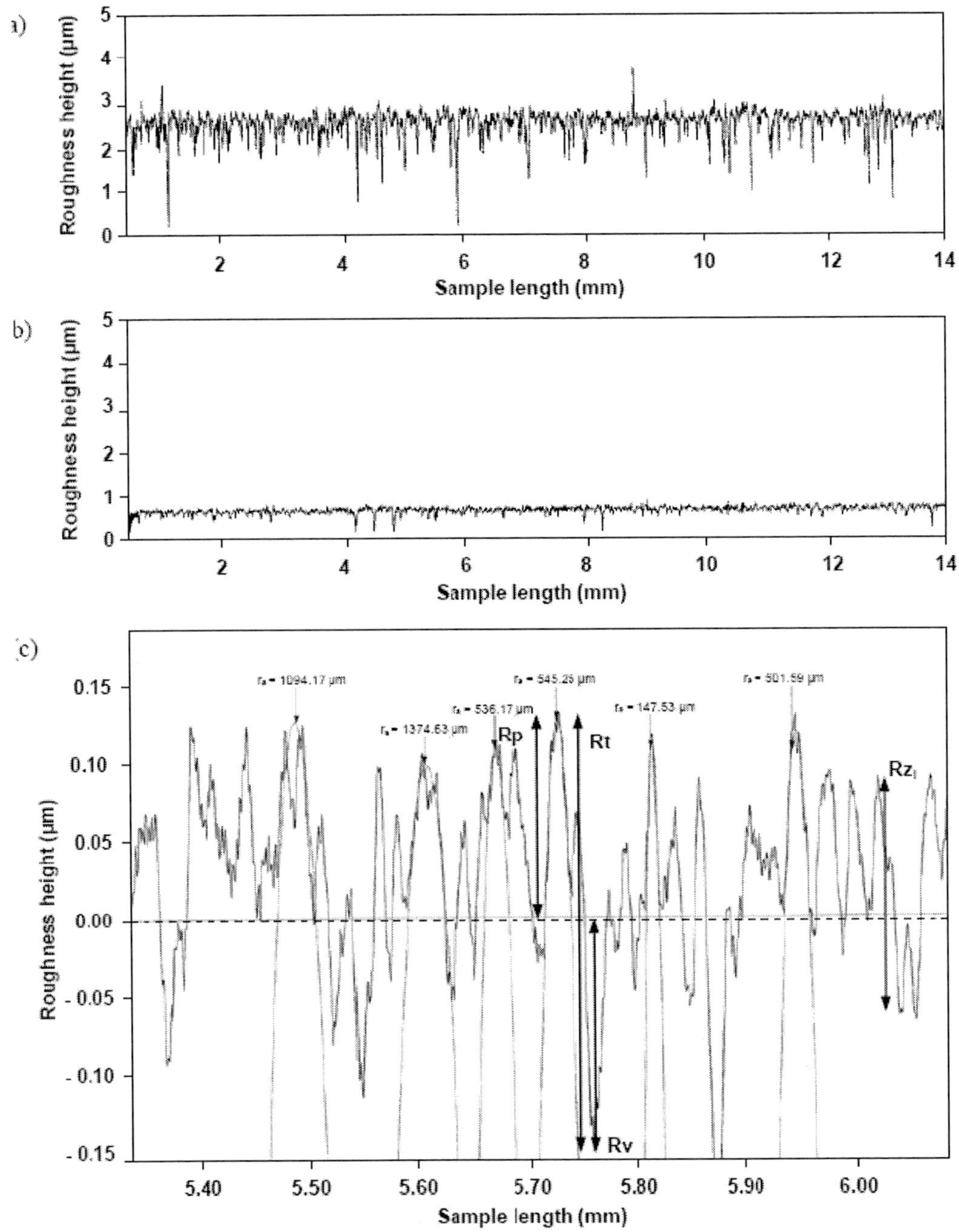

Figure 11. Normalised roughness profiles for steel measured perpendicular to the sliding direc-tion, (a) ground Ra = 0.20 µm, (b) polished Ra = 0.05 µm, (c) detail for polished Ra = 0.05 µm.

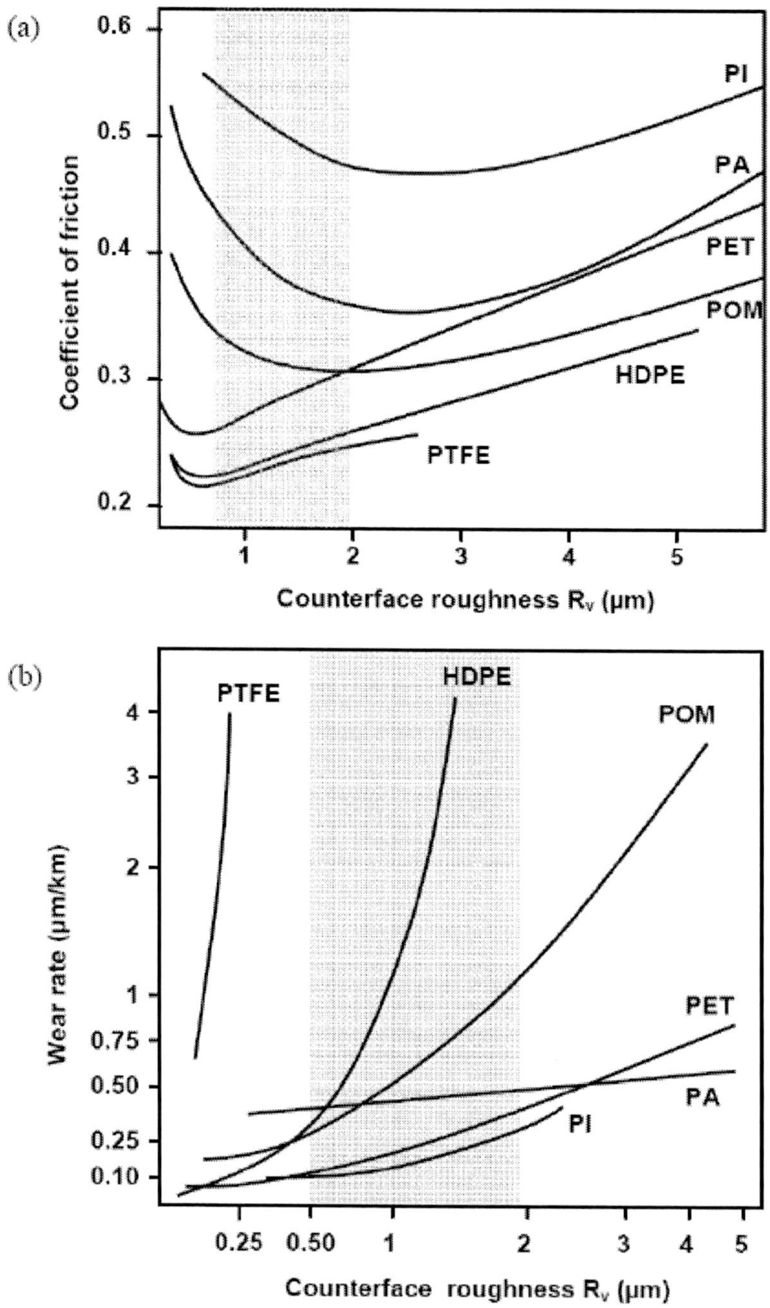

Figure 12. Theoretical relation between counterface roughness and tribological properties, with an indication of present working range: (a) polymer friction, (b) polymer wear, PI = polyimide; PA = polyamide, PET = polyethylene-terephtalate, POM = polyoxymethylene, HDPE = high-density polyethylene, PTFE = polytetrafluoroethylene.

Wear rates often increase continuously at higher roughness due to a combined effect of abrasion and fatigue. The lack of a transition in wear rates at low roughness can possibly be related to the effect of transfer film formation, which is promoted at low roughness but has a rather lumpy morphology and act as an abrasive due to little mechanical locking points on the counterface [21]. Smooth transfer is also inhibited at very high roughness and does not protect against wear. Wear debris was mainly observed on top of the roughness peaks rather than in the valleys [22]. The counterface roughness for small-scale tests of polyimides is chosen at Ra = 0.05 μm because Tewari and Bijwe [23] reported that polyimides are not suitable for abrasive wear applications, while the incorporation of fillers even deteriorates the abrasive wear resistance. This results from the Lancaster-Ratner equation [24] indicating that abrasive wear rate is proportional to a toughness parameter $1/(R_m.e)$ with R_m the ultimate tensile strength and e the elongation at break: sintered polyimides are rather brittle, mainly due to low elongation, and easily fracture (see Table 1). Due to the huge difference in hardness between steel and polymer contact bodies, small abrasive action of some metallic protrusions remains even at low roughness. Roughness parameters for large-scale testing and comparative small-scale testing of thermoplastics are chosen at Ra = 0.20 μm in parallel to industrial surface finishing of bearing elements. Counterfaces are ground and afterwards polished with GRID 600 abrasive SiC paper towards the required roughness.

5. Thermal Testing Conditions

The mechanical energy input from friction force and sliding velocity is transformed into internal energy or heat, which causes the temperature of the sliding bodies to increase. The mechanisms and exact location for this energy dissipation are not exactly known. Some investigators as Landman et al. [25] contend that these processes occur by atomic-scale interactions in the top surface layer while, e.g., Rigney et al. [26] believe that most energy dissipation or frictional heating originates from plastic deformation in the bulk solid beneath the contact region. For polymers, it is mainly the visco-elastic hysteresis loss resulting in heating of the sliding interface. Experimental work by Kennedy [27] has shown that at least 95 % of the energy dissipation happens within the top 5 to 10 μm of the contact bodies. The energy dissipation is theoretically concentrated within the real contact area and in absence of lubricant, heat is conducted into the two members through small contact spots. The temperature distribution at the sliding interface thus depends on the contact geometry, the coefficient of friction, the sliding parameters, the thermal material properties and the roughness conditions. As the real contact surface A′ is only a small fraction of the apparent contact area A, the local energy density is reasonably high. For low contact stress situations, asperity interaction results in numerous temperature flashes [28] over areas of a few microns, possibly causing local flow or melting. For highstress situations, the real area of contact A' is close to the apparent area of contact A and essentially a single contact occurs during sliding, levelling the local flash temperatures. Also if one of the sliding members is compliant, in casu a polymer, the local temperature rise is smaller than for, e.g., metal or ceramic contacts where flash temperatures of 700°C can occur. The real area of polymer contacts increases through creep and deformation, which reduces the heat generated per unit of macroscopic area. Besides the asperity flash temperature, the nominal temperature rise or bulk temperature becomes significant in sliding contact. The latter requires to consider the finite size of the body, the repeating feature of the sliding motion and total thermal equilibrium of the bodies. The different contributions to the ‘surface temperature’ and their significance in relation to present sliding conditions are detailed within this paragraph.

It is clear that frictional heating and the resulting contact temperature significantly alter the polymer properties and their sliding behaviour so that a meaningful prediction and/or measurement of the surface temperature should be made. It was also suggested by Kalin [29] that high temperatures existing in the sliding interface can cause tribochemical reactions that would not occur under normal conditions. Bulk temperatures can be experimentally measured over large contact areas, while different techniques to measure the transient temperatures at individual asperity contacts have been used with limited success [30]. The latter values are generally calculated [31], while measurements over large contact areas are performed, e.g. by infrared thermography or thermocouples. Numerical techniques have been developed and used by Varadi et al. [32] to evaluate the contact temperature evolution between steel and composite surfaces in sliding contact. Transient heat conduction problems were analysed by Gecim and Winer [33] or Yevtusenko and Ukhanska [34], concluding that almost 87 % of the steady-state value was immediately attained. Wang and Komvopoulus [35] developed a fractal theory by assuming spherical asperity tips. The computation is however tedious and it is, for two reasons, rarely necessary: first because approximate analytical methods are usually sufficient to give insight into the degree of heating relatively to the polymer's softening or melting point and, secondly, the physical assumptions about contact size and heat transfer coefficients are the real limits to accuracy. An experimental-analytical approach might be preferred and estimated temperatures might be correlated to physical transitions in polymer structure, surface analysis and wear [36].

The following questions rise: "Which temperature rise should be taken into account for determination of the homogeneous polymer surface temperature during reciprocating sliding ?", and "What is the significance of the temperature rise to evoke variations in the polymer sliding behaviour?" :

- If a single concentrated heat source does not repeat the same sliding path over the contact surface and if the counterface body is very large, all heat generated at a contact spot is transferred into the three-dimensional heat sink and the counterface bulk temperature levels out without any significant nominal bulk temperature rise. The contact temperature in that case can be found from the work done by Blok [28]. However, if the heat source passes repeatedly over the same point on the surface, there still exists a nominal counterface bulk temperature rise before the next sliding step regenerates heat at the same contact spot. The frictional heat generated during one sliding pass needs some time to flow away from the contact area and the bulk temperature does not completely level out. The temperature conditions of the counterface for the second sliding stroke seem different compared to the first sliding stroke and an extra surface temperature rise relatively to the initial bulk temperature should be taken into account.
- The maximum steady-state temperature T* at the polymer surface during reciprocating sliding has three contributions, being (i) the environmental temperature T_{env} or initial bulk temperature, (ii) the temperature rise $\Delta T_{b,steel}$ due to progressive heating of the steel counterface during reciprocating sliding and (iii) the temperature $\Delta T_{b,polymer}$ created through heat that is generated on the polymer surface during each new sliding event, making its surface temperature higher than the nominal steel counterface temperature. The summation of the environmental temperature T_{env} and

the temperature rise of the steel counterface $\Delta T_{b,steel}$ is considered as a 'nominal bulk temperature' or the 'actual steel surface temperature', or from Formula (18):

$$T^* = T_{env} + \Delta T_{b,steel} + \Delta T_{b,polymer} \tag{18}$$

An appropriate expression for each temperature component can be selected from the available theoretical models and should be verified against the experiments, as done in our previous work [36]. For a rectangular testing geometry and considering whether the contact partner is moving or stationary, the Loewen and Shaw temperature model [37] can be applied for the $\Delta T_{b,steel}$ component and the Jaeger temperature model [38] can be used for the $\Delta T_{b,polymer}$ component. As such, the following analytical temperature model apply for the presented testing geometry, according to Formula (19):

$$T^* = 23°C + A\frac{\mu p v \ell}{k} + 4{,}2.10^{-4}\frac{\mu F_N\sqrt{v}}{b\sqrt{\ell}} \tag{19}$$

with A the geometry factor, μ the steady-state coefficient of friction, p the steady-state contact pressure, v the sliding velocity, ℓ the semi-length, k the steel conductivity, F_N the normal load, and b the semi-width. As such, the rectangular testing geometry $2b$ x 2ℓ is taken into account by considering a geometry factor. The aspect ratio b/ℓ is important for the heat flow distribution over the steel plate (long-range), while the aspect ratio is not considered for the polymer surface temperature as an isolated polymer surface fraction in the sliding interface does not 'feel' its environmental geometry (short-range). The maximum polymer surface temperature T* was found to be relavant for correlations with the intrinsic polymer transition temperatures [36]. It governs homogeneously over the entire substrate and for longer sliding times than the *flash*-temperatures around local contact asperities.

CONCLUSION

The friction and wear tests of polymers in Hertz contacts was evaluatesd. This tribotesting configuration is important for simulation of line-contacts (e.g. gears, …) and requires effective control of the contact conditions and thermal sliding regimes in order to provide reliable data. The purpose of this study was to investigate the reliability of sliding experiments of a polymer in line contact during reciprocating motion with large strokes. For oscillating sliding tests, the wear can be quantified more accurately compared to continuous sliding, because only a small material part is involved in the wear process. Polymer contacts are also influenced by visco-elastic deformation, being an additional parameter affecting the repeatability of tribological data. An in-depth discussion on the repeatability of test results and influencing parameters will contribute to the development of better tribotesting procedures.

The statistical repeatability of sliding tests for sintered polyimide in line contact was studied. There is a statistical variation of about 7 to 9% on dynamic friction, about 10 to 12% on static friction, about 8 to 9% on wear rates from weight loss and about 12 to 15% from wear rates from dimensional measurements. Compared to literature, coefficients of friction show higher variation and the wear rates show lower variation than steel/steel pairs. The interpretation of wear-data from on-line and off-line techniques allows to make difference between real wear, creep and thermal expansion of the polymer specimens.

The contact conditions and transition in macroscopic contact situation into a more conformal contact during sliding, has an important influence on the sliding behaviour. The experimental deformation during static loading of polyimide is up to a factor 10 larger than the calculated Hertz deformation, due to the additional influences of visco-elastic deformation and sub-surface stresses. Most deformation occurs at running-in, in combination with high wear and coefficients of friction. The limit between running-in and steady-state sliding happens at a wear depth of about 250 μm corresponding to leveling of the contact pressures. A transition between rather time-dependent recoverable deformation at 50 to 100 N and permanent plastic deformation at 100 to 150 N is measured during static loading, creep and recovery tests after sliding. On-line and offline wear rates are in agreement when corrected for creep and thermal expansion. Another influencing contact parameter is the counterface roughness, which can be narrowly controlled by secure preparation of the steel counterfaces. The initial counterface roughness is selected within an optimum working range mentioned in literature and according to practical application: the average steel roughness Ra = 0.05 μm is applied for polyimides.

An analytical expression for calculating the maximum polymer surface temperature T* has been proposed for the present sliding configuration. Under reciprocating sliding, the temperature T* has three contributions, i.e. the environmental temperature, the nominal steel bulk temperature and the polymer surface temperature. The nominal steel bulk temperature corresponds to progressive heating of the steel bulk and is calculated from the Loewen and Shaw model, agreeing to measurements with a thermocouple distanced at 10 mm from the sliding stroke. The polymer surface temperature results from additional heating at the polymer surface during each sliding motion and is estimated from the Jaeger model. The Jaeger model used to calculate the temperature T* considers the integrated effect of flash temperatures. The calculated maximum polymer surface temperature T* is 10 to 15°C higher than measurements with a thermocouple placed immediately near the sliding stroke, but it is physically not possible to measure temperatures in the contact itself. Measured and calculated values represent homogeneous surface temperatures, acting over for longer times the entire polymer surface. Therefore, the changes in tribological properties for polymers can be explained by relating the maximum polymer surface temperature T* to intrinsic polymer transition temperatures.

ACKNOWLEDGMENTS

Pieter Samyn acknowledges the Research Foundation Flanders (F.W.O.) for financial support and assignment of a post-doctoral research grant.

REFERENCES

[1] Kar, M.K., Bahadur, S. (1974). *Wear*, 30, 337–348.
[2] Neale, M.J., Gee, M.: Guide to wear problems and testing for industry, Professional Engineering Publishing Limited, St. Edmundsbury Press Ltd, Suffolk (2000).
[3] Seabra, L.C., Baptista, A.M. (2002). *Wear*, 253, 394–402.
[4] Samyn, P., De Baets, P., Schoukens, G., Hendrickx, B. (2003). *Polym. Eng. Sci.*, 43, 1477–1487.
[5] Samyn, P., De Baets, P., Kerestes, R., Zsidai, L., Kalacska, G., Kislinder, E., Kozma, M. (2007). *Tribotest*, 13, 83-100.
[6] Fusaro, R.L. (1988). *Tribol. Trans.*, 31, 174–181.
[7] Fusaro, R.L. (1978). *ASLE Trans.*, 21, 125-133.
[8] Sheiretov, T., Van Glabbeek, W., Cusano, C. (1995). *Tribol. Trans.*, 38, 914–922.
[9] Iwabuchi, A., Hori, K., Sugawara, Y. (1988). *Wear*, 125, 67–81.
[10] Zsidai, L. De Baets, P., Samyn, P., Kalacska, G., Van Peteghem, A.P., Van Parys, F. (2002). *Wear*, 253, 673-688.
[11] Samyn, P., Van Paepegem, W., Leendertz, J.S., Suister, E., Van Schepdael, L., Degrieck, J., De Baets, P. (2007). *Eng. Struct.*, 29, 2673-2691.
[12] Lhymn, C. (1987). *Wear*, 114, 223-239.
[13] Klaffke, D. (1995). *Wear*, 189, 117-121.
[14] Yamaguchi, Y.: Tribology of plastic materials. Elsevier, Amsterdam (1990).
[15] Devore, J.L.: Probability and statistics for engineering and the sciences, Brooks/Cole, Pacific Groove, CA (2009).
[16] Hamrock, B.J., Jacobson, B.O., Schmid, S.R.: Fundamentals of machine elements, WCB/McGraw-Hill Higher Education, Boston (1999).
[17] Stachowiak, G.W., Batchelor, A.W.: Engineering tribology, Butterworth-Heinemann, Boston (2001).
[18] Bhushan B.: Modern tribology handbook, Vol. I, CRC Press, Boca Raton (2001).
[19] Horng, J.H., Len, M.L., Lee, J.S. (2002). *Wear*, 253, 899-913.
[20] Erhard, G. (1983). *Wear*, 84, 167-181.
[21] Uetz, H., Wiedmeyer, J.: Tribologie der Polymere, Carl Hanser Verlag, Munchen (1985).
[22] Ovaert, T.C., Ramachandra, S. (1995). *Int. J. Machine Tools and Manufacture*, 35, 311-316.
[23] Tewari, U.S., Bijwe, J. (1991). *Tribol. Internat.*, 24, 247-254.
[24] Lancaster, J.K. (1969). *Wear*, 14, 219-223.
[25] Landman, U., Luedtke, W.D., Ringer, E.M. In: Fundamentals of Friction: macroscopic and microscopic processes, Singer, I.L., Pollock, H.M. (eds.), Kluwer Academic Publishers, Dordrecht (1992).
[26] Rigney, D.A., Hirth, J.P. (1979). *Wear*, 53, 345-370.
[27] Kennedy, F.E. (1982). *J. Lubr. Technol.*, 104, 582-588.
[28] Blok, H. (1963). *Wear*, 6, 483-494.
[29] Kalin, M. (2004). *Mater. Sci. Eng.*, A 375, 390-397.
[30] Hooke, C.J., Mao, K., Walton, D., Breeds, A.R., Kukureka, S.N. (1993). *Tribol. Trans.* 115, 119-124.

[31] Matysiak, S.J., Yevtushenko, A.A., Ivanyk, E.G. (2002). *Int. J. Heat and Mass Transfer* 45, 193-199.
[32] Varadi, K., Neder, Z., Friedrich, K., Flöck, J. (1998). *Tribol. Internat.*, 31, 669-686.
[33] Gecim, B., Winer, W.O. (1985). *J. Tribol.*, 107, 333-342.
[34] Yevtusenko, A., Ukhanska, O. (1994). *Wear*, 176, 19-23.
[35] Wang, S., Komvopoulos, K. (1994). *J. Tribol.*, 116, 812-823.
[36] Samyn, P., Schoukens, G. (2008). *Polym. Eng. Sci.*, 48, 774-785.
[37] Bhushan B.: Principles and Applications of tribology, Wiley-Interscience, New York (1999).
[38] Jaeger, C.J. (1942). *Proc. R. Soc. NSW*, 76, 1107-1121.

In: Advances in Polymer Research
Editors: Losif Sidorov, Fedor Nikitin, pp. 185-258
ISBN: 978-1-61942-223-0

APPLICATION OF LATTICE CLUSTER THEORY TO THE CALCULATION OF MISCIBILITY - AND INTERFACIAL BEHAVIOR OF HYPERBRANCHED POLYMER CONTAINING SYSTEMS

***S. Enders*[1] *and T. Zeiner*[2]**
[1] TU Berlin, Fachgebiet Thermodynamik und Thermische Verfahrenstechnik, Straße des 17. Juni 135, D-10623 Berlin.
[2] TU Dortmund, Lehrstuhl Fluidverfahrenstechnik, Emil-Figge-Str. 70, D-44227 Dortmund.

ABSTRACT

Newly developed hyperbranched polymers possess a compact, highly branched, three-dimensional structure, which has a high density of functional end groups and inherently low viscosity. The combination of these two properties, low viscosity and high reactivity, makes them attractive candidates for an overwhelming variety of applications. The experimental and theoretical investigation of the phase behavior of hyperbranched polymer systems is a crucial requirement for a successful introduction of new applications to highly competitive markets. In this context, thermodynamic models, which accurately account for the impact of polymer branching on the phase behavior of polymer systems, play a very important role.

The lattice cluster theory (LCT) is an extension of the well-known Flory-Huggins theory, especially in the calculation of the entropy of the lattice. Whereas the Flory-Huggins theory is limited to linear chains the LCT can be applied to arbitrary chain architecture. This situation permits the incorporation of the architecture in the thermodynamic functions useful for phase equilibrium calculations. The polymer architecture plays an important role in the physical properties of hyperbranched polymers. Additionally, the combination of the LCT with the density gradient theory allows the theoretical investigation of the interfacial properties between the demixed phases.

INTRODUCTION

The development of linear polymers and their impact on all aspects of modern life is one of the major achievements of the last century. As this field has matured it is increasing apparent that further developments will likely arise, not from the synthesis of totally new linear polymers, but from more accurately controlling the architecture of polymers from currently available monomers. The production and processing of polymers are influenced by the presence of phase separation and segregation, which may be either necessary or highly undesirable. For example, proper orientation and crystallization conditions are needed to secure useful fibers and films; on the other hand, segregation of highly viscous phases during a polymerization process may lead to catastrophic consequences like plugged lines or overheated reactors [1,2]. Partial miscibility also plays important roles in biology and medicine, as for example in the formation of cataracts in eye lenses [3], protein separations [4] and fibril formation related to Alzheimer's disease3. Polymer solutions, which are mixtures of high-molecular weight compounds (solutes) and low-molecular weight solvents, have been examined experimentally, since the early 1930s. These studies revealed some abnormalities when the results were compared with those for low-molecular weight solutions e.g. an unexpectedly small vapor depression, small boiling point elevation, small osmotic pressure and extremely high solution viscosities. These abnormalities became a strong motivating force in the search for a thermodynamic theory for polymer solutions. For polymer solutions, where the difference in molecular size of the components is very large, miscibility gaps become highly asymmetric. A useful approximate model describing the thermodynamic properties, including the phase behavior, in polymer systems is the well-known Flory-Huggins lattice theory [5]. In the lattice model the mixture is represented by a number of regularly arranged lattice sites, each of the same size. The lattice is thought to have a definite coordination number z, which however; does not remain as a relevant variable in the simplest version of the theory. Each polymer molecule is considered to be composed of a number of segments, and the entropy of mixing is evaluated by counting the number of distinguishable ways of placing the molecules on the lattice. Flory and Huggins[5] in their initial works were mainly concerned with the effects of varying molecular size on the thermodynamic properties.

In the framework of this theoretical approach no information about the architecture of the monomers and the resulting polymers are involved. The architecture of the polymer is often determined by the functionality of the monomers from which it is formed. This property of a monomer is defined as the number of reaction sites at which may form chemical covalent bonds. The basic functionality required for forming even a linear chain is two bonding sites. Higher functionality yields branched or even crosslinked or networked polymer chains. Modern polymerization strategies such as dendritic macromolecular chemistry involve the formation of large multiples of covalent bonds between homogeneous monomers to produce large molecules or infinite networks with a broad range of structure control [6].

Historically, each of the three macromolecular architectural classes, i.e. (I) linear, (II) crosslinked, and (III) branched, has spawned rich polymer science (Figure 1). These architectural discoveries have been characterized by the emergence of new syntheses, structures, phenomena, properties, and products that have dramatically improved the human condition. Nanotechnology initiatives have focused on new synthesis strategies, structures, phenomena, and properties associated with length scales of 1-100 nm. These dimensions encompass biological building blocks (protein, DNA, RNA, etc.) and abiotic application areas (nanophotonics and nanoelectronics) [7]. Dendritic polymers are recognized as the fourth major class of polymeric architecture consisting of three subsets that are based on degree of structural control, namely: a) random hyperbranched polymer, b) dendricraft polymers and c) dendrimers [7,6].

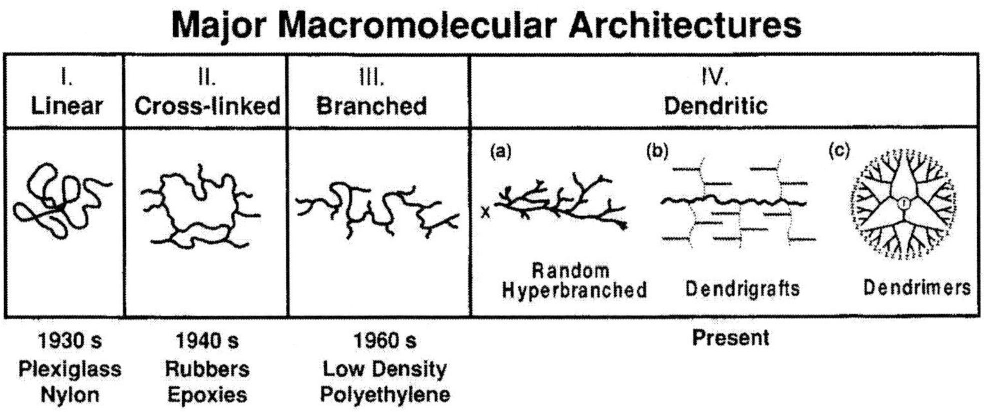

Figure 1. Representation of the four major classes of macromolecular architectures [6].

The outstanding polymer chemist Flory [8] was the first to hypothesize concepts, which are now recognized to apply to statistical, or 'random hyperbranched' polymers. However; the first synthesis of hyperbranched polymer could by realized 35 years later [9] Webster and Kim [10] coined the popular term 'hyperbranched polymer' that has been widely used to describe this type of dendritic macromolecules. In theory, all polymer-forming reactions can be utilized for the synthesis of hyperbranched polymers, however; in practice some reactions are more suitable than others [6]. Hyperbranched polymers are typically prepared by polymerization of AB_X monomers, where x is two or more. They are produced by the one-pot polymerization of AB_X monomers or macromonomers involving polycondensation [11,12,13,14,15] ring opening [9,16,17,18] or polyaddition [19,20] reactions hence the product usually consist of broad statistical molecular weight distribution [21,22,23,24] Frechet et al. [25] presented the first example of a hyperbranched vinyl polymerization initiating the birth of a 'second generation' of hyperbranched polymers. Frechet and Chang [26] reported proton-transfer polymerization as a versatile route to hyperbranched polymers. Remarkable progresses have been made in recent years in the exploration of metal-mediated and metal-free click polymerization [27,28] systems and in the syntheses of linear and hyperbranched polytriazoles with regioregular molecular structures and advanced functional properties [29]. Hyperbranched polymers are often modified to tailor their properties for a specialized purpose. Five modification methods have been developed: 1) end-capping with short chains or organic molecules; 2) terminal grafting via living polymerization; 3) growing hyperbranched polymers on the surface, or grafting from/onto the surface; 4) hypergrafting to obtain hyperbranched polymers with a linear macromolecular core; 5) blending or crosslinking.

Star polymers are three-dimensional hyperbranched structures in which linear arms of the same or different molecular weights emanate from a central core. The existence of numerous functional groups in a small volume makes these polymers important for use in biological and pharmaceutical applications. Biologically active molecules can be immobilized on the surface of the polymer gel or incorporated into the network.

Unique architecturally driven properties that may be expected from hyperbranched polymers will be largely derived from their a) amplified number of terminal functional groups, b) new rheological properties based on less chain entanglement, c) new architectural arrangements that may modulate crystallinity, flow characteristics and glass transition properties in designed systems and d) the formation of micelles.

The degree of branching is one of the most important molecular parameters of hyperbranched polymers since it determines many physical properties including the phase behavior of these polymers. This contribution focuses on the incorporation of the architecture of the polymers in the thermodynamic equations necessary for phase equilibrium calculations.

PROPERTIES AND APPLICATIONS OF HYPERBRANCHED POLYMERS

The unique properties of hyperbranched polymers are mainly manifested by their intrinsic globular structure and large number of terminal functional groups. Unlike dendrimers, however; hyperbranched polymers have elements of conventional polymers, namely molecular weight dispersity, isomerism, and geometrical shapes. Because the addition of each monomer takes place randomly, a large number of geometrical isomers can be formed even for a given molecular weight and branching degree. The polydispersity of the hyperbranched polymers is expected to increase to infinity at infinite polymer molecular weight. Besides the complex structure, also molar mass determination for hyperbranched polymers is far from trivial. It is obvious that molar mass determination by GPC lacks the fact that linear standards are not suitable for calibration. The application of light scattering and viscosity detector in the GPC as mostly done for branched polymers improves the results but still the broad molar mass distribution and the large number of polar end groups might cause problems [30,31].

Some of these polymers exist as unusual colloid-like aggregates and form unimolecular micelles. Unimolecular micelles are defined as a class of macromolecules, wherein an interior hydrophobic core is surrounded by a hydrophilic surface layer. These structures closely resemble the shape of classical micelles except that they are static, in contrast to the dynamic nature of micelles, with all end-groups attached to the central core. The ability of guest molecules to penetrate the lipophilic interior can be used as drug or gene nano-carriers [10,32,33,34,35,36] In addition dendrimers can be surface engineered to release the drug at desired site, that is, as targeted drug delivery. This property along with the solubilisation behavior could improve the bioavailability of drugs. Hyperbranched polymers should be an attractive candidate for a gene- or drug-delivery system in aqueous media and could provide the phase-transfer carriers between water and organic media. Recently [34,37,38] biodegradable unimolecular reversed micelle consisting of a hyperbranched hydrophilic core and hydrophobic shell, were introduced. Liang et al. [39] demonstrate the possibility that a unimolecular micelle can simultaneously deliver both polar and apolar guests. In a biphasic water/chloroform mixture, the nanocapsule can transfer anionic, water-soluble guest from an aqueous phase to the chloroform phase; while when dissolved in water, the nanocapsule can efficiently capture both ionic and apolar solutes. Release of the guest can occur under the stimulus of pH or the switch of medium.

Above their glass transition temperature, these hyperbranched polymers are considered to minimize their free energy by lowering their free volume via a conventional nematic mesophase which is generated by a conformational change of their structural units from gauche to anti [40]. Percec et al. [41] were the first to report on flexible, non-spherical polymers with AB_2 mesogens in the branches that exhibit nematic and smectic behavior. On the other hand, aromatic polyamides exhibit polymer aggregation in the absence of a complexing salt resulting in a mesomorphic phase [42].

In general, hyperbranched polymers and dendrimers are more soluble in the same solvents than their linear analogous [43]. For higher generation dendritic polymers, solubility characteristics depend predominantly on the properties of their functional endgroups. As an example, dendrimers with hydrophobic interiors such as polyethers and polycarbosilanes can be made water soluble by introducing hydrophilic groups. Oppositely, water soluble dendrimers can be generated to be hydrophobic by converting their functional groups into hydrophobic units [44]. Furthermore, end group modification allows ideally to optimize their properties for special applications and to fine tune e.g. miscibility, melt rheology, surface and optical properties as well as biocompatibility.

Numerous applications in the field of coatings and blends, nanoscience, microelectronics, information technology, optics, lithography, organic light-emitting diodes, solid electrolytes, photoresponsive materials, chemical- and biosensors, and medicine (drug delivery systems, tissue engineering, magnet resonance contrast agents) have been suggested for hyperbranched polymers but only a few have yet reached full commercial exploitation. Several reviews [6,7,15,45,46,47,48,49,50,51,52,53,54] about this issue were publicized in the past. In this contribution only a few of them will be highlighted.

Application in Medicine

After drug administration, the drug may pass through different physiologic barriers and/or pathways, decreasing the actual amount of drug that reaches the site. Tissue specificity, product stability and solubility all desirable characteristics of drug, but are not always attained. Therefore, the need of develop a drug carrier system with such characteristics is of great importance. In the last 25 years [55], there have been numerous efforts focused on the development of the drug carrier systems. Investigators have made attempts to develop a specific drug carrier system, which can maintain continuous drug levels in a desired range, reduce side effects by improved tissue or organ specificity.

Many polymeric carriers have been investigated for therapeutic applications [55]. However; only a few polymers such as linear poly (lactide-co-glycolide), polyethylene glycol, and acrylic-based polymeric carriers have been introduced at commercial scale for controlled-release applications. Müller et al. [56] indicated that most polymeric drug delivery systems suffer from two major drawbacks, i.e., the cytotoxicity of polymers and the lack of processability. Hence, the successful formulation of a polymeric drug delivery system requires a system solution, i.e. a) a carrier with a narrow molecular mass distribution that fulfils nontoxicity and controlled-release criteria, b) sufficient loading capacity for the desired applications and c) processability by means of a commercial encapsulation technique. Even after considering the aforementioned system solution, factor such as the bioavailability, biodegradability, biodistribution, and drug efficacy can still limit the effectiveness of a polymeric drug delivery system. The processing of many therapeutic formulations has been suggested with commercial encapsulation techniques employ volatile organic compounds for the processing of polymeric carriers, where the residuals leading to the major health concern. However; most polymeric carriers such as polyesters and polyamides are very difficult to process in the absence of volatile organic compounds, because of high melting points, limited solubilities in supercritical gases, or high solution and melt viscosities are nonprocessable in most encapsulation techniques. Furthermore, harsh operating conditions and the cost effectiveness of the encapsulation technique may also limit the commercialization of the polymeric drug delivery system.

In the search for an ideal carrier system, the hyperbranched polymers may have significant potential. Hyperbranched polymers and their substitutes can be used as nanomaterials for host-guest encapsulation for several molecules such as dyes, pharmaceuticals, cosmetics, fragrances, catalysts and pollutants as well as the fabrication of organic – inorganic hybrids, and even directly as nanoreactor [37] for some reactions. As carriers, hyperbranched macromolecules can offer their interior or peripheral functional groups to covalently fix bioobjects, or depending on their core-shell architecture, to sequester guest molecules. The information stored at a molecular level plays a key role in this process. For a controlled release application, a change in pH, temperature, pressure, or a bacterial, enzymatic, or catalytic activity disintegrates the carrier system, leading to the release of the encapsulated active substance. The loading capacity and the release kinetics of carriers based on hyperbranched polymers are dependent on the polymer backbone, the number and type of functional groups, the molar mass, the polydispersity, and the amphiphilicity of the macromolecule. In controlled-release applications, hyperbranched polymers are functionalized in such a way that the shell of the hyperbranched carrier not only protects the encapsulated guest molecules, but also responds to the target environment via specific physiological change such as temperature, pH, chemical, and/or enzymatic reactions [57,58]

Well-characterized, hyperbranched polymers were subjected to functionalization for preparing drug delivery systems of low toxicity, high loading capacity, ability to target specific cells and transport through their membranes [57] They open new routes for the development of controlled-drug delivery systems with the potential to inhibit microbial adhesion to host tissues. In past years, the scope of encapsulation has expanded from pharmaceutical applications to cosmetics and agrochemical products[51]. The common goal for these applications is a high degree of control over the release mechanism for the encapsulated active substances.

As hydrophobic drug substances are difficult to introduce in the human body, many studies focus on increasing the solubility of drugs in the aqueous phase. As an example, tamoxifen, which is a hydrophobic breast anticancer drug, was encapsulated by Tziveleka et al. [59] in hyperbranched polyglycerols modified with polyethylene glycol. The results demonstrate that molecular encapsulation, based on hyperbranched nanocarriers, allows transporting 12 times more tamoxifen into an aqueous phase, compared to the solubility of the pure drug. The release of tamoxifen observed upon addition of sodium chloride is, in most of the cases, significant only at concentrations exceeding the physiological extracellular concentration [59].

Hyperbranched polymers such as polyesters, polyethyleneimine, polyglycerol, poly(ethylene glycol) and different copolymers have been investigated as potential polymeric vehicles for drug delivery applications and have proven to be highly biocompatible, thermally and chemically stable[57]. Moreover, although most dendritic polymers show cytotoxicity, low-molecular mass hyperbranched polyglycerols have proven to be nontoxic. The encapsulation of therapeutic agents such as ibuprofen [60,61,65,63,64], acetaminophen [65], tamoxifen [59,66], ketoprofen [63], diflunisal [63] naproxen [63], paclitaxel

[67,68,69,70,71,72,73], doc [73], glimepiride [7], doxorubicin hydrochloride [75,76], methotrexate [75], sodium ibandronate[75] cisplatin[77], chlorambucil [78], cytochrome c [79], amphotericin B [80], nimodipine [81], indomethacin [82], doxorubicin [76,83], and docetaxel [70] were studied. The in vitro release of ibuprofen from drug-dendrimer complex is appreciably slower compared to pure ibuprofen. The complex drug enters A549 cells much more rapidly than pure drug suggesting that dendrimers may be able to carry the complex drug inside cells efficiently [64]. Hyperbranched Polyol with 128 OH end groups appears to encapsulate approximately 24 drug molecules [60].

The use of oral antidiabetic drugs for management of type 2 diabetes increases rapidly caused by the discovery and approval of several new types of oral antidiabetic drugs with different mechanism of pharmacological action [74]. Many of these drugs show poor solubility, slow dissolution rate in water, pH-dependent solubility and high permeability. Several approaches to improve water solubility include prodrugs, complexation, cosolvency, solid state modifications, surfactants were investigated. Among these the addition of cosolvents, the formation of cyclodextrins or micellar inclusions and the preparation of solid dispersions are the most commonly used. Many of these solubilisation techniques have their own limitations, toxicity, nephrotoxicity. The transformation of the drug to its amorphous form is often desirable since the solubility increases from a few to many-fold. The presence of hydrophilic compounds in close contact with the drug molecules increases the solubility by maintaining the drug in a molecular state and maximizing the surface area of the compound. The polymeric molecules also act as crystallization inhibitors and preserve the drug in its amorphous state [7]

Polymeric biomaterials are also of particular interest for the parenteral administration of peptides and proteins [84]. In an initial phase, release occurs predominantly by pore diffusion through an interconnecting network formed by the dissolving drug substance itself. The second release phase is governed by polymer degradation. Polyphasic drug release profiles can be overcome either by formulation approaches or by modification of the biodegradable polymers. While the release properties of biodegradable microspheres can be modified only in a limited sense, polymer modifications provide a broader spectrum of possibilities. Hydrophilic multi-arm polyethylene oxides seem to be promising candidates for parenteral protein delivery systems, allowing a synchronization of both pore diffusion and polymer erosion for the controlled release [84].

Boltorn hyperbranched dendritic polymers functionalized with mannose have been used to inhibit DC-SIGN-mediated infection in an Ebola-pseudotyped viral model [85]. Their physiological solubility, lack of toxicity and especially their low price suggest the application of these glycodendritic polymers for possible formulation as microbicides [85].

Gene delivery is a relatively new discipline of drug therapy and is defined as the incorporation of foreign nucleic acid that can be mediated by viral and nonviral methods into the host cells. An ideal gene delivery method must meet three major criteria: a) it should protect the transgene against degradation by nucleases in intercellular matrices, b) it should bring the transgene across the plasma membrane and into the nucleus of target cells, and c) it should have no detrimental effects. Nucleic acids are conventionally bound to viral vectors, but polymeric carriers such as polyethylenimine (PEI) have also been used. Linear PEI shows high transport efficiency, because of the relatively better loading capacity. However; an increase in the loading capacity of PEI occurs with an increase in the molar mass, which, in turn, increases cytotoxicity. Moreover, factors such as nondegradability, cytotoxicity, and short circulation time in bloodstream hinder the desired effects of the therapy. A common feature of the hyperbranched polymers related to gene therapy is the exhibition of polyvalent interactions, while for multifunctional derivatives, a number of targeting ligands determine specificity, and another type of group secures stability in biological milieu and prolonged circulation, while others facilitate their transport through cell membranes [58,86]. Furthermore, polymers employed for gene delivery should be or become cationic in the biological environment for the formation of complexes with the negatively charged genetic material [87]. Hyperbranched polyethylenimine and their modifications [72, 88,89,90,91,92,93], polyethylene glycol based cationic hyperbranched polymers [94,95], hyperbranched polysiloxysilane [96], hyperbranched Boltorn H® with introduced tertiary amines [97], hyperbranched poly(amido amine)s [98,99,100,101] and poly(amino ester)s [90,102] have been investigated as carriers for gene delivery applications. It was found that hyperbranched polyglycerols were nontoxic to mice, even after the dosage injection of 1g per kg body weight [103]. Hyperbranched poly(amido amine) a highly efficient non-viral vector for gene delivery into numerous cell lines in vitro and in vivo [90,104]. However; results obtained from in vitro studies do not always correlate to similar observations obtained in the in vivo experiments [6,87].

The products for skin, hair or body care are generally referred to as "personal care products". Many of them contain biologically active substances, such as vitamins, supplements, or fragrances. In a recent study, it was shown that hyperbranched polymer molecules having primary amine groups react with aldehyde-based fragrances to form a fragrance delivery system [105]. Such fragrance moieties can e hydrolyzed to release fragrance molecules such as aldehydes and/or ketones via a reverse Schiff base reaction mechanism.

Dendritic substances have made crucial advances and have already been tested in preclinical studies, particularly in the field of contrast media for magnetic resonance. The relaxation time of the water protons are significantly shortened by application of the contrast media, such as Magnevist (gadolinium salt of diethylenetriamine pentaacetic acid) into the organ that is to be visualized and leads to a picture with a very good signal-to-noise ratio. However; clinically used contrast media reveal a disadvantage. Immediately after intravenous application they diffuse into the extravenous area as low-molecular compounds. Do prevent the diffusion higher-molecular weight compounds have been developed in which several gadolinium complexes are covalently bound to albumin, dextran, and polylysine. Unfortunately, no preparation is on its way to clinical studies, because of bad renal elimination of the compounds. The most promising solution of this problem consists in the use of a new compound [106] consists of a trimesinic acid central building block, to which second generation lysine dendrons with a total of 24 complexed gadolinium ions are anchored. Animal tests showed the quantitative renal elimination and the higher intravascular retention time, which ensures an excellent signal-to-noise ratio [106]

Tissue engineering has recently attracted a great deal of attention for its potential in repairing, reconstructing, regenerating, or replacing tissues in damaged or diseased organs. A number of tissues and organs have already been investigated using this approach, including cartilage, liver, skin, bone, tendon, ureter, intestine, pancreas, and blood vessels. One approach in tissue engineering employs a hybrid system combining biomaterials as structural scaffolds to organize living cells into a desired structure in vitro or in vivo. Polymers for these scaffolds require good biocompatibility, suitable biodegradability, and the ability to interact specifically with appropriate cells. The control of protein adsorption is a key issue in many tissue engineering. All this desired properties can be found for hyperbranched polymers [107,108,109,110,111,112,113]

The escalating global incidence of bacterial infection, particularly in chronic wounds, is a problem that requires significant improvements to existing therapies. For example, Shepherd et al. [114] have developed hyperbranched poly(N-isopropylacrylamide) functionalized with the antibiotics vancomycin and polymyxin-B that are sensitive to the presence of bacteria in solution. When the polymers were removed from the infected skin, either in a polymer gel solution or in the form of hydrogel membranes, they removed bound bacteria, thus reducing the bacterial load in the infected skin model[114]. Because poly(ethylene glycol) is a nontoxic, water-soluble polymer that resists recognition by the immune system, hydrogels prepared from poly(ethylene glycol) star polymers are excellent candidates as biomaterial.

Ever increasing demands on the healthcare industry offer challenging opportunities for producing products that are not only more cost effective and outperform existing products, but also provide ready access to a variety of new biological and analytical reagents. Recently, Xu et al.[115] synthesized a new dendritic core-shell architectures with pH-labile linkers based on hyperbranched polyglycerol cores and biocompatible poly(ethylene glycol) shells which encapsulate the anticancer agent doxorubicin and a dye for near-infrared imaging.

Application in Chemistry

Metal nanoparticles continue to evoke great current interest due to their tremendous potential in designing smart materials for a wide variety of applications (for instance silver [116,117,118,119] or gold [120,121] with antibacterial activity [122,123] for drug delivery purpose or antibacterial coating, platinum [124,125], gold [126,127]' or titania [128,129] for catalysis, iron for introduction of magnet properties [130]). Much emphasis has been placed lately in developing methodologies that could modulate the size and shape of these metal particles. Dendrimers that are monodisperse in nature with a regular and highly branched three-dimensional architecture, provide a useful platform to accomplish this goal. These hyperbranched macromolecules have been widely explored as templates in the construction of metal nanoparticles [131,132].

In general, these metallodendrimers are thought to be macromolecular materials that can combine the advantage of both homo- and heterogeneous catalysts [133,134,135,136] Because of their 'pseudo'-spherical nature and their resultant conformations the metal sites in these well-defined polymeric catalysts should be easily accessible for substrate molecules and reagents, and therefore exhibit characteristics usually encountered in homogeneous catalysis such as fast kinetics, specificity and solubility. Owing to their precise persistent nanoscale size they may be easily removed from product streams, e.g. by means of ultrafiltration techniques.

Gladitz et al. [137] produced a long-term-stable silver nanoparticle dispersions with narrow size distribution using amphiphilic-modified hyperbranched polyethyleneimines and investigated the antibacterial performance and morphology of thin silver-loaded hyperbranched polymer coatings on poly(ethylene terephthalate) prepared by different wet coating techniques. Thermosensitive gold nanoparticles with tunable lower critical solution temperature have been prepared by coating the nanoparticles with a thermo- and pH-responsive hyperbranched polyelectrolyte [120]. The globular, polyfunctional structure of the hyperbranched polymers will offer new options to control phase-transition temperatures of the nanoparticles.

Membrane-based gas separations have attracted much attention in the past because they offer many significant advantages over traditional separation processes such as low energy consumption, low capital investment cost, simple and easy operation.

A large number of polymeric materials have been studied for this application. Hyperbranched polymers [138] show an average cavity size of 0.5 nm, which is larger than that of common gas molecules such as H_2, O_2, CO_2, N_2 and CH_4. Rölker et al. [139] examined the application for gas separation such as CO_2 absorpt ion from flue gas. For example, one amine-terminated hyperbranched polyimide membrane crosslinked with terepththaldehyde exhibited fairly high CO_2/N_2 separation performance and good selectivity [49].

The gas permeability of different gases such as CO_2, O_2 and N_2 for hyperbranched polyimide containing trifluoromethyl groups [140], for star-like poly(ethylene oxide)s [141], for polyimide - silica hybrid membranes [142,143] for polyimide [144,145], and Boltorn-modified polyimide [146] was investigated. Sterescu et al. [147] describes the preparation, characterization and the permeation properties of poly(2,6 dimethyl-1,4-phenylene oxide) dense polymer films containing aliphatic hyperbranched polyesters, Boltorn (H20, H30, and H40). The gas permeability of poly(2,6 dimethyl-1,4-phenylene oxide) with 1.0 wt % of Boltorn is 2-3 times higher than the pure polymer, while at higher concentration (9.1 wt %) of Boltorn the permeability becomes almost 50% of the pure polymer [147]. The gas pair selectivity, however; stays constant.

The increase in permeability at low concentration of Boltorn is due to the increase of the free volume, probably due to hydrogen bonds between Boltorn and the oxygen of poly(2,6 dimethyl-1,4-phenylene oxide) backbone, whereas the decreased permeability containing higher concentration of Boltorn (9.1 wt %) is due to two reasons: decrease in free volume as determined by positron annihilation lifetime spectroscopy as well as phase separation [147]. The hyperbranched polyesters form aggregates that migrate to the top surface of the membranes.

The applications of highly crosslinked thermoset polymers are limited by their intrinsic brittleness. To increase their toughness, different solutions, most often based on the addition of toughening or plasticizing components, have been developed. When the resulting blended system is homogeneous, the increase in toughening is generally provided by plastification of the thermoset matrix or reduction of the crosslink density. Moreover, other properties such as the Young's modulus or glass-transition temperature are usually strong affected.

To limit the effect of these drawbacks, the pure resin system can be blended with phase separating rubbery modifiers, which will remain only partially dissolved in the thermoset. Phase separation may be promoted during processing either by temperature change or by chemically induced phase separation. In both cases, one resin-rich phase and a second phase rich in modifier are formed. The second phase appears as small particles whose diameter may range from a few nanometers to some tens of microns. Boogh et al. [148] suggested the application of hyperbranched polymers as reactive modifiers. They [148] have been shown to have a marked toughening effect without loss of thermomechanical properties and processability.

The large number of terminal functional groups allows the use of hyperbranched polymers as crosslinking agents.

The peculiarities of dendritic structures offer new possibilities to manipulate the properties of surface by coating them. In the formation of these macromolecules at interfacial areas, either gas-water or gas-solid transitions, a change in conformation compared to that in solution has been noticed [47].

Hyperbranched polymers can also be utilized in nanoimprint lithography. A fascinating application of the nanoimprint lithography technique is to fabricate so-called quantum magnetic disks, comprising single magnetic domain bits surrounded by a non-magnet material.

Pattering of polymer films at micron or even submicron resolutions is of critical technological importance in the microelectronics industry. Because hyperbranched polymers have many functional groups, moieties with interesting optical, electrochemical, biological, and mechanical properties can be incorporated into the hyperbranched polymer films. Related techniques including template-based approach and photolithography have been developed by Crooks and coworkers [149].

Based on their unique properties, hyperbranched polymers can be applied as toughener for thermosets, curing, cross-linking or adhesive agents, dye-receptive additives for polyolefins, compatilizers, dispersers, processing aids, and rheology modifiers or blend component. The Boltorn^{TM} three-generation hyperbranched polyesters can act as outstanding toughener in epoxy matrix composites, and they can include more than a two-fold increase in the critical strain energy release rate for both low and highly cross-linked matrices, without affecting either the viscosity of the uncured resins or the thermomechanical properties of the cured material [49].

In the tubular film blowing process of linear low density polyethylene (LLDPE), Boltorn H30 hyperbranched polyester successfully acted as a processing aid, and sharkskin was eliminated with no significant influence on the other physical properties of LLDPE films [150,151]. In processing, the hyperbranched polymer has a tendency to migrate to the surface, leading to a hyperbranched polymer-rich surface, which creates the opportunity to tailor the surface properties for various potential applications.

Polypropylene (PP) is a versatile and widely used polymeric material, however; PP does not accept dyes, owing to its non-polar structure as well as its high crystallinity, resulting from the high stereo-regularity of PP. The dyeability of PP with Dispere Blue 56 can be markedly enhanced through the incorporation of hyperbranched macromolecules (Hybrance PS 2550) into PP prior to fibre spinning [152,153].

Perstorp demonstrated that hyperbranched aliphatic polyols can be used, for example, for hardcoat applications, because they allow for a unique combination of coating properties such as high hardness, scratch resistance, and flexibility while ensuring at the same time low viscosity and low shrinkage.

Several pharmaceutical, electronic and optical manufacturing application require "ultrapure" water that meet stringent limits concerning the dissolved and suspended solids, organic carbon, dissolved gases, and biological organisms.

In this context, hyperbranched polymers have been suggested for the extraction of model impurities, such as polycyclic aromatic hydrocarbons via molecular encapsulation. Recently [154,155], it was suggested unimolecular micelles based on hyperbranched polymers as "nanosponges" to extract pollutants from an aqueous phase followed by their subsequent release in an organic phase.

Application in Energy Technique

A solid polymeric electrolyte should meet the requirements of 1) being amorphous, 2) having a high solvating power for appropriate ions, 3) good ion transport, and 4) electrochemical stability. It is well known that polyglycerols segments satisfy the last three requirements, and hyperbranched polymers are usually amorphous. Thus, hyperbranched macromolecules possessing ethylene glycol chains have been designed, prepared and used as novel polymeric electrolytes or ion-conducting elastomers [156,157,158]. Novel polymer electrolytes have been prepared with hyperbranched polymers as the host and lithium trifluoromethanesulfonimide, $LiN(CF_3SO_2)_2$, as the ion source. Analysis of the salt concentration dependence of the glass transition temperature indicates that intermolecular interactions occur in the polymer electrolytes and that these hyperbranched polymers can function as a 'solvent' for the lithium salt [159].

Hyperbranched polymers can also be used to developed fuel cell membrane material [160,161,162,163]. It was found that the molecular weights of the hyperbranched polymers do not affect significantly the ionic conductivity, but the molecular weight distribution might affect it, and also further branching at the terminals of the hyperbranched polymers led to a decrease in the ionic conductivity [160]. Additionally, it was demonstrated that applying the concept of dry polymer system to proton conduction is one possible approach toward high-temperature fuel cells [161,164,165,166].

In recent years, the search to develop large-area solar cells at low cost has led to research on photovoltaic systems based on nanocomposites containing conjugated polymers. These composite films can be synthesized and processed at lower costs and with greater versatility than the solid state inorganic semiconductors that comprise today's solar cells [167,168,169] However; the best nanocomposite solar cells are based on a complex architecture, consisting of a fine blend of interpenetrating and percolating donor and acceptor materials. Cell performance is strongly dependent on blend morphology, and solution-based fabrication techniques often result in uncontrolled and irreproducible blends, whose composite morphologies are difficult to characterize accurately. Here we incorporate three-dimensional hyperbranched colloidal semiconductor nanocrystals in solution-processed hybrid organic-inorganic solar cells, yielding reproducible and controlled nanoscale morphology.

Phase change materials have attracted lots of interest because of their high storage density and constant temperature during phase change process. Their applied field is extensive, such as solar energy storing, smart air-conditioning, buildings, agricultural greenhouse, temperature-regulating textiles, heat management of electronics, telecommunications and microprocessor equipment. Solid-solid phase change materials have some advantages, for instance no liquid or gas generation and small volume change. Cao and Liu [170] prepared hyperbranched polyurethane for thermal energy storage as a novel material. The phase transition enthalpy was more than 100J/g with a transition point at 67°C. Doped hyperbranched polyurethane leads to a material with a transition enthalpy of 125.0J/g and the main decomposition temperature of 437°C [171,172].

However; one should recognize that extensive research of hyperbranched polymers is still in its infancy, and the main object to apply hyperbranched polymers in industry fields is still a dream with few exceptions. DSM introduced hyperbranched polyesteramides to oil field applications, because of their ability to suppress the crystallization of gas hydrated from water-hydrocarbon mixtures [51]. BASF demonstrated that hyperbranched polymers can be used as cross-linkers for coatings [51]. Evonik Degussa uses hyperbranched polymers as performance additives and enzymatically degradable carriers [51].

THERMODYNAMICS OF HYPERBRANCHED POLYMERS

As polymer dissolution plays a crucial rule in many applications. The paper describes theoretical approaches to quantify this complex phenomenon. The phase behavior of concentrated solutions of hyperbranched polymers is of great importance for their application in the field of chemical engineering (membranes, extractive distillation, extraction, absorption, adsorption, etc.) or in the field of medicine (drug delivery systems, tissue engineering, development of sensors, etc.). Kim [173] to point out "Better understanding on physical properties of these polymers, such as solubility and miscibility of these polymers in solvents or with polymers, and functional group dependency to the thermal relaxation process are needed for further development of the subject."

Role of Branching

Branching of polymer chain plays an important role in liquid-liquid phase equilibria (LLE) in polymer solutions and influences quantities like separation temperature and/or separation pressure of a polymer solution. Kleintjens at al. [174] showed for the system polyethylene + diphenyl ether that the two-phase region of a branched polyethylene solution may be shifted by more than 10°C compared with that of a linear polyethylene sample of about equal number and mass average molar mass.

De Loos et al. [175] measured cloud-point curves and critical curves in the system ethylene + branched polyethylene and concluded that the cloud-point curves of a system with a branched polyethylene are at significantly lower pressure (10 – 40 MPa) that the cloud-point curves of a system with a linear polyethylene with comparable molecular-weight distribution.

Striolo and Prausnitz [176] measured the osmotic second virial coefficients for linear and star poly(ethylene oxide) in water and propan-1-ol. They [176] figured out that at fixed temperature, the osmotic second virial coefficient for star poly(ethylene oxide) is lower than that for the linear polymer of comparable molecular weight.

Kennedy et al. [177] generalized the Flory-Huggins free enthalpy of mixing expression to solution of branched polymers, assigning different interaction energies to solvent-end segment and solvent-middle segment contacts. In 1980, Kleintjens and Koningsveld [178] realized "To data, polymer solution theories have failed to describe quantitatively the effect of polymer chain branching on liquid-liquid phase relationships in solutions". They tried to improve this situation using an adopted lattice treatment to incorporate this phenomenon [178]. A branched polymer molecule has been presented as a chain, composed of end, linear middle, and branched middle segments, each interacting through a specific surface area with the immediately surrounding molecules or segments. The expression obtained for the energy of mixing contains five parameters. The model described spinodal, critical, and cloud-point data within experimental error over practically the full measured concentration range. Phase compositions, however; could only be reproduced qualitatively.

On the basis of experimental batch-cell data, increasing short-chain branch density was found to reduce the cloud-point pressure of poly(ethylene-1-butene) solution in propane, by as much as a factor of 3 [179]. This trend was captured by a SAFT (Statistical Association Fluid Theory) approximation via an effective segment energy that depends on the branch density. However; it was not possible to accomplish quantitative agreement with the performed experiments.

Blas and Vega [180] presented an extension of the SAFT for branched chain molecules using Wertheim's first- and second-order thermodynamic perturbation theory with a hard-sphere reference fluid. Molecules are formed by hard spherical sites which are tangentially bonded. Linear chains are described as freely jointed monomeric units, whereas branched molecules are modeled as chains with a different number of articulation points, each of them formed by three arms. The theory was applied to the critical temperature of light branched alkanes, whereas the results show that the theory qualitatively predicts the experimental data, although the proposed molecular description was too simple to account for the finer details of the molecular architecture of branched chains [180]. Additionally, the branching term accounts for the branching units that are given as rigid tetramer units within a hyperbranched macromolecule and hence, only branching points with four bonds can be described.

A similar SAFT-approach was used by Kozlowsky et al. [181] for the calculation of phase behavior of polymer solutions containing hyperbranched polymers. The vapor-liquid equilibrium could be modeled very close to the experimental data over a large pressure range [181]. Unfortunately, the calculation of the liquid-liquid phase split was not successful [181]

The experimental investigation of the phase behavior of hyperbranched polymer solutions is a crucial requirement for a successful introduction of new applications to highly competitive markets. Several studies dealing with the liquid-liquid equilibria in binary [181,182,183,184,185,186,187,188] and ternary [181,182,184,189,190,191] systems can be found in the literature. For instance, Boltorn U-3000 is completely soluble in hydrocarbons (hexane, heptane, benzene, and toluene) and ethers (methyl-tert-butyl ether or ethyl-tert-butyl ether) but only partially miscible with alcohols, where the solubility in alcohols increases with an increase of the chain length of the solvent [183]. Boltorn H3200 shows a liquid-liquid phase split with lower critical solution temperature that interferes with the solidification surface of Boltorn H3200 [185].

Lieu et al.[192] as well as Mio et al. [193] presented vapor-liquid equilibrium data for binary solutions of dendrimers in a variety of organic solvents. These authors figured out that the generation number and chemical composition of the dendrimers have a significant effect on solvent-dendrimer compatibility.

Domanska and Zolek-Tryznowska [194] measures the solid-liquid and the liquid-liquid phase diagrams at ambient pressure for the system Boltorn W3000 with alcohols, or with ethers, where the liquid-liquid equilibrium shows a LCST behavior. Ko et al. [195] investigated the influence of salt on the solid-liquid phase behavior of hyperbranched polymers.

Using inverse gas chromatography the infinite-dilution activity coefficients were measured [196,197,198,199,200,201]. Solvent activity coefficients of amine-terminated poly(propylenimine) of generation 2 to 5 in different solvents (methanol, ethanol, ethyl acetate, toluene, and THF) change with respect to the dendrimer generation number, reaching a minimum at generation 4. Because the dendrimers are basic, the solvent quality is higher for slightly acidic solvents, such as methanol and ethanol [196].

Other properties, like association [202], densities and viscosities [203,204] or osmotic compressibility [205], osmotic virial coefficients [176] of hyperbranched polymer solution were also measured.

Recently, Cheng et al. [206] reports for the first time the influence of degree of branching (DB) on the thermoresponsive phase transition behaviors of hyperbranched multiarm copolymers. Two series of PEHO-star-PEOs (series A) and PEHO-star-PDMAEMAs (series B) with the hydrophobic DB-variable PEHO core and different kinds of linear arms (PEO arms or PDMAEMA arms) were synthesized. It was found these two series demonstrate thermoresponsive phase transitions with the LCST [206]. The studies on the LCST transition mechanism indicate that series A belongs to the thermoresponsive polymer system with LCST transition based on hydrophilic-hydrophobic balance, while series B belongs to the thermoresponsive polymer system with LCST transition based on coil-to-globule transition [206]. For series A, the LCST phase transition is highly dependent on the DB of the PEHO core in copolymers, whereas for series B, the LCST phase transition is independent of the DB but dependent on solution pH [206]. Compared with linear thermoresponsive polymers hyperbranched ones having spheroid-like structure exhibited an unusual salt effect: a non-linear LCST decrease upon increasing the concentration of various salts such as NaCl, KCl or Na_2SO_4 has been observed [207].

Using small-angle neutron scattering Martter et al. [208] measured bulk thermodynamic interaction parameters for blends of anionically polymerized star with different number of arms and linear polybutadienes of well-defined architecture and molecular weight. Comparison of these measured values with results from comparable polystyrene blends suggests the existence of nonuniversal aspects in the thermodynamic interaction due to entropic contributions [208]. While the interaction parameter for polystyrene star/linear blend increases monotonically with number of arms in the star, the value in the polybutadiene star/linear blends does not [208]

Massa et al [209] provide a survey of the phase behavior of blends of hydroxyphenyl- and acetoxyphenyl-terminated hyperbranched polyesters with polymers such as polycarbonate, polyester, and polyamides. The hydroxyterminated hyperbranched polyester blend miscibility was identical to that of poly(vinylphenol); this suggests that strong interactions due to hydrogen bonding, more so than chain architecture, dominate the blend miscibility [209]. Miscibility was found in blends of the two hyperbranched polyesters and in those of poly(acetoxystyrene) with poly(vinylphenol) but none of the blends of linear analog with hyperbranched polymer was single phased [209].

All these experimental observations demonstrate clearly the role of the architecture of the polymer and the interaction between different segments present in the system.

Thermodynamic Model – Lattice Cluster Theory

Classic theories of polymer systems, as typified by the widely used Flory-Huggins (FH) theory [1,5], treat individual monomers as single entities, devoid of any chemical structure. In the framework of the FH theory the polymers are taken to have random configurations, which arise from placing individual monomers at single lattice sites, FH-theory subject to strict constraint applies to solvent molecules when they are also present. Flory and Huggins [5] employed a very simple mean-field approximation that effectively ignores the details of the polymer chain connectivity and, therefore, that cannot distinguish between linear, star, branched, comb polymer architecture. The FH theory [5] is based on the polymer chain cohesive interaction energy parameter. Because most of the interaction between the hyperbranched polymers resides at their chain ends, rather than the chain interaction, the conventional Flory-Huggins theory may not be able to predict the miscibility of hyperbranched polymers.

While the lattice cluster theory (LCT), originally developed by Freed and coworkers [210,211,212,213,214,215], extends well beyond FH theory first by endowing individual monomers with explicit molecular structures that occupy several lattice sites. Moreover, the LCT develops a vastly superior solution of the lattice model to include important contributions to thermodynamic properties from packing and interaction induced local correlations. The LCT describes the thermodynamic consequences of the existence of short-range correlation that arise from packing constraints and the different monomer-monomer interactions. Because only short-range correlations are included, the LCT technically is also a mean-field theory. The accuracy of the LCT has been tested against Monte Carlo simulation of various lattice model polymer systems [216,217,218,219,220,221,222] demonstrating that the LCT provides an excellent estimation of all thermodynamic properties in the mean-field regions, with the exception of the specific heat at lower temperatures.

Our application of the LCT to hyperbranched polymers are designed to probe the limitations of the LCT as a guide to future efforts for rectifying these limitations, to determine the qualitative features governing the thermodynamic properties of these fascinating systems [188,223,224,225,226,227]

The partition function for the extended lattice model of polymer systems is evaluated within the LCT by introducing approximations for the bonding constraints and for the excluded volume and van der Waals interactions [215]. We assume that all chains are fully flexible. The theoretical description of polymer thermodynamics at a fixed pressure requires treating the system as compressible. The compressibility of the system can be introduced into the extended lattice models by allowing for the presence of empty lattice sites, called voids. The voids do not represent a species of particles in the thermodynamic sense, nor do they have interactions; they are merely a simple bookkeeping device for keeping track of the excess free volume. When computations are performed for constant pressure systems, the excess free volume fraction is determined numerically from the equation of state, derived us usual by defining the pressure as the derivative of the free energy with respect to total volume.

The structure of the polymer enters the LCT free energy through a series of geometrical indices $N_{j,i}$ that enumerate the number of distinct sets of j sequential bonds in a single chain of species, i.

$$N_{1,i} = M_i - 1 \tag{1}$$

where M_i is the number of monomers in a polymer chain of type i (segment number).

$$N_{2,i} = M_i - 2 + b_{3,i} + 3b_{4,i} \tag{2}$$

where $b_{3,i}$ is the number of branching points of degree 3 of the polymer chain of type i in which 3 bonds meet, and hence $b_{4,i}$ is the number of branching points of degree 4 of the polymer chain of type i in which 4 bonds meet. The quantity $N_{2,i}$ describes the number of two consecutive bonds in a polymer chain, i. The number of three consecutive bonds in a polymer chain, i, $N_{3,i}$ is given by:

$$N_{3,i} = M_i - 3 + 2b_{3,i} + 6b_{4,i} \tag{3}$$

The number of ways in which three bonds meet at a lattice site for a polymer chain, i, $N_{\perp,i}$ can be calculated using:

$$N_{\perp,i} = b_{3,i} + 4b_{4,i} \tag{4}$$

The number of distinct ways of selecting two non-sequential bonds on the same chain, i, can be expressed by $N_{1,1;i}$:

$$N_{1,1;i} = \frac{1}{2} N_{1,i} \left(N_{1,i} - 1 \right) - N_{2,i} \tag{5}$$

Finally, the number of distinct ways of selecting two sequential bonds and one non sequential bonds on the same chain, i, reads:

$$N_{1,2;i} = N_{1,i} N_{2,i} - 2N_{2,i} - 2N_{3,i} - 3N_{\perp,i} \tag{6}$$

The geometrical coefficients, used in Eqs. 1 - 6, can be calculate from the chemical structure of arbitrarily chosen hyperbranched polymer knowing the quantities M_i, $b_{3,i}$ and $b_{4,i}$. Let us assume the system consists of two different components. For simplification we introduce the quantities K_i for component 1 and L_i for component 2 in the following manner:

$$\begin{aligned} &K_1 = N_{1;1} / M_1; \quad K_2 = N_{2;1} / M_1; \quad K_3 = N_{3;1} / M_1; \\ &K_4 = N_{\perp;1} / M_1; \quad K_5 = N_{1,1;1} / M_1; \quad K_6 = N_{1,2;1} / M_1; \end{aligned} \tag{7}$$

and

$$\begin{aligned} &L_1 = N_{1;2} / M_2; \quad L_2 = N_{2;2} / M_2; \quad K_3 = N_{3;2} / M_2; \\ &L_4 = N_{\perp;2} / M_2; \quad L_5 = N_{1,1;2} / M_2; \quad L_6 = N_{1,2;2} / M_2; \end{aligned} \tag{8}$$

The lattice theory of a binary blend represents each component related to LLE of hyperbranched polymer solutions. ($i = 1,2$) as n_i monodisperse polymer chain placed on a regular array of N_L lattice sites and coordination number z. In LCT the Helmholtz free energy is expanded in a double power series of $1/z$ and $\varepsilon_{LCT}/(k_B T)$, where k_B means the Boltzmann constant. The interaction energy ε_{LCT} is defined by

$$\varepsilon_{LCT} = \varepsilon_{ii} + \varepsilon_{jj} - 2\varepsilon_{ij} \tag{9}$$

where ε_{ii} is the energy of a non-bonded polymer-polymer of polymer type i, ε_{jj} is the same quantity for polymer type j, and ε_{ij} is the energy of a contact between polymer of type i and polymer of type j. ε_{LCT} is related to the usual χ-parameter of the Flory-Huggins theory by $\chi = z\varepsilon_{LCT}/(2k_B T)$. To obtain an expression for the Gibbs energy of mixing from the partition function, we assume that, at the pressure of interest here, the change in molar volume upon mixing is negligible. In this limit of an incompressible lattice the Helmholtz free energy is equal to the Gibbs free energy. The Gibbs energy of mixing for a binary blend emerges from the LCT as [223,227]:

$$\frac{G_{Mix}}{N_L k_B T} = \frac{\phi_1}{M_1} \ln \phi_1 + \frac{(1-\phi_1)}{M_2} \ln(1-\phi_1) + \sum_{i=1}^{6} \alpha_i (\phi_1)^i \tag{10}$$

where ϕ_1 is the concentration of component 1, expressed in segment fraction, and the α_i-values are:

$$\alpha_1 = \frac{\varepsilon_{LCT}}{k_B T}\left(\frac{z}{2} - K_1\right) + \frac{(K_1 - L_1)^2}{z} + \frac{\varepsilon_{LCT}}{z k_B T}\begin{pmatrix} 2K_2 + K_3 + 3K_4 - 4L_1K_5 + K_6 - 2K_1L_1 \\ -4L_1K_2 + 2L_1K_1^2M_1 - K_1K_2M_1 \end{pmatrix}$$

$$+\frac{1}{z^2}\begin{pmatrix} -4L_1^4M_2 - 4L_1L_2 + 2L_1^2L_2M_2 + 2K_1^4 + K_2^2 + 8K_1^3/3 + L_2^2 - 4K_1K_2 - 2K_1K_3 \\ -2K_1K_6 - 2K_1^4M_1 + 4K_5K_1^2 - 6K_1K_4 + 2K_1^2K_2M_1 + 4L_1K_2 + 4K_1L_2 - 8L_1^2K_1 \\ +6L_1^4 + 16L_1^3/3 + 2K_1^2L_1^2M_1 + 4L_1^3K_1M_2 + 2L_1K_3 + 2L_1K_6 - 2K_2L_2 - 8L_1^3K_1 \\ -2K_1L_1L_2M_2 - 2K_1L_1K_2M_1 - 4K_5L_1^2 + 6L_1K_4 + 2K_1L_3 - 6L_1L_4 - 8K_1L_5L_1 \\ +8L_5L_1^2 + 2K_1L_6 - 2L_1L_3 - 2L_1L_6 + 6K_1L_4 \end{pmatrix} \tag{11}$$

$$\alpha_2 = \frac{1}{z^2}\begin{pmatrix} -L_2^2 - 8L_1^3 - 2L_1^2L_2M_2 + 6L_1^4M_2 - 8K_1^2L_1 + 2K_1^2L_1^2M_2 + 2L_1K_1L_2M_2 \\ +2L_1K_1K_2M_1 - 12K_1^2L_1^2 + 8K_5L_1^2 + 4L_1L_2 + 2K_2L_2 - 8K_1K_5L_1 \\ +16K_1L_1^2 + 4K_1^3L_1M_1 - 4K_1L_2 - 4L_1K_2 - 2L_1K_3 - K_2^2 - 6L_1K_4 \\ -2L_1K_6 - 12L_1^4 - 2K_1^2K_2M_1 + 24K_1L_1^3 + 4K_1K_2 + 2K_1K_3 + 2K_1K_6 \\ +6K_1K_4 - 4K_1^2L_1^2M_1 - 8K_1L_1^3M_2 - 2K_1L_3 + 6L_1L_4 + 16K_1L_5L_1 \\ -12L_5L_1^2 - 2K_1L_6 + 2L_1L_3 + 2L_1L_6 - 4L_5K_1^2 - 6K_1L_4 \end{pmatrix}$$

$$+\frac{\varepsilon_{LCT}}{zk_BT}\begin{pmatrix} -2K_6 - 6K_4 + 12L_1K_5 + 10K_1L_1 - 4L_1K_1^2 - 4L_1^3 + 2K_1^3M_1 - 4K_1^2 \\ +2L_2 - 2K_3 + 8L_1^2K_1 + 2L_1^3M_2 + 2K_1K_2M_1 + 8L_1K_2 - 4K_1K_5 - 4K_2 \\ -6L_1K_1^2M_1 - 4L_1^2 - L_1L_2M_2 - 4K_1L_2 + L_3 + 3L_4 + L_6 - 4L_1L_5 \end{pmatrix} \quad (12)$$

$$+\frac{\varepsilon_{LCT}}{k_BT}\left(2K_1 - L_1 - \frac{z}{2}\right) + \left(\frac{\varepsilon_{LCT}}{k_BT}\right)^2\left(-\frac{z}{4} - K_1^2(1+2M_1) + 3K_2 + 4K_5 + 3K_1 - \frac{L_1}{2}\right)$$

$$-\frac{(K_1 - L_1)^2}{z}$$

$$\alpha_3 = \frac{1}{z^2}\begin{pmatrix} K_1L_1\left(8K_1 - 8K_1^2 + 24K_1L_1 - 24L_1^2 - 8L_1\right) - 8\left(K_1^3 - L_1^3\right)/3 \\ +(K_1 - L_1)^2\left(-2L_1^2M_2 + 2M_1K_1^2 - 4K_5\right) - 8\left(K_1^3 - L_1^3\right)/3 + 4L_5K_1^2 \\ -8K_1L_1L_5 + 4L_1^2L_5 + 8L_1^4 + 4L_5K_1^2 - 8K_1L_1L_5 + 4L_1^2L_5 \end{pmatrix}$$

$$+\frac{\varepsilon_{LCT}}{zk_BT}\begin{pmatrix} 20L_1K_1^2 + 8K_5K_1 - 12L_1K_5 + 8K_1^2 - 4K_1^3 + 2K_2 + K_3 + 3K_4 + K_6 - 2L_2 \\ +L_1M_2(L_2 + 2K_1L_1 - 4L_1^2) - 16K_1L_1 + 12L_1^3 - 28L_1^2K_1 + 4\left(-L_1K_2 + K_1L_2\right) \\ +K_1M_1\left(6L_1K_1 - 4K_1^2 - K_2\right) - L_3 - 3L_4 - L_6 + 8L_1L_5 - 4K_1L_5 + 8L_1^2 \end{pmatrix} \quad (13)$$

$$+\left(\frac{\varepsilon_{LCT}}{k_BT}\right)^2\left(\frac{z + 5L_1 - 13K_1}{2} - 3K_2 - 6K_5 + 3K_1^2M_1 - L_2 + 6K_1^2 + 2L_1^2 - 6K_1L_1\right)$$

$$+\frac{\varepsilon_{LCT}(L_1 - K_1)}{k_BT}$$

$$\alpha_4 = -2\frac{(K_1 - L_1)^4}{z^2} + \frac{\varepsilon_{LCT}}{zk_BT}\begin{pmatrix} -2K_1^2L_1M_1 - 12L_1^3 - 4K_1K_5 + 4L_1K_5 + 8K_1^3 - 2K_1L_1^2M_2 \\ -28L_1K_1^2 - 4\left(K_1 - L_1\right)^2 + 2K_1^3M_1 + 2L_1^3M_2 + 32K_1L_1^2 \\ -4L_1L_5 + 4K_1L_5 \end{pmatrix} \quad (14)$$

$$+\left(\frac{\varepsilon_{LCT}}{k_BT}\right)^2\begin{pmatrix} -\frac{z}{4} + L_2 + 4K_5 - 2K_1^2M_1 + K_2 + 6K_1 - 4L_1 + L_1^2M_2/2 - 7L_1^2 - 12K_1^2 \\ +18K_1L_1 - L_5 \end{pmatrix}$$

$$\alpha_5 = -4\frac{\varepsilon_{LCT}(K_1 - L_1)^3}{zk_BT} + \left(\frac{\varepsilon_{LCT}}{k_BT}\right)^2\begin{pmatrix} \frac{K_1^2M_1 - L_1^2M_2}{2} - 18K_1L_1 - 2K_1 + 2L_1 + 8L_1^2 \\ +10K_1^2 + L_5 - K_5 \end{pmatrix} \quad (15)$$

$$\alpha_6 = -3\left(\frac{\varepsilon_{LCT}}{k_BT}\right)^2(K_1 - L_1)^2 \quad (16)$$

Using standard thermodynamics Eq. 10 can be applied to liquid-liquid phase equilibrium calculations of a binary system. In principle, the binary system can be made from a) two solvents, b) two linear polymers, c) a linear polymer and a solvent, d) two linear polymers, e) hyperbranched polymer and a solvent, f) two hyperbranched polymers, g) a linear and a hyperbranched polymer. The parameters ($M_i, b_{3,i}, b_{4,i}$) occurring in Eqs. 11 - 16 estimate the type of the component. Using Eq. 10 the spinodal condition can be derived, resulting in:

$$0=\frac{1}{\phi_1 M_1}+\frac{1}{(1-\phi_1)M_2}+2\alpha_2+6\alpha_3\phi_1+12\alpha_4\phi_1^2+20\alpha_5\phi_1^3+30\alpha_6\phi_1^4 \tag{17}$$

Additionally, the critical condition based on Eq. 10, reads:

$$0=-\frac{1}{\phi_1^2 M_1}+\frac{1}{(1-\phi_1)^2 M_2}+6\alpha_3+24\alpha_4\phi_1+60\alpha_5\phi_1^2+120\alpha_6\phi_1^3 \tag{18}$$

Liquid-liquid phase separation can occur either upon heating or cooling, corresponding to lower (LCST) and upper critical solution (UCST) phase boundaries, respectively. The principle shape of the LLE behavior depends on the temperature dependency of the Gibbs energy of mixing. Caused by the above mentioned series expansion until the second order with respect to the interaction energy, Eq. 10 can also be written in the following manner:

$$\frac{G_{Mix}}{N_L k_B T}=\delta_0+\frac{\delta_1}{T}+\frac{\delta_2}{T^2} \tag{19}$$

Using the following abbreviations:

$$v_1=b_{31}+3b_{41} \qquad v_2=b_{32}+3b_{42} \tag{20}$$

δ_0, δ_1 and δ_2 - values in Eq. 19 are:

The quadratic temperature dependence in Eq. 19 ensures that all principle types of phase behavior in a binary system, namely a) only UCST, b) only LCST, c) closed loop miscibility gaps and d) UCST + LCST for the same system (hour-glass phase diagrams), can be calculated. However; in the case where $\delta_2=0$ only one miscibility gap with UCST or LCST behavior can be modeled, where the sign of δ_1 determine, if a UCST or a LCST phase behavior is present.

$$\delta_0 = \frac{-1}{3M_1^4 M_2^4 z^2}\left[\begin{array}{l} 6\phi_1^4\left(M_2-M_1\right) \\ +2\phi_1^3 M_1\left(\begin{array}{l} 3M_2\left(M_2-M_1\right)\left[\begin{array}{l} -12M_1+M_1M_2 \\ +2\left(M_2-M_1\right)\left(v_2M_1-v_1M_2\right)\end{array}\right] \\ +\left(M_1^3-M_2^3\right)\left(M_2-M_1\right)\end{array}\right) \\ +3\phi_1^2M_1^2\left[\begin{array}{l} M_2^2\left(M_2v_1-M_1v_2\right)^2+4v_2M_2\left(-M_2^2-3M_1^2+4M_1M_2\right) \\ +2\left(M_1-M_2\right)\left[\begin{array}{l} \frac{zM_2^2}{2}\left(M_1-M_2\right)+4v_1M_2^2 \\ +\left(2M_2^2-M_2+6\right)\left(M_1-M_2\right)\end{array}\right]\end{array}\right] \\ +\phi_1\left[\begin{array}{l} +2M_1^4\left(-6M_2^2+2M_2-9\right)+2M_2^2\left(-6M_1^2+M_1-3\right) \\ +M_1M_2\left[\begin{array}{l} -3M_1^3\left(-8v_2+M_2v_2^2\right)+6M_1^2M_2\left(\left(M_2v_1-4\right)v_2-2v_1\right) \\ -3M_2^3v_1\left(M_1v_1-4\right)\end{array}\right] \\ +6M_1^3M_2\left(4+4M_2^2-M_2\right)-3zM_1^2M_2^2\left(M_1-M_2\right)^2\end{array}\right]\end{array}\right] + \frac{\phi_1\ln\left(\phi_1\right)}{M_1}+\frac{\left(1-\phi_1\right)\ln\left(1-\phi_1\right)}{M_2} \tag{21}$$

$$\delta_1 = \frac{-\varepsilon_{LCT}}{2k_B M_1^3 M_2^3 z}\left[\begin{array}{l} 8\phi_1^5\left(-3M_2M_1\left(M_1-M_2\right)+M_1^3-M_2^3\right) \\ +4\phi_1^4\left\{\begin{array}{l} +4M_2^3-6M_1^3 \\ +M_1M_2\left[\begin{array}{l} \left(M_1-M_2\right)\left(M_1v_2-M_2v_1\right) \\ -\left(M_1-M_2\right)^2+16M_1-14M_2\end{array}\right]\end{array}\right\} \\ +2\phi_1^3\left\{\begin{array}{l} -4M_2^3+12M_1^3 \\ +M_1M_2\left(\begin{array}{l} -2\left(M_2M_1+4M_1-4M_2\right)\left(M_1v_2-M_2v_1\right) \\ +M_1M_2\left(\begin{array}{l} \left(z+3\right)M_1M_2\left(M_1-M_2\right) \\ +4\left(M_1-M_2\right)^2-28M_1+20M_2\end{array}\right)\end{array}\right)\end{array}\right\} \\ +\phi_1^2M_1\left(\begin{array}{l} 2\left(M_1M_2\left(-M_1M_2^3-3M_1M_2+6M_2^2-2M_1+4M_2\right)-2M_2^3\right) \\ +z^2M_1^2M_2^3-2zM_1M_2^2\left(M_1M_2+M_1-2M_2\right)-8\left(M_1-M_2\right)^2 \\ +8M_2^2v_1\left(M_2-M_1M_2-M_1\right) \\ +4M_1M_2v_2\left(M_1M_2-2M_2+2M_1\right)\end{array}\right) \\ +\phi_1M_1^2M_2^2\left(-2zM_2\left(1-M_1\right)-6M_2+2M_1M_2+4M_2v_1-M_1M_2z^2\right)\end{array}\right] \tag{22}$$

$$\delta_2 = -\frac{\varepsilon_{LCT}^2 \phi_1^2}{4k_B^2 M_1^2 M_2^2} \left[\begin{array}{l} 12\phi_1^4 (M_1 - M_2)^2 \\ +2\phi_1^3 \left(-16(M_1 - M_2)^2 - 4M_2^2 + M_1 M_2 \begin{pmatrix} 4 - 5(M_1 - M_2) \\ +2M_1 v_2 - 2M_2 v_1 \end{pmatrix} \right) \\ +\phi_1^2 \begin{pmatrix} +M_1^2 M_2^2 (z+6) + 2M_2 M_1^2 (9 - 4v_2) - 4M_1 M_2^2 (8 - 3v_1) \\ +48M_2^2 - 72M_2 M_1 + 28M_1^2 \end{pmatrix} \\ +\phi_1 \begin{pmatrix} -24M_2^2 - 8M_1^2 + 2M_2 M_1 (12 - 3M_1 + 17M_2 - M_2 M_1 (6+z)) \\ +4M_1 M_2 (M_1 v_2 - 3M_2 v_1) \end{pmatrix} \\ +M_1 M_1 (M_1 M_2 (6+z) + 4M_2 (v_1 - 3) - 2M_1) + 4M_2^2 \end{array} \right] \tag{23}$$

Let us discuss two special cases. The first case A should by characterized by

$$M = M_1 = M_2 \qquad b_{i,j} = 0 \tag{24}$$

The conditions in Eq. 24 imply that the two components having the same molecular weight and no branching points occur in the molecules. Eqs. 21 - 23 in combination with Eq. 24 lead to:

$$\delta_{0,\mathrm{A}} = \frac{\phi_1 \ln(\phi_1) + (1-\phi_1)\ln(1-\phi_1)}{M} \tag{25}$$

$$\delta_{1,\mathrm{A}} = \frac{\varepsilon_{LCT} \phi_1 (1-\phi_1)\left(6 + 2z + M\left(z^2 - 2z - 2\right)\right)}{2k_B M z} \tag{26}$$

Assuming for the coordination number $z >= 6$ then Eq. 26 tell us that only the sign of the interaction energy of the considered mixture in the limit of Eq. 24 determines if LCST or UCST behavior can be found.

$$\delta_{2,A} = -\frac{\varepsilon_{LCT}^2 \phi_1^2 (1-\phi_1)^2 \left(M^2 (z+6) - 14M + 4\right)}{4k_B^2 M^2} \tag{27}$$

According Eq. 27 the quantity $\delta_{2,A}$ can only become zero if the second bracket in the numerator of Eq. 27 vanish. The coordination number z can have value between 6 and 12. In this range the root of the bracket leads to segment number smaller then 1 or the complex numbers without any physical sense. These findings mean, that the LCT can describe all types of phase diagrams. This is in contrast to the classical FH-theory, where LCST-behavior or closed miscibility gaps or hour-glass phase diagrams can only be modeled, if the classical χ is assumed to be function of temperature. The solution of the critical condition (Eq. 18) in combination with Eq. 24 provide the critical temperatures:

$$T_{C1,2,A} = \pm\sqrt{3}\frac{\varepsilon_{LCT}}{k_B M}(1-\phi_1)\phi_1\sqrt{M\left((z+6)M^2 - 14M + 4\right)} \tag{28}$$

The last equation demonstrate that one of the critical temperatures is located at negative temperatures and hence without any physical sense.

The next special case B which we can discuss reads:

$$M = M_1 = M_2 \qquad v_1 > 0 \quad v_2 = 0 \tag{29}$$

Hence, the mixture consists of two molecules having the same molecular weight, but component 1 have a branching structure and component 2 is linear. In this case the critical temperatures are given by:

$$T_{c1,2,B} = \frac{(1-\phi_1)\phi_1\varepsilon_{LCT}}{Mz(1-2\phi_1)k_B}\left[\begin{array}{l} -6Mv_1\phi_1(1-\phi_1) \\ \pm\sqrt{\begin{array}{l} 3Mz^2\left(4+M(6v_1-14)+M^2(6+z)\right) \\ -12\phi_1 Mz^2\left(4+M(9v_1-14)+M^2(6+z)\right) \\ +12\phi_1^2 M\left(z^2\left(4+M(17v_1-14)+M^2(6+z)\right)+3Mv_1^2\right) \\ -24\phi_1^3 M^2 v_1\left(5z^2+3v_1\right)+36\phi_1^4 M^2 v_1^2 \end{array}} \end{array}\right] \tag{30}$$

According Eq. 30 only one positive critical temperature for $\varepsilon_{LCT} > 0$ and $\phi_1 < 0.5$ can be calculated. The presence of two positive critical temperatures require a critical segment fraction above 0.5 for $\varepsilon_{LCT} > 0$. The introduction of the architecture in the theoretical framework can lead not only to a quantitative change in the phase diagram, but also to qualitative difference in comparison with the FH-theory.

The architecture and the size asymmetry also affect the range of T over which mean-field and Ising-type critical behaviors are observed [215,228]. These temperature range are expressed in terms of the Ginzburg number Gi, which provide a rough estimate of the magnitude of the reduced temperature $\tau \equiv (T - T_C)/T$ at which the crossover from mean-field to Ising-type behavior occurs. Mean-field theory holds for $\tau \geq 10Gi$, while the Ising critical behavior corresponds to $\tau \leq Gi/10$. The range $Gi/10 < \tau < 10Gi$ describes a crossover domain with $\tau \approx Gi$ [228]

Important contribution emerge also from the so-called "equation of state effects", the variation of polymer properties with thermodynamic state and, in particular, with pressure [229]. This polymer fluid compressibility is introduced into the lattice model by allowing lattice sites to be empty. The void volume fractions were determined from the appropriate equation of state for a given system [229].

The LCT is extended to study all the possible nematic orderings of rectangular mesogens, namely, rodlike, discotic, and biaxial nematics for oblong mesogens, and rodlike and discotic nematics for square mesogens [230]. The model employed so far treats each polymer as a completely flexible entity, subject only to the excluded volume constraints against multiple occupancy of a lattice site by any two united atom groups. However; bond angle constraints and steric interaction of, e.g., hydrogen atoms and side groups to local chain stiffness and the semiflexibility of polymer molecules. For example, torsional motions about single C-C backbone bonds in many olefins are limited to a trans and a pair of gauche conformations with differing energies. The generalization of the LCT to include explicit trans-gauche energy differences was applied to study the combined influences of chain stiffness disparities, monomer molecular structures, energetic asymmetries, and nonrandom mixing on the miscibilities of binary polymer blends [217,231,232]. The LCT was also used to determine the essential microscopic parameters that influence the phase separation in binary blends of linear semiflexible lattice chains with equal polymerization indices [231,233,234]. A generalized entropy theory of glass formation was developed by merging the LCT for the thermodynamics of semiflexible polymer melts at constant pressure with the Adam-Gibbs relation between the structural relaxation time and the configurationally entropy [235,236,237,238]. Characteristic temperatures and structural relaxation times for different classes of glass-forming polymer liquids are computed using a revised entropy theory of glass formation that permits the chain backbone and the side groups to have different rigidities [239]. The theory is applied to glass formation at constant pressure or constant temperature, where the calculations provide new insights into physical factors influencing the breadth of the glass transition and the associated growth of relaxation times [239].

The LCT was used to study the microscopic molecular factors affecting the miscibilities of A_xB_{1-x}/C binary mixtures (where the homopolymer C is either different or identical to the A_xB_{1-x} random copolymer species) [240,241,42,243,244,245,246]. The LCT was also used to study the influence of monomer structure, i.e., short chain branching, on the miscibility of binary polymer blends [216,234], where compressible blend model with a single interaction energy for all united atom groups were applied.

A further extension of the LCT theory is the development of a thermodynamic model able to describe the influence of monomer structure and local correlations on the free energy of strongly interacting and self-assembling polymer systems [247,248]. This extension combines a systematic high dimension (1/d) and high temperature expansion (that is appropriate for weakly interacting systems) with a direct treatment of strong interactions. The general theory is illustrated for a binary polymer blend whose two components contain "sticky" donor and acceptor groups, respectively. The free energy is determined as an explicit function of the donor-acceptor contact probabilities that depend, in turn, on the local structure and both the strong and weak interactions.

THERMODYNAMIC MODEL – WERTHEIM THEORY

Another characteristic feature of the hyperbranched polymers is the large number of functional groups, which can have special interactions, like association sites. These association sites allow the formation of cross-associates as well as self-associate. The LCT can be combined with a model for association, like the Wertheim-perturbation theory [188,226] or the extended chemical association model [227]

In Wertheim's approach [249,250,251,252], the overall density of the particles in the fluid is split in two parts. For a fluid consisting of particles with one attractive spot the formalism is in terms of two densities: the overall particle density ρ is divided in a density of non-bonded particles ρ_0 and in a density of particles ρ_1 that did form an attractive bond with another particle [249]:

$$\rho = \rho_0 + \rho_1 \tag{31}$$

In a fluid of particles with only one attractive spot we can thus recognize non-bonded and bonded particles. The total 2-particle distribution function $G(1,2)$ is constituted from contributions arising from the correlations between two particles which have not formed an attraction bond $g_{00}(1,2)$, two particles of which one has formed an attraction bond $g_{10}(1,2)$ and $g_{01}(1,2)$, and two particles that both have formed an attraction bond $g_{11}(1,2)$. The coordinates of particles 1 and 2 are given by (r_1,Ω_1) and (r_2,Ω_2), where r_i denotes the position and Ω_i the orientation of particle i. To get a multiple density approach, the pair interaction potential $u(1,2)$ depending on the orientation of particle 1 and 2 must be split in two parts, an isotropic repulsive part $u_R(r_1,r_2)$ and a directional attractive part $u_{att}(1,2)$. The Mayer function $f(1,2)$ can be divided in an attractive $F(1,2)$ and a purely repulsive part $f_R(r_1,r_2)$ [249]

$$f(1,2) = f_R(r_1,r_2) + F(1,2) \tag{32}$$

with

$$f_R(r_1,r_2) = \exp\left(\frac{-u_R(r_1,r_2)}{k_B T}\right) - 1 \tag{33}$$

and

$$F(1,2) = \exp\left(-\frac{u_R(r_1,r_2)}{k_B T}\right)\left(\exp\left(-\frac{u_{att}(1,2)}{k_B T}\right) - 1\right) \tag{34}$$

It has been proven [249,250] that this division of the Mayer function allows a diagrammatic expansion of ρ in terms of the activity z, f_R - and F -bonds similar to LCT. According the suggested expansion, the overall density ρ can be split into densities of bonded and non-bonded particles. The ρ_0 and ρ_1 -values, are then both classified by a different part of the set of diagrams that constitutes ρ. Starting from the grand canonical partition function and using these expansions of ρ_0 and ρ_1, Wertheim [249,250] derived an exact diagrammatic expansions of the structural correlations $g_{00}(1,2)$, $g_{01}(1,2)$, $g_{10}(1,2)$ and $g_{11}(1,2)$ in terms of ρ_0 and ρ_1, f_R- and F-bonds. He then defined, along the same lines as the direct correlation function is defined, for fluids consisting of hard spheres, partial direct correlation functions $c_{00}(1,2)$, $c_{01}(1,2)=c_{10}(1,2)$ and $c_{11}(1,2)$ and derived their diagrammatical expansions in terms of ρ_0, ρ_1 and $f_R(1,2)$ and $F(1,2)$.The partial correlation could then be related to an Orstein-Zernicke matrix equation. This procedure bears strongly resemblance to the derivation of the Orstein-Zernicke equation for hard spheres [253,254].

To obtain the fluid structure, the Orstein-Zernicke matrix equation has to be combined with an appropriate radial distribution function $g(r_1,r_2)$ as a closure equation and a self consistency relation based on Eq. 31. The self consistency relation is a mass balance equation determining division in bonded and non-bonded. For fluids without directional forces, the Orstein-Zernicke matrix equation and a closure equation determine the fluid structure in terms of particle density, ρ. For a fluid with directional forces, the Orstein-Zernicke matrix equation and a closure equation generate the correlations for $g_{00}(1,2)$, $g_{10}(1,2)$ and $g_{11}(1,2)$ in terms of ρ_0 and ρ_1.

These correlations define the distribution of the particles over bonded and non-bonded particles. For fluids consisting of hard spheres this distribution is not possible, because there is only one type of particles. For fluids with directional attractive forces, the ρ_0 and ρ_1 functions assign $g_{00}(1,2)$, $g_{10}(1,2)$ and $g_{11}(1,2)$ with help of Orstein-Zernicke matrix equation and on the other hand the $g_{00}(1,2)$, $g_{10}(1,2)$ and $g_{11}(1,2)$ determine the values of ρ_0 and ρ_1. This last step is necessary for internal consistency and is provided by the self consistency relation. The Wertheim theory has three constituent parts, an Orstein-Zernicke matrix equation, closure equation and a self consistence mass balance equation. Wertheim [251,252] extended this formalism to a multi-component mixture with different interaction sites.

Caused by the incompressibility assumed in the LCT the Wertheim theory needs to be modified. One possibility is a lattice Wertheim theory suggested by Nies et al. [255,256]. The Wertheim theory was transferred on a fully occupied lattice, so an incompressible fluid is regarded [255,256]. The following expression for the Helmholtz energy of association was derived [255,256]:

$$\frac{f_{asso}}{RT}=\sum_i \phi_i \left[\sum_{A_i}\left[\ln\left(X_{A_i}\right)-\frac{X_{A_i}}{2}\right]+\frac{1}{2}M_i^A\right] \tag{35}$$

where X_{A_i} is the segment molar fraction of the non-bonded polymer segments and M_i^A is the number of association sites at one molecule. The notion “Non-bonded polymer segments” means in this case that the segments do not contribute in association. The value X_{A_i} is calculated as follows:

$$X_{A_i} = \left[1 + \sum_j \sum_{B_j} \phi_j X_{A_j} \Delta^{A_i B_j}\right]^{-1} \qquad (36)$$

where the summation over B_j runs over all molecules and association sites. The association strength $\Delta^{A_i B_j}$ is given by:

$$\Delta^{A_i B_j} = K_i \left(\exp\left(\frac{\varepsilon_{ij}^{Asso}}{k_B T}\right) - 1\right) \qquad (37)$$

with K_i being the ratio of the nearest-neighbour positions with the proper orientation to all possible orientations of the component i and ε_{ij}^{Asso} is the association energy. The difference between the lattice Wertheim association model and the association model used in the SAFT equation of state lies in the association volume, K_i, as in the SAFT model the density is used to calculate the association volume.

Using this version of association theory, the association sites have to be distributed on a molecule. For example a water molecule has four association sites, one at each proton of the hydrogen and one at each lone electron pair of the oxygen. After the localization of the association sites, the possible interactions of the association sites have to be defined; for instance that only a proton and a lone electron pair can interact with each other and not two proton sites or two electron pair sites.

This leads for multi-component cross associating systems to a non linear system of equations of type Eq. 36, which has to be solved numerically to calculate the Helmholtz energy or the Gibbs energy, respectively. In opposite to this, in the case of only one associating molecule, the association can be determined analytically.

To consider the cross-association between different molecules mixing rules for the association volume and the association energy are applied to calculate the parameters for the cross-association [188]:

$$K_{ij} = \frac{K_i + K_j}{2} \qquad (38)$$

and:

$$\varepsilon_{ij}^{Asso} = \sqrt{\varepsilon_{ii}^{Asso} \varepsilon_{jj}^{Asso}} \qquad (39)$$

By combination of the LCT with the lattice Wertheim approach both important effects occurring in systems built up from hyperbranched polymers can be take into account for the calculations of thermodynamic properties and phase equilibria.

Phase Equilibria Calculation – Examples

The general equation for the Gibbs energy of mixing (Eq. 10) in combination with Eqs. 11 to 16 provided in the last section allows the calculation of phase diagrams of various binary mixtures. The application to a special type of mixture can be realized with the help of Eqs. 7 and 8. In the following section chapters the use of these equations are discussed for polymer solutions as well as for polymer blends.

Hyperbranched Polymer + Solvent

The thermodynamic description of the LLE of hyperbranched polymer solutions presents a significant challenge to any polymer theory concerned with modeling the molecular factors controlling the mixing behavior. This analysis is designed to understand factors influencing the phase behavior and to aid in sorting out some of the puzzling data related to LLE of hyperbranched polymer solutions.

The incompressibility assumption implies that the LCT for binary systems (Eq. 10) made from hyperbranched polymer (component 1) and a solvent (component 2) contains only one adjustable parameter, the interaction energy, ε_{LCT} , defined in Eq. 9.

All remaining quantities of the LCT are determined from the architecture of the polymer and the solvent. The first step in the LLE-modeling is the determination of the parameters (M_i, b_{3i}, b_{4i}) for the molecules present in the mixture. Hyperbranched macromolecules generally exhibit a much less perfect structure than dendrimers, due to the presence of not only dendritic and terminal units, but also a large fraction of linear segments (Figure 2)

This can be characterized by the degree of branching. A first definition of the degree of branching (DB) was introduced by Frechet et al. [25]. This DB-value is given by:

$$DB = \frac{D+T}{D+T+L} \tag{40}$$

where D,L, T represent the relative abundances of the dendritic, linear and terminal units in the polymer, respectively. In Eq. 40 the linear segments are also counted as branching points, which leads to an overestimation of the DB in the case of low molecular weight polymers. Wurm and Frey [54] suggested the following expression:

$$DB = \frac{2D}{2D + L} \tag{41}$$

that gives a more accurate value for DB, even for the low molecular weight regime. It is important to point out that a dendrimer is limited to distinct generations that are less accessible in high generations, while with hyperbranched polymers it is possible to access a broad range of molecular weights and degree of branching [54]. However; in order to apply the LCT to hyperbranched polymers three characteristic parameters are necessary. The quantities given in Eqs. 7 and 8 reflect the internal architecture and size of the molecules, subject to the limitations imposed by the lattice morphology. The structure is chosen, whenever possible, as most closely corresponding to a united atom model in which united atom groups, such as CH_n reside at single lattice sites. In this chapter solutions of three hyperbranched polyesters with the generation numbers $g = 2$ (Boltorn H20), $g = 3$ (Boltorn H30) and $g = 4$ (Boltorn H40) are studied.

A detailed experimental characterization of these hyperbranched polymers can be found in the literature [30,31]. All these molecules possess the same core: $O\left[CH_2C(C_2H_5)(CH_2O-)_2\right]_2$.

Depending on the generation number g the molecules additionally include a different number of groups A: $COC\left(CH_3\right)\left(CH_2O-\right)_2$ and B: $COC\left(CH_3\right)\left(CH_2OH\right)_2$. The general formulae of the polymers are for $g = 2$: $(core)A_4B_8$, for $g = 3$: $(core)A_{12}B_{16}$ and for $g = 4$: $(core)A_{28}B_{32}$.

According to these formulae (neglecting polydispersity) the molar masses take the values 1642 g/mol ($g = 2$), 3498 g/mol ($g = 3$) and 7210g/mol ($g = 4$). The number of OH- end groups equals the number of B-units.

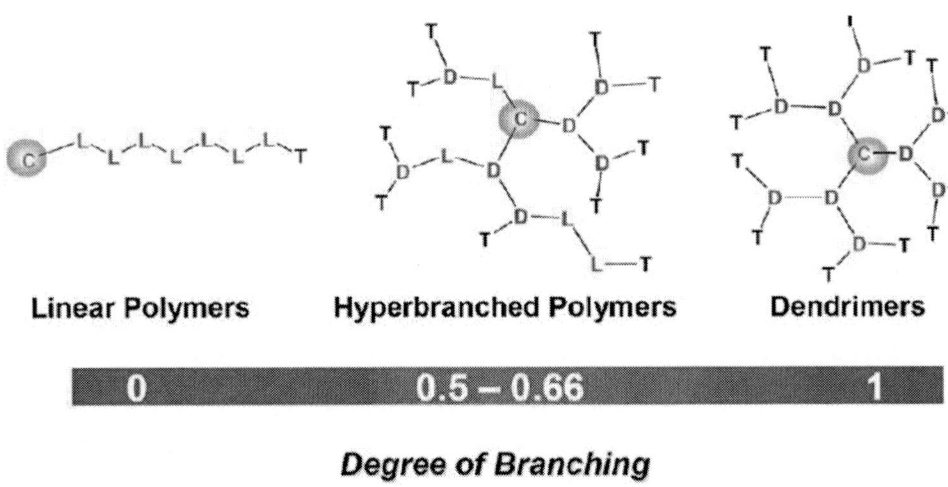

Figure 2. Schematic representation of the degree of branching for linear and hyperbranched polymers compared to dendrimers [54].

Figure 3 shows schematically a hyperbranched polyester with the generation number, $g = 3$. To describe the architecture of a hyperbranched polymer two specification in addition to the generation number g will be used. The separator length n is the number of segments between two branching points.

It denotes the number of segments of an A unit or a B unit. The number of core segments is given by the quantity n_0. It is assumed that one water molecule occupies one lattice site, so the number of core segments and the separator length can be determined by dividing the core, A and B units in groups that have a molar mass comparable to a water molecule.

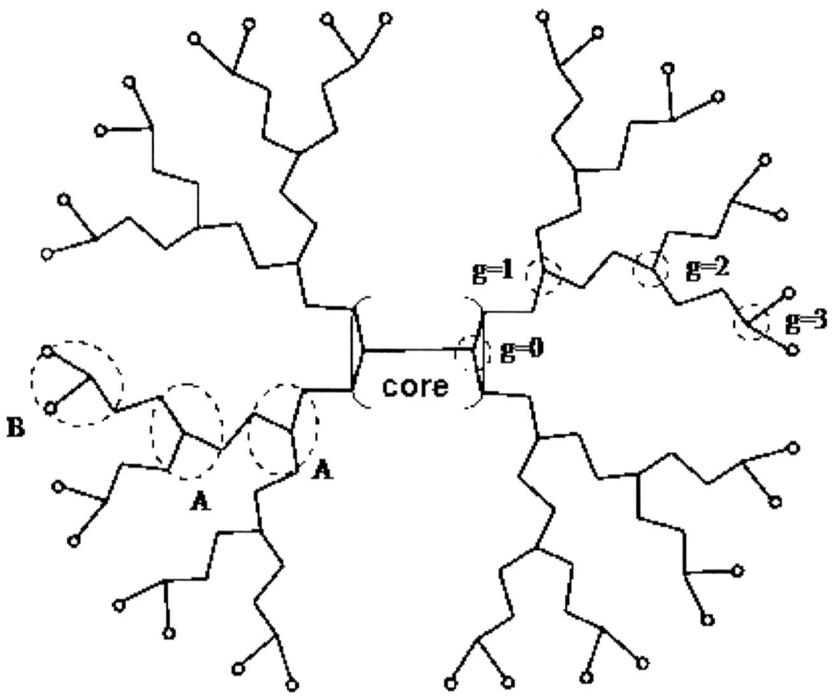

Figure 3. Schematic presentation of a hyperbranched polymer of generation number g=3 [187].

The solvents (water, propan-1-ol) were taken as linear molecules, where water occupies one lattice place and propan-1-ol three of them. The values of the separator length, number of core segments and generation number are collected in Table 1.

Using these architectural parameters, the topological coefficients of the LCT can be calculated, where the number of segments of a hyperbranched polyester molecule can be calculated as follows [223]:

$$M = 4\left(2^g - 1\right)n + n_0 \tag{42}$$

Each A-unit and each B-unit possesses one branching point of degree 3. The core contains two such branching points. Branching points of degree 4 and higher do not exist in these polymer molecules. So the number of branching points is calculated as follows [223]:

$$b_3 = 4\left(2^g - 1\right) + 2 \tag{43}$$

In contrast to the experimental procedure, where the concentration is usually given in mass fraction, w_i, in the theoretical framework the concentrations of the components are measured by the segment fraction, ϕ_i. Both quantities are connected via:

$$\phi_1 = \frac{M_1 N_1}{M_1 N_1 + M_2 N_2} = \frac{\dfrac{M_P}{M_{M,1}} w_P}{\dfrac{M_1}{M_{M,1}} w_1 + \dfrac{M_2}{M_{M,2}} w_2} \tag{44}$$

where the indices 1 and 2 indicate that the specific value ($M_{M,i}$ is the molar mass of the component i) belonging to the polymer or to the solvent.

Table 1. Architectural parameters describing hyperbranched polyester

Separator length n	4
Number of core segments n_0	7
Generation number g	$g = 2$ (Boltorn H20) $g = 3$ (Boltorn H30) $g = 4$ (Boltorn H40)

In order to study the influence of the association phenomena on the phase equilibrium the self-association of the solvents, the self-association of the polymer and the cross-association between solvent and polymer are modeled using the Wertheim lattice perturbation theory. This Wertheim theory needs a location of the association sites on each component in the mixture. Water is modeled as a 4C-modell, meaning that one association place is located at each lone electron pair of the oxygen and one association site is at each hydrogen atom. To consider the association phenomena of solutions containing hyperbranched polyester, the OH-groups are modeled by an association site on one lone electron pair of the oxygen and one association site on the hydrogen. The other lone electron pair of the oxygen is assumed to be hidden, because of steric considerations. Propan-1-ol is modeled using the same assumptions.

First, some model calculations for the system composed of hyperbranched polymer and a solvent occupying only on lattice site will be performed.

Figure 4 demonstrated different binodale lines obtained with different numbers for the coordination number, z. The energy parameter ε_{LCT} can be fitted to critical temperature. Within the FH-theory only the well-known χ-parameter occur. However; within the LCT this χ-parameter contains the coordination number z as well as the interaction energy, ε_{LCT}. Hence, both parameters are connected. If one chose the coordination number, z, then the interaction energy ε_{LCT} must also be changed in order to represent the same critical temperature.

Figure 4 shows the influence of the coordination number on the liquid-liquid phase split of hyperbranched polyester solutions calculated by the LCT using different interaction energies to align the critical temperatures because the shape of the three cloud point curves should be compared. The cloud-point curve calculated with $z = 6$ shows a sharp decline of the high concentrated polymer branch and a slow rise on the diluted side of the curve. This behavior is not typical for polymer solutions. Using higher coordination numbers leads to more realistic cloud-point curves. Therefore, in the following calculations of LLE the coordination number $z = 12$ is used. Moreover, the LCT is a series expansion of the partition function in the inverse coordination number and the interaction energy, $\varepsilon_{LCT} / k_B T$.

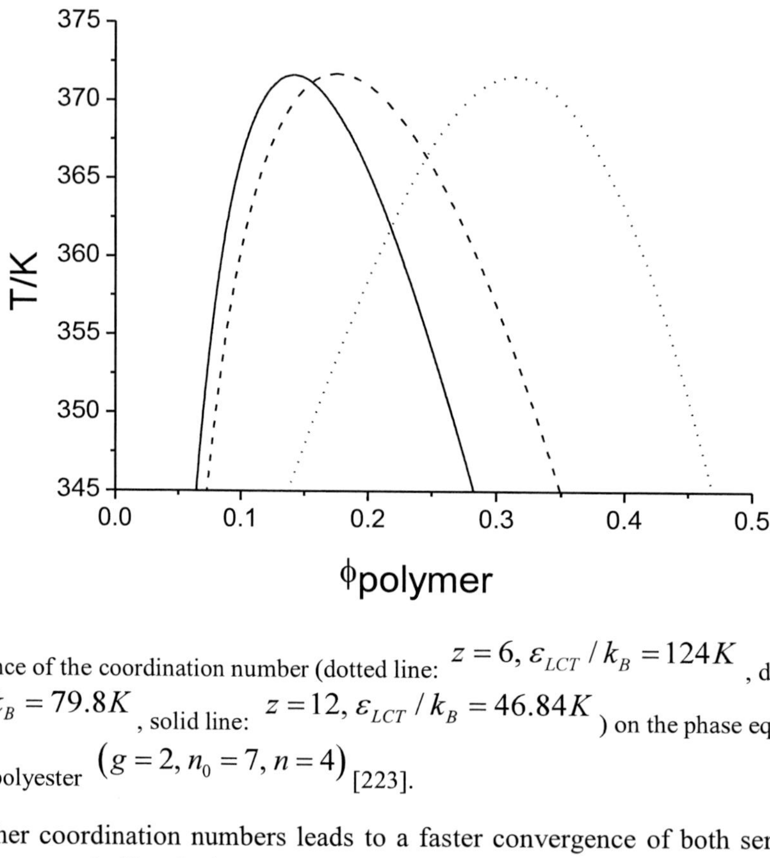

Figure 4. Influence of the coordination number (dotted line: $z = 6, \varepsilon_{LCT} / k_B = 124K$, dashed line: $z = 8, \varepsilon_{LCT} / k_B = 79.8K$, solid line: $z = 12, \varepsilon_{LCT} / k_B = 46.84K$) on the phase equilibria on hyperbranched polyester $(g = 2, n_0 = 7, n = 4)$ [223].

Using higher coordination numbers leads to a faster convergence of both series as also the interaction energy decline for larger coordination numbers.

Comparing the three cloud point curves calculated with $z = 12$, but different values for ε_{LCT} , leads to the conclusion, that the interaction parameter ε_{LCT} has no impact on the critical composition or on the shape of the cloud point curve (Figure 5). By increasing or decreasing the interaction parameter ε_{LCT} the critical temperature is shifted to higher or lower values. That means, that the critical composition and the shape of the cloud point curve is determined by coefficients such as the separator length or the number of core segments, which are given by the architecture of the polymer. By growing separator length, n, the critical point is shifted to higher critical temperatures and lower critical segment fractions[223]. Increasing the number of core segments, n_0, leads to higher critical temperatures, lower critical composition and a narrower cloud point curve [223]. This effect can also be traced back to the fact that an increase of the number of core segment also increases the number of segments. But this influence is not as strong as the influence of the separator length, because the increase of the number of segments is equal to the increase of the number of core segments. Increasing the generation number of hyperbranched polymers, g, leads to higher critical temperatures, lower critical segment fractions and to narrower LLE [223] (Figure 6). This effect can be explained by the fact that the number of segments increases in power of the generation number, g. The critical temperature of the hyperbranched polyester solution is lower compared to their linear analogous and the critical composition of hyperbranched polyester solutions is higher than the critical temperatures of the linear analogous [223].

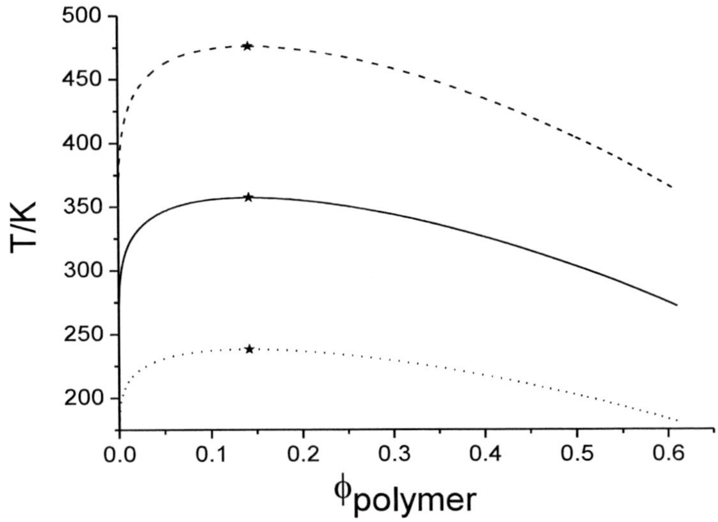

Figure 5. Dependence of the liquid-liquid equilibria (stars: critical points, dashed line: $\varepsilon_{LCT} / k_B = 60K$, solid line: $\varepsilon_{LCT} / k_B = 45K$, dotted line: $\varepsilon_{LCT} / k_B = 30K$) on the interaction energy for the hyperbranched polyester $(g = 2, n = 4, n_0 = 7)$ [223].

Beside the architecture of a hyperbranched polymer also the number as well as the kind of functional groups has an important impact on the liquid-liquid equilibria of hyperbranched polymer solutions.

Applying the Wertheim association theory two parameters, namely the association volume, K_i, and the association energy, ε_i^{Asso}, must be specified. This theory uses two parameters, which have to be fitted on the LLE. For water these parameters can be obtained using the vapor pressure curve resulting in $\varepsilon_{water}^{Asso} / k_B = 1800K$ and $K_{water} = 0.01$. However; for the polymer no independent physical quantities are available. Therefore, these parameters must be fitted to the LLE.

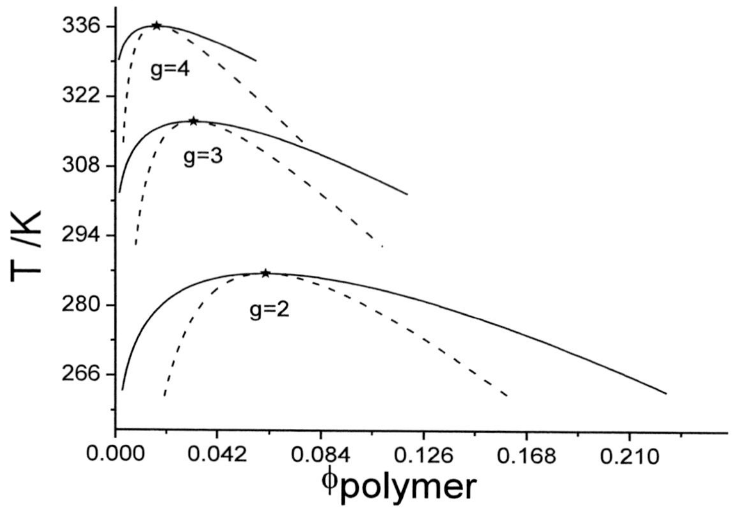

Figure 6. Influence of the generation number, g, on the phase equilibria (solid line: binodal, broken line: spinodal, star: critical point) on hyperbranched polyester solutions $\left(\varepsilon_{LCT} / k_B = 30K, n_0 = 2, n = 3, z = 12\right)$.

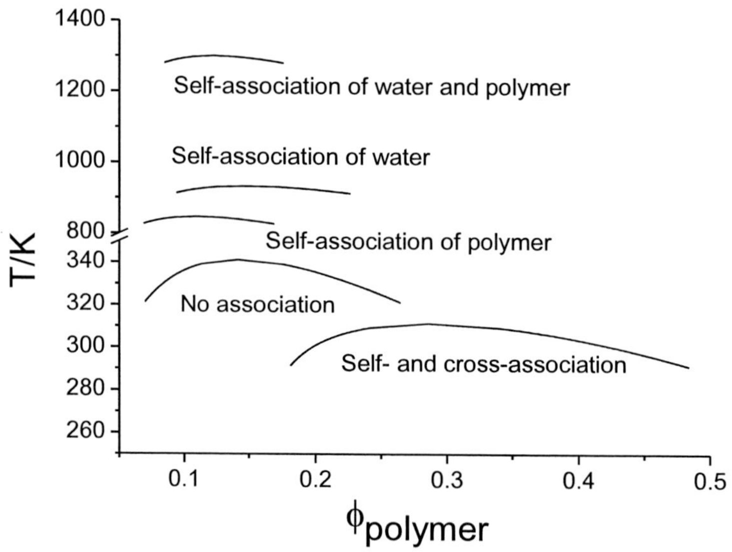

Figure 7. Impact of different kinds of association (self- and cross-association) on the spinodal curve of aqueous solutions of hyperbranched polyester $(g = 2)$ and the association parameter $\left(K_1 = 0.0673,\ \varepsilon_1^{Asso} / k_B = 1100K,\ K_2 = 0.01,\ \varepsilon_2^{Asso} / k_B = 1800K\right)$ [188,223].

To analyze the influence of association, the spinodal curves of the system hyperbranched polyester ($g = 2$) and water are calculated with different kinds of association (Figure 7). The parameters used for this model calculation are the same parameters used to calculate the liquid-liquid equilibrium of Boltorn H20. If only self-association of the polymer and water is considered, the critical point is shifted to unrealistic high critical values. Just the consideration of the self- and cross-association together leads to realistic critical temperatures. With help of Figure 7 the influence of cross- and self-association compared to the non-polar polymer in a non-polar solvent is investigated. By considering the self-and cross-association the critical point is shifted to higher segment fraction of the polymer and towards lower critical temperatures. Regarding these results it has to be pointed out that if the association phenomena are considered in the calculation both the self- and cross-association have to be included. These findings are contradictory to the results of Jang and Bae [257], who concluded that the self-association of water has the greatest influence on the liquid-liquid equilibrium of aqueous hyperbranched polyester solutions.

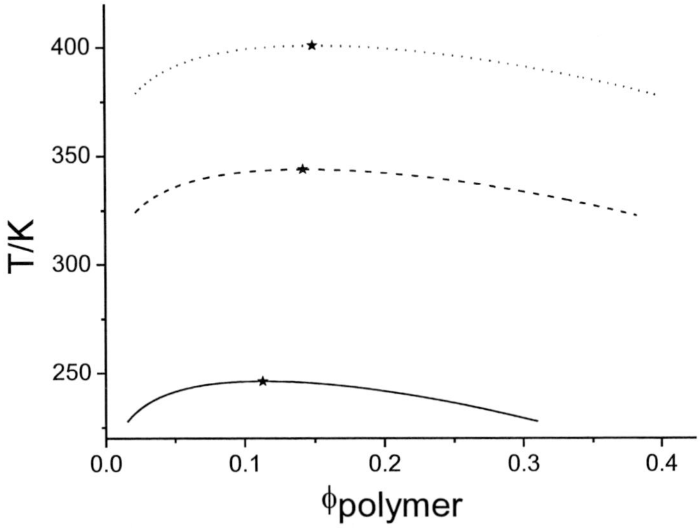

Figure 8. Influence of the association energy on the binodal curves (dotted line: $\varepsilon_1^{Asso}/k_B = 1100K, \varepsilon_2^{Asso}/k_B = 2000K$, dashed line: $\varepsilon_1^{Asso}/k_B = 1100K, \varepsilon_2^{Asso}/k_B = 1800K$, solid line: $\varepsilon_1^{Asso}/k_B = 1500K, \varepsilon_2^{Asso}/k_B = 1800K$) and the critical point (stars) of aqueous hyperbranched polyester solution using LCT ($\varepsilon_{LCT}/k_B = 30K$) in combination with Wertheim theory ($K_1 = 0.045, K_2 = 0.01$) [223].

The influence of the association energy on the liquid-liquid phase equilibria of aqueous hyperbranched polyester solutions is shown in Figure 8. By increasing the association energy of water the critical temperature and the critical segment fraction of the polymer are shifted to higher values. On the other hand, the critical temperature and the critical polymer segment fraction gets smaller by increasing the association energy of the polymer end groups. Another effect of changing the association energy is the broadening of the miscibility gap by increasing the association energy of water and the narrowing of the cloud point curve by rising up the association energy of the polymer. This behavior of the cloud point curve by increasing the association energy of water can be explained by the fact that water tends to association cluster of water molecules. Due to reasons the miscibility gap broadens. By increasing the association energy of the polymer molecules, water tends to build up association cluster with polymer molecules and the polymer molecules are hindered to build large clusters with each other because of steric effects. So the miscibility gap gets closer and the critical temperature becomes lower.

Figure 9 shows the influence of the association volume on the phase equilibria of hyperbranched polyester + water system. Increasing the association volume of water or the association volume of the polymer has a similar effect on the miscibility gap as changing the association energy of water and the polymer. By increasing the association volume of water the critical temperature and the critical segment fraction is shifted to higher values. On the other hand, an increase of the association volume of the polymer leads towards lower critical temperatures and lower critical composition.

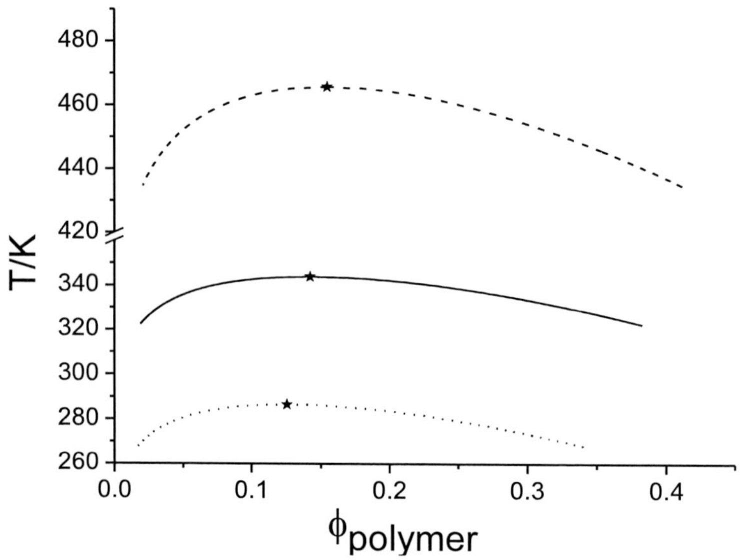

Figure 9. Influence of the association volume on the LLE (stars: critical point, solid line: $K_1 = 0.045, K_2 = 0.01$, dashed line: $K_1 = 0.045, K_2 = 0.02$, dotted line: $K_1 = 0.06, K_2 = 0.01$) of aqueous hyperbranched polyester solution using LCT ($\varepsilon_{LCT} / k_B = 30K$) in combination with Wertheim theory ($\varepsilon_1^{Asso} / k_B = 1500K, \varepsilon_2^{Asso} / k_B = 1800K$) [223].

The broadening of the cloud point curve by increasing the association volume of water can be explained in the same way as by increasing the association energy because the association strength is proportional to the association volume. Moreover, there is also a narrowing of the immiscibility gap by increasing the association volume of the polymer.

For the binary system composed of Boltorn H20 and water two data sets related to the LLE could be found in the literature [187,188] (Figure 10). However; both data sets differ strongly from each other, where the data given in Ref. [188] are located at much higher temperatures. The demixing temperatures were estimated during heating and during cooling. Both temperatures were very close together, within maximal 2 K [188]. The data in Ref. [188] do not show the shoulder found by Yang and Bae [187]. Having in mind the melting temperature of Boltorn H20 is above 323K maybe the data found by Yang and Bae [187] are rather connected to the solid-liquid equilibria instead the liquid-liquid equilibria.

The branch of the cloud-point curve for the diluted polymer solution could by descript in excellent agreement with the experimental data, if the LCT either alone or in combination with the Wertheim theory is utilized (Figure 9). However; deviations between the experimental and theoretical cloud-point curve in the concentrated range occur. Taking the association forces in terms of self- and cross association into account leads to a slightly improvement of the calculated results by shifting the polymer volume fraction in the concentrated phases to larger values and hence the application of the Wertheim approach improves the calculation results in the right direction. Having in mind that the increased number of the used adjustable parameter it can be concluded the higher effort is not warrantable.

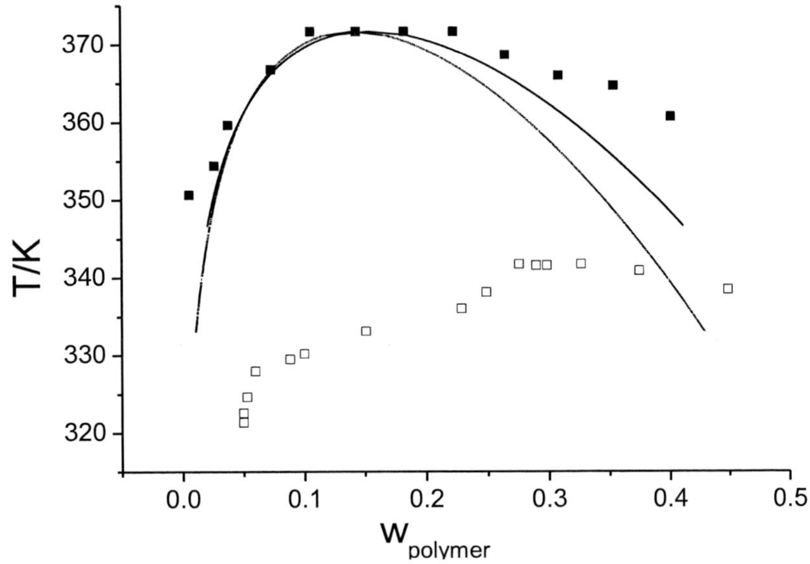

Figure 10. Experimental (solid symbols [188], open symbols [187]) and calculated (solid line: LCT + Wertheim theory, broken line: LCT) LLE for the system Boltorn H20 + water.

Including only self association and not cross association in the theoretical calculations leads to extreme high demixing temperatures (Figure 7), were the polymer is hardly stable. Only if both effects (self- and cross association) are taken into account the calculation results agree with the experimental findings. This statement is contrary to the results found by Yang and Bae [257]. Using another version of the LCT in combination with another association model for the self association of water Yang and Bae [257] concluded the dominant impact is the self association of water on the LLE. However; like discussed in the literature [224] the reason for this conclusion can be errors in the applied version of the LCT. Similar results could be obtained for the system Boltorn H20 + propan-1-ol [188,223]. These binary phase equilibrium investigations form the bases for the study of the ternary system made from Boltorn H20 + water + propan-1-ol [226].

Other hyperbranched polymer from the Boltorn family is Boltorn U3000. Domańska et al. [183] measured phase equilibria of polymer solutions in propan-1-ol and butan-1-ol. Boltorn U3000 is fatty acid modified polyester of generation number 3. The molecule consists also of the core $C(CH_2O-)_4$, of 12 separator groups A: $COC(CH_3)(CH_2O-)_2$ and 16 groups B: $COC(CH_3)(CH_2OH)(CH_2OR)$ with the end groups OH and OR. Here, R origins from the 16-carbon-long alkyl acid (R: $CH_3-(CH_2)_{14}CO-$). Again, the core is divided into 5 segments and the separator group A into 3 segments. Considering the 16-

carbon-long group R the group B has 20 segments. All together a Boltorn U3000 molecule may be described by $M_1 = 373$ segments, where molar mass is $7192\, g/mol$. Furthermore, there are 28 branching points of degree 3 ($b_{3,1} = 28$) and one branching point of degree 4 ($b_{4,1} = 1$). The segment numbers of the linear solvents propan-1-ol and butan-1-ol were assumed to be $M_2 = 3$ and $M_2 = 4$. Caused by 16 OH groups of Boltorn U3000, similar to the solution containing Boltorn H20 self- and cross-association occurs. The association phenomena can be treated with a chemical association model[227] or with the Wertheim perturbation theory. In this contribution we focus our attention to the combination of the LCT with the Wertheim perturbation theory. The relevant parameters within the LCT are: $g = 3, n = 3, n_0 = 5$.

In order to minimize the number of adjustable parameter within the Wertheim perturbation theory we take to parameter describing propan-1-ol from the above discussed system propan-1-ol + Boltorn H20[188], namely $K_2 = 0.11$ and $\varepsilon_2^{Asso} / k_B = 1745K$. Using the experimental data from the literature[183] the three remaining parameters ($\varepsilon_{LCT}, K_1, \varepsilon_1^{Asso}$) must be fitted. The phase split for the polymer solution Boltorn U3000 + butan-1-ol were calculated keeping the parameter K_1, ε_1^{Asso} from the system Boltorn U3000 + propan-1-ol. For this system the parameter $\varepsilon_{LCT}, K_2, \varepsilon_2^{Asso}$ must be adjusted to the experimental data given in the literature [183]. The calculated LLE together with the experimental data [183] are depicted in Figure 11. The adjusted parameters are listed in Table 2.

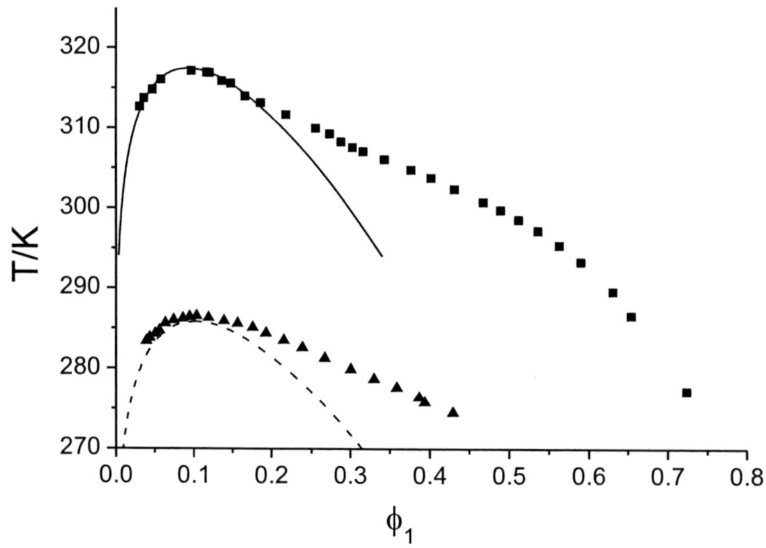

Figure 11. Experimental (squares: Boltorn U3000 + propan-1-ol [183], triangles: Boltorn U3000 + butan-1-ol) and calculated (solid line: Boltorn U3000 + propan-1-ol, broken line: Boltorn U3000 + butan-1-ol) LLE.

Table 2. Model parameters for solution of Boltorn U3000 in propan-1-ol or in butan-1-ol

Parameter	Boltorn U3000 + propan-1-ol	Boltorn U3000 + butan-1-ol
ε_{LCT} / k_B	10.9 K	8.35 K
K_1	0.023	0.023
$\varepsilon_1^{Asso} / k_B$	1200 K	1200 K
K_2	0.011	0.01
$\varepsilon_2^{Asso} / k_B$	1745 K	1710 K

Similar to the results obtained for the system Boltorn H20 in water (Figure 10) or in propan-1-ol [188], the LCT in combination with the Wertheim approach describe the branched of the cloud point curve related to the diluted polymer solution very close to the experimental data. However; the branch of the cloud point curve describing the composition of the polymer-rich phase some derivation occur. For the system Boltorn U3000 + propan-1-ol the experimental cloud point curve shows a characteristic shoulder. Our theoretical framework was not able to describe this shoulder.

POLYMER BLENDS

Polymer blends are formulated by mixing polymers with different chemical structures to create new materials with beneficial properties of the individual components. While FH-theory explains some basic trends in blend miscibility, the theory completely neglects the dissimilarity in monomer structures that is central to the fabrication of real blends. The application of the LCT for mixtures made from two linear polymers or copolymers was intensively studied by Freed et al [211-215,221]. An excellent review could be found in the literature [215], where the influence of the monomer structure on the compatibility of the blend is discussed in great detail. Several illustrative applications of the LCT provide new molecular-scale interpretations for many nontrivial phenomena occurring in polymer systems [215]. The applications also demonstrate the predictive ability of the theory and its usefulness in analyzing thermodynamic data over a wide variety of polymer mixtures, ranging from binary homopolymer blends of various types of copolymer systems [215].

Recently [258], several new experimental and theoretical phase equilibrium data of different polymer blends were published. The theoretical calculations were performed using two different equations of state (Sanchez-Lacombe EOS and Perturbed Chain-Statistical Associated Fluid Theory - EOS [PC-SAFT]), where a nice agreement between the experimental data and the modeling result could be achieved [258].

Both equations of state require three pure component parameters and one binary interaction parameter, if no association is modeled. Even, if polar groups able to form associates are present, no associations were taken into account [258]. Now we would like to investigate, if the LCT is also able to model the experimental phase diagrams. For comparison we use only the LCT and neglect also the association forces.

This approach means that only one parameter must be fitted to experimental data, and no pure-component physical properties, like density data, are involved. For this purpose we select two mixtures, namely polybutadiene (PBD) + polystyrene (PS) and polyethylene glycol mono -methyl ether (PEGE) + 2,6-dimethyl phenylene oxide (PPO).

The first mixture, composed from PDB und PS, five different blends, differing in the molecular weight of the polymers (Table 3) experimental cloud point curves are available. Within the LCT we assume every polystyrene monomer occupy one lattice site and every butadiene monomer two lattice sites.

Table 3. Studied PBD/PS blends

blend	$M_{M,PBD}[g/mol]$	$M_{M,PS}[g/mol]$	M_{PBD}	M_{PS}	$\varepsilon_{LCT}/k_B\,[K]$
1	1100	1340	10	13	29.95
2	1100	1670	10	16	31.1
3	1100	4370	10	42	21.5
4	2350	1900	22	18	23.0
5	2350	3300	22	32	21.6

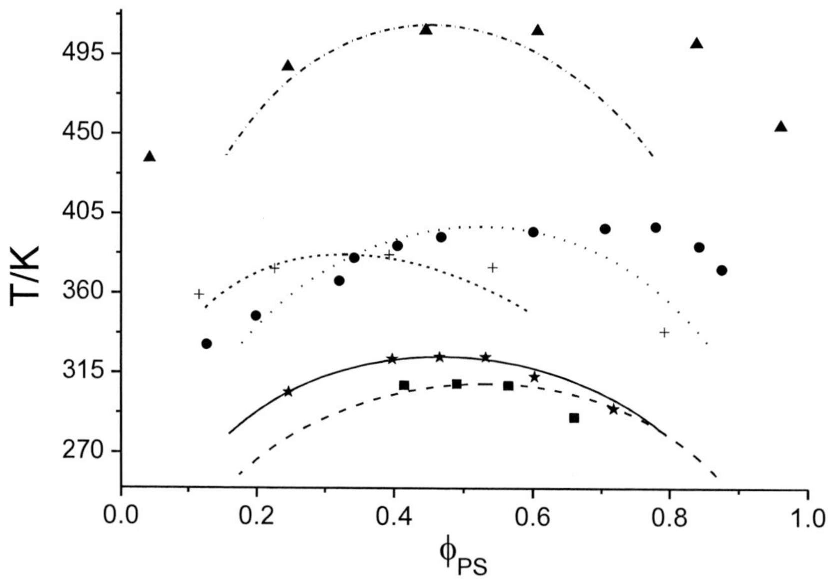

Figure 12. Experimental [258] (squares: blend 1, stars: blend 2, crosses: blend 3, circles: blend 4, triangles: blend 5) calculated LLE using the parameters given in Table 3.

The calculated phase diagrams together with the experimental data [258] are plotted in Figure 12. First, from this figure it can be concluded that the LCT does a good job in modeling the experimental phase diagram. The results obtained with the LCT are very similar then the results obtained with the Sanchez-Lacombe EOS [258]. However; one must have in mind the different number of adjustable parameters. The calculations performed with the PC-SAFT-EOS are closer to the experimental data [258]. Maybe the LCT calculation can be improved by the incorporation of empty lattice sites.

The second example deals with blend made from PEGE and from PPG, where the monomers of both polymers carry polar groups. In the first approach the association will be neglected. We assume the PPG monomer occupy four lattice sites and the PEGE three of them. The parameters are collected in Table 4 and the phase diagrams are demonstrated in Figure 13.

Table 4. Studied PEGE/PPG blends

blend	$M_{M,PEGE}[g/mol]$	$M_{M,PPG}[g/mol]$	M_{PEGE}	M_{PPG}	$\varepsilon_{LCT}/k_B[K]$
1	750	2000	33	78	8.55
2	550	2000	24	78	9.35

The theoretical results depict in Figure 13 show clearly the limits of the applied theoretical framework. The calculated miscibility gaps are too small in comparison with the experimental findings. One possible reason for this situation can be the polar interaction. However; the equations of state (SL-EOS and PC-SAFT) were able to model the experimental phase diagram, without to take the polar interaction into account. PPG can also show typical surfactant properties, like micelle formation. However; micelle formation is not covered by the used theory.

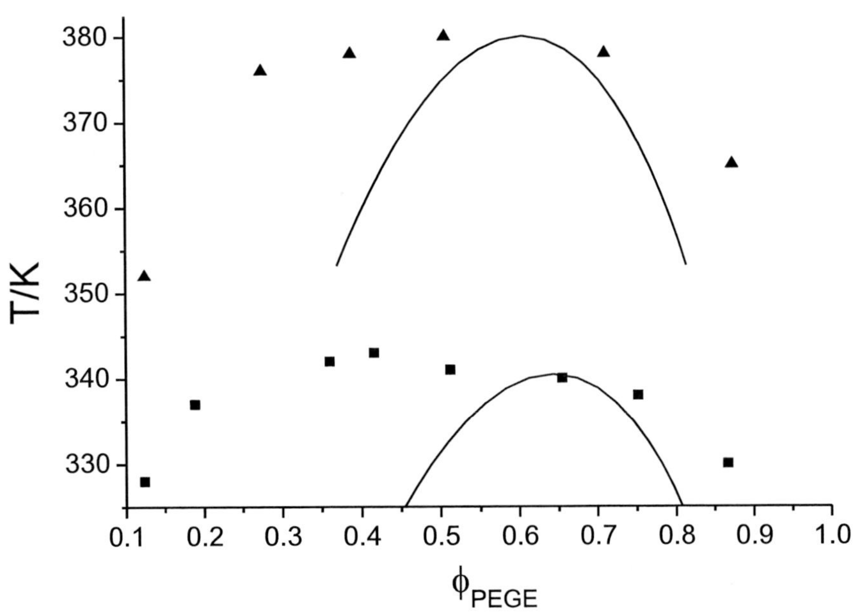

Figure 13. Experimental [258] (triangles: blend 1, squares: blend 2) calculated LLE using the parameters given in Table 4.

Unfortunately for blends built up from linear polymers and hyperbranched polymers no experimental data could be found in the public literature. Therefore, we are limited to model calculations. Let us study the phase behavior of a mixture of a linear polymer having different molecular weights with a hyperbranched polymer, for example Boltorn H20. The architectural parameters of the hyperbranched polymer are $g = 2, n = 4$ and $n_0 = 7$. The assumed segment-number and the energy parameter are listed in Table 5, where the energy parameters were chosen to ensure a miscibility gap with the identical critical temperature. The calculated cloud-point curves are plotted in Figure 14.

Table 5. Studied blends made from Boltorn H20 (1) + linear polymer (2)

blend	M_1	$\varepsilon_{LCT} / k_B [K]$
1	1	61
2	10	10
3	50	3.42
4	250	1.88
5	700	1.4

With increasing molecular weight of the linear polymer the critical concentration shifts to lower values of ϕ_2 (Figure 14). Additionally, with increasing molecular weight of the linear polymer a smaller value of ε_{LCT} is required to generate a miscibility gap. The dependency of ε_{LCT} on the molecular weight is plotted in Figure 15. In the limit of high molecular weights only very small interaction energy is sufficient to create a demixing behavior.

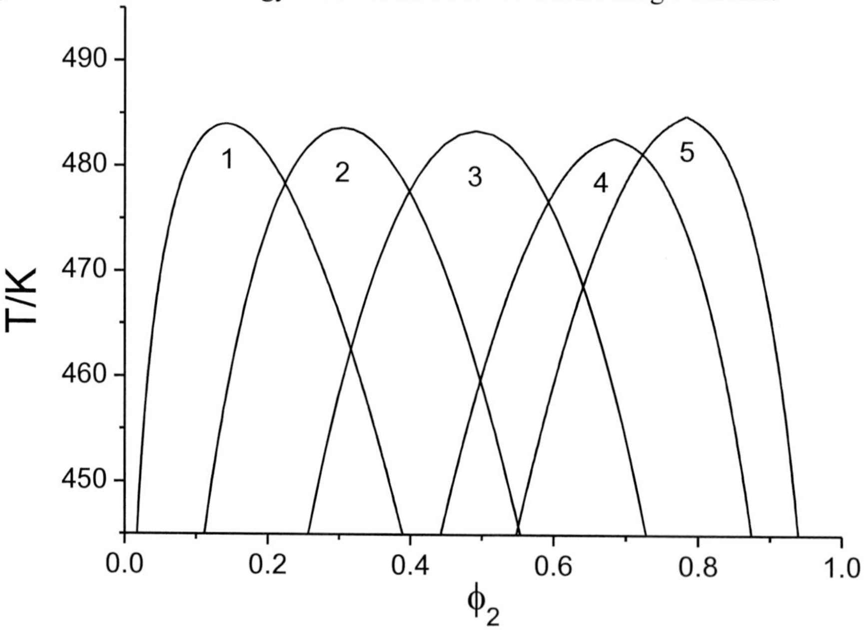

Figure 14. Calculated demixing curves for mixtures made from linear polymers with different molecular weights (1) and hyperbranched polymer Boltorn H20 (2) using the parameters given in Table 5. The numbers indicate the number of blend..

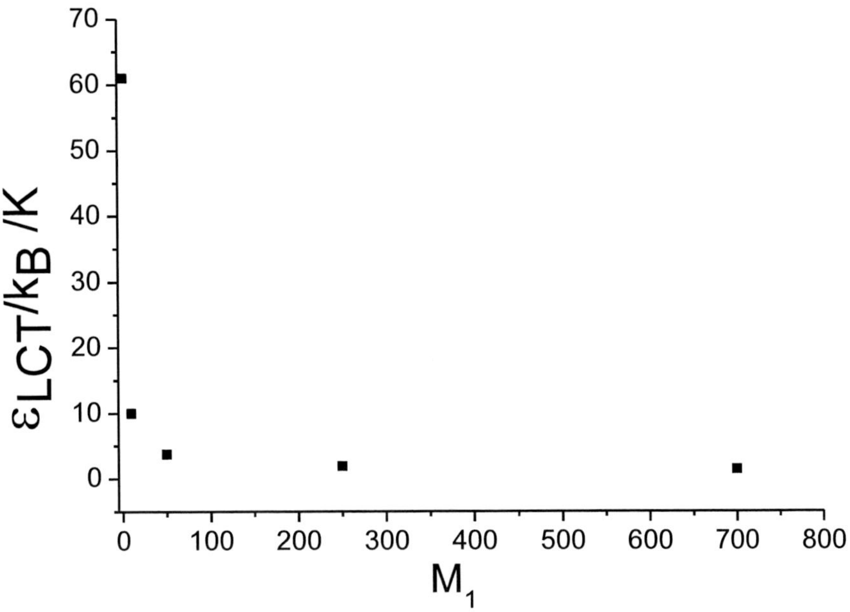

Figure 15. Interaction energy ε_{LCT} as function of the segment number for the system linear polymers with different molecular weights (1) and hyperbranched polymer Boltorn H20 (2).

This finding is very similar to the calculation results of the FH theory for the mixtures of two linear polymers [1,5]. In order to calculate the phase composition the improved description of the entropy of mixing by the LCT in comparison with FH theory plays an important role, because the entropic effect prevail the enthalpic effect.

Next, we would like to study the phase behavior for the mixture of two hyperbranched polymers. Unfortunately, no experimental data could be found in the public literature. Therefore, we can perform only model calculations without comparison with experimental data. For this purpose we chose the mixture of Boltorn U3000 and Boltorn H20 or Boltorn H30 or Boltorn H40. The architecture of both hyperbranched polymers were explained in the chapter above. We assume a very low value for the interaction energy ε_{LCT} in order to see if a miscibility gap can be created by the different molecular mass and the different polymer architecture.

For the mixture composed of Boltorn H20 + Boltorn U3000 a miscibility gap exist (Figure 16), but the corresponding demixing temperatures are below the melting temperature of the polymers. For the practical application of the polymer blend a high compatibility can be expected. This is in contrast to the mixture Boltorn H30 + Boltorn U3000, where the miscibility gap is located in the temperature range suitable for several applications. At constant temperature (i.e. 300K) up to 10% Boltorn H30 can be blended with Boltorn U3000 without phase separation. This can be useful to control several physical properties. On the other hand, only a very limited amount of Boltorn U3000 can be added to Boltorn H30. A mixture made from Boltorn H40 and Boltorn U3000 is only thermodynamic stabile at small concentrations. The calculated demixing temperature can hardly be found in the experiment, because the polymer shows some degradations at temperatures over 400K.

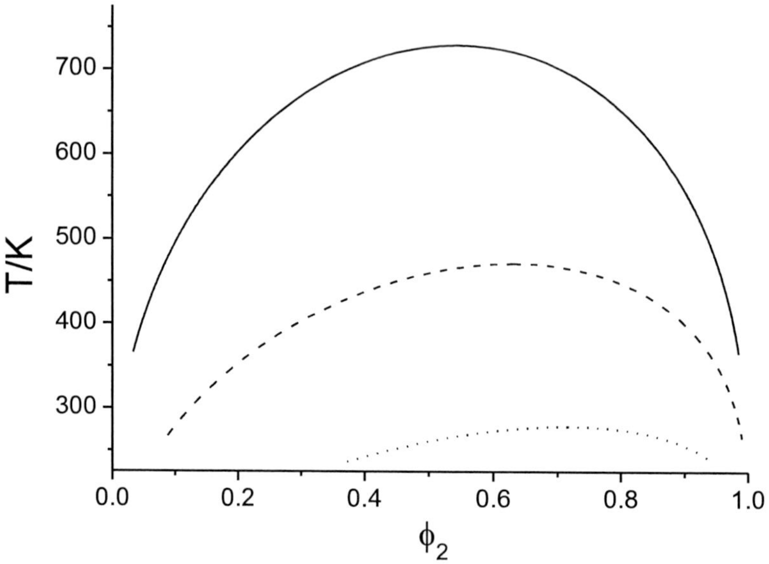

Figure 16. Calculated LLE for a mixture of Boltorn U3000 (1) with different Boltorn H-polymers (2) (solid line: Boltorn H40, dashed line: Boltorn H30, dotted line: Boltorn H20) using LCT with $\varepsilon_{LCT} / k_B = 1K$.

If one assumed that the interaction energy, ε_{LCT} , do not depend on the architecture of the considered hyperbranched polymer, then $\varepsilon_{LCT} = 0$ for a mixture made from Boltorn H20 and Boltorn H40 is valid[224]. In this case our model predicts a homogeneous mixture of the whole temperature range. However; if a slight dependence of the interaction energy on the polymer architecture, maybe on the segment number or the connected quantity the generation number, is permissible then a miscibility gap can be predicted. For instance, if $\varepsilon_{LCT} / k_B = 1K$ a critical temperature of $T_C = 247K$ can be calculated using the LCT approach. This critical temperature is well below the melting temperature of both polymers. However; if the interaction energy is slightly increased, maybe to $\varepsilon_{LCT} / k_B = 1.5K$, a critical temperature of $T_C = 371\,K$ and a critical segment fraction of $\phi_C = 0.32$ can be calculated. This result can be interpreted, that also hyperbranched polymers of the same chemistry, but different architecture can show phase separation behavior. Using FH-theory for linear polymers the well-known interaction parameter χ can also be a function of the molecular weight. This was often found for oligomers [1,2,5].

In comparing lattice cluster theory predictions and experimental data, it must be borne in mind that the theory is based upon an admittedly oversimplified lattice model. Nevertheless, the theory provides a description of the gross influences on thermodynamic properties of polymer architecture. The improved theories might be constructed by further developments of the LCT to incorporate more aspects of reality, or they could require use of much more sophisticated atomistic non-lattice theories. The later approach would necessitate very heavy numerical computations, especially for the coexistence curves, and they are to be contrasted with the almost trivial computational requirements for applications of the analytical formulas of the LCT. Beside the architecture of the hyperbranched polymer covered with the LCT the number and the type of the terminal functional groups must be taken into account in the thermodynamic equations. These specific interactions can be modeled by self- and cross association phenomena with the help of the Wertheim perturbation theory.

THERMODYNAMICS OF INTERFACIAL PROPERTIES

One of the major applications of hyperbranched polymers is their use in polymer blends or as additives in linear polymers. The effectively depends strongly on the surface properties, especially, on the tendency of the hyperbranched polymer to migrate to the surface to form a lubricating layer. The surface energy plays a key role in polymer processing and blending. For the application of hyperbranched polymer for encapsulation purposes of different guest molecules in medicine [96,259] or dyeing [81] the surface energy plays also a dominant role. The surface energy can be controlled by the type of functional terminal groups [260] or charges. The behavior of hyperbranched polymers on the surface is substantially dependent on the formation of micellar-like structures [261]. For this reason several experimental investigations were performed in order to characterize the surface properties.

The surface properties of neat and modified Boltorn® hyperbranched polyesters were studied by water-contact-angle measurements and the micro-Wilhelmy wetting technique to determine the solid and melt surface tensions, respectively [262, 263]. The surface tensions of the melts are extremely high for hydroxyl-terminated hyperbranched polyesters and approach that of water, indicating the preferential surface segregation of hydroxyl terminal groups [262]. The substitution of hydroxyl groups with alkyl chains dramatically reduces the melt surface tension to the point where a large degree of substitution produces the surface tension equivalent of pure alkanes [262]. The molecular adsorption of third (HBP3) and fourth (HBP4) generations of hyperbranched polyesters with 32 and 64 hydroxyl-terminal groups on a bare silicon surface was studied with scanning probe microscopy and was described in the terms of the Langmuir isotherm [264]. The shape of hyperbranched polymers with 32 hydroxyl-terminal groups within an adsorbed layer evolved from a pancake with a thickness less than 1 nm for very low surface coverage to densely packed wormlike bilayer structures with a thickness of about 3 nm for the highest surface coverage [264]. The molecules of the fourth generation hold a stable, close-to-spherical shape with a diameter of 2.5 nm throughout the entire range of surface coverage including both dense monolayer's and isolated molecules [264]. High intramolecular flexibility of HBP3 molecules as compared with constrained mobility of bulkier branches of HBP4 is considered to be responsible for different surface behavior [264].

Mikhaylova et al. [265] applied a combination of different surface-sensitive techniques to obtain better understanding of adsorption processes of model proteins on hydrophilic surfaces of hydroxyl (phenolic group) terminated hyperbranched aromatic polyesters (HBP-OH) in comparison to hydrophobic polystyrene (PS) surface. Surprisingly, the adsorbed amount of lysozyme and human serum albumin on the hydrophilic surface of HBP-OH having weak negative surface charge was larger than on the hydrophobic PS surface. The adsorption kinetics of both proteins on the two types of surfaces showed very fast adsorption process [265]. Axisymmetric drop shape analysis was applied to measure in situ the changes in the contact angle and interfacial tensions as a result of the adsorption of protein molecules from a sessile solution droplet [265]. The observed effects are very complex since not only the presence of proteins, including the arrangement and conformation of the adsorbed molecules, but also the properties of the polymer surfaces play a role [265].

Peleshanko et al. [261] reported the synthesis of novel hyperbranched amphiphilic poly(ethylene oxide)-polystyrene $(PEO\text{-}PS)_n$ copolymers obtained by controlled radical polymerizations. Langmuir monolayers displayed reversible amphiphilic behavior at the air-water interface [261]. The random, mixed character of short hydrophilic and hydrophobic fragments results in peculiar surface behavior: unlike regular linear and star block copolymers, the amphiphilic hyperbranched macromolecules with higher PEO content are spread at the air-water interface and short PEO fragments are not submerged into the water subphase even at high compression [261].

The nature of the solid-liquid interface of hydroxyl-terminated hyperbranched aromatic polyester (HPB-OH) thin films was studied by Mikhaylova et al. [266]. Contact angle measurements, atomic force microscopy, zeta potential measurements and spectroscopic ellipsometry were used to characterize the physical and chemical structure of the HBP-OH thin films as well as the adsorption-relevant surface properties [266]. All surface-sensitive methods used indicate marked swelling behavior of HBP layers in aqueous phosphate buffer saline [266]. Moreover, the degree of swelling, zeta potential as well as surface free energy, strongly depend on the time the films had been annealed above their glass transition temperature [266,267].

The influence of hyperbranched polymer grafted polypropylene (PP-HBP) on the interfacial adhesion between fusion bonded bilayers of polypropylene (PP) and polyamide 6 (PA6) and on the properties of PP/PA6 blends was investigated [268]. The interfacial adhesion between PP-HBP compatibilized bilayers was ten times higher compared to maleic anhydride grafted PP (PP-MAH) compatibilized bilayers [268]. This is attributed to the higher diffusivity and functionality of PP-HBP leading to the formation of more PP-PA6 copolymers at the interface [268]. The high diffusivity of PP-HBP is an asset for multilayer film extrusion, while for blends; the high functionality permits the use of less compatibilizer for similar property improvements [268].

Recently, dendritic core-shell architectures which are based on hyperbranched polyglycerol for the solubilisation of hydrophobic drugs have been synthesized and characterized [81]. The core of hyperbranched polyglycerol has been modified with hydrophobic biphenyl groups or perfluorinated chains to increase the core hydrophobicity of the macromolecules. These amphiphilic core-shell type architectures were then used to solubilize pyrene, nile red, and a perfluoro tagged diazo dye, as well as the drug nimodipine in water. Surface tension measurements and scanning electron microscope were used to reveal the aggregation properties of these complexes. The formation of supramolecular aggregates with diameters of $20\,nm$ and critical aggregate concentrations of $2\cdot 10^{-6}\,mol/L$ have been observed [81].

Besides the thermodynamics of the surface properties, the kinetics, which is also strongly depended on the polymer architecture [269], is an essential feature for several applications. Díez-Pascual et al. [269] reported a comparative study between Langmuir and Gibbs monolayers of a hyperbranched polyol, poly(propylene glycol) homopolymers, and poly(propylene glycol)-poly(ethylene glycol) copolymers with different structure and molecular weight. With the help of dynamic surface tension and surface pressure measurements the adsorption kinetics were studied, where the kinetics were consistent with a rapid diffusion stage followed by a slow reorganization at the air-water interface [269].

Recently, the surface tension of the pure hyperbranched polymer, Boltorn U3000, and binary mixtures of Boltorn U3000 with alcohols (butan-1-ol, hexan-1-ol) have been measured at atmospheric pressure in the range of temperatures from $298.15\,K$ to $328.15\,K$, or of Boltorn U3000 with methyl-tert-butylether (MTBE) in the range of temperatures from $298.15\,K$ to $318.15\,K$ [270]. The influence of hyperbranched polyesters with different functional end groups on the surface tension of mixtures with an oligo(ester diol) was investigated [271] using Wilhelmy balance technique. The results indicate that the surface tension of the pure hyperbranched polyesters strongly depends on the functionality of the end groups, where the functionalization of the hydroxyl end groups by short alkyl chains (methyl, tert-butyl) reduced the surface tension depending on the degree of substitution [271]. The surface tension of the mixtures with the hydroxyl-terminated hyperbranched polyester was slightly increased at higher concentrations of the hyperbranched polymer compared to the surface tension of the pure ester diol [271]. On the other hand, the surface tension of mixtures could be considerably decreased using 1% of hyperbranched polyester polyols partially substituted with short alkyl chains [271], whereas the modified hyperbranched polyesters act as surface active agents. Peleshanko et al. [272] presented a review focused on the recent developments in the field of the highly branched molecules (excluding dendrimers) with emphasis on their surface behavior, microstructure, surface morphology, and properties.

Unfortunately, no experimental data related to the miscibility of system containing hyperbranched polymers could be found in the literature.

Theories of polymer interfacial properties have been developed along two fundamentally different lines [273], the self consistent field techniques and free energy functional methods, which have been designed originally for describing the strong and weak segregation limits, respectively. However; only a limit amount of work dealing with hyperbranched polymers can be found in the literature. For example, Qian et al. [274] used self-consistent (SCF) mean-field lattice simulations to study entropy-driven segregation of various branched and hyperbranched polymeric additives in chemically similar linear polymer hosts. The simulations account for the effect of molecular architecture on local configurationally entropy in the blends, but ignore the effect of architecture on local density and blend compressibility. Star, dendrimer, and comb-like additives are all found to be enriched at the surface of chemically identical linear host polymers [274]. The magnitude of their surface excess increases with increased number of chain ends and decreases with increased segmental crowding near the branch point [274]. Provided the number of arms and molecular weight of the branched additives are maintained constant, Qian et al.[274] found that the simplest branched architecture, the symmetric star, exhibits the strongest preference for the surface of binary polymer blends.

In this contribution we focus our attention to a special version of free energy functional methods, namely the modified density gradient theory, originally developed by van der Waals [275] and Lord Rayleigh [276] and later rediscovered by Cahn-Hilliard [277]. The theory of van der Waals [275] and Lord Rayleigh [276] leads to a description of the state of matter within the vapor-liquid interface. The goal of the theory was to elucidate the density profiles and to evaluate the associate surface tension. Van der Waals [275] used in his description a local free energy density consisting of a uniform term, evaluated at the local density, and a kind of gradient energy density term which is proportional to the square of the density gradient. The interfacial density profiles can be obtained by minimizing this free energy density under the constraint that the total number of molecules is constant, and resulting in an expression that allows evaluation of surface tensions. In version provides by Cahn-Hilliard [277] the density gradient approach leads to a general expression of Helmholtz energy of an inhomogeneous system, in which the homogeneous portion of the free energy can be taken from any equation of state or activity coefficient model, while the inhomogeneous contribution is truncated at the leading square gradient approximation. Several examples using this approach for the calculation of surface properties, like interfacial tension, density profiles within the interface and selective enrichment in the interface can be found in the literature [278,279,280,281,282,283,284,285,286]

In 1986, Wang et al. [287] presented a continuum model for polymer chains near an asymmetric (A-B) liquid-liquid interface where each side of the interface can have different polymer-surface interactions. For example, one side can attract the macromolecule, while the other can repel it. The model contains different monomer free energies and different excluded volume interactions for the macromolecule in the two solvents [287].

Later [288,289,290] the de Gennes modified square gradient approximation as arising from the entropic contributions associated with chain connectivity were used. McMullen et al. [288] employed density functional methods to derive the free energy and grand potential functionals appropriate to homopolymers and blends based on the FH theory. Later on, this theoretical framework was extended to block copolymers [289] permitting the treatment of compressible systems and therefore of more strongly first-order microphase separations. Freed and coworkers [291 292] discussed the influence of the concentration dependent interaction parameter, χ, within the FH limit on the interfacial properties, especially on the interfacial width and tension. In order to study the surface tension of dilute polymer solutions a renormalization group approach was applied [293]. An interpolation function extends the calculations to intermediate values of the polymer-surface interaction strength and provides the surface pressure as a function of both polymer-polymer and polymer-surface interactions [293].

Using a compressible version of FH theory within the density functional treatment for the liquid-liquid interface of phase separated binary polymer blends [294] involved several numerical problems, because the coupled, nonlinear Euler-Lagrange equations for the interfacial profile and tension are shown to become numerically unstable and permit analytic solution in the asymptotic profile wings. This analysis was limited to symmetric binary blends because this restriction permits to compute numerically the central portion of the interfacial profile [294].

Freed [295] provide a bridge between the density functional and self-consistent-field formulations for inhomogeneous polymer systems by deriving the self-consistent-field equations from a density functional approach. The final density functional emerges in the form of a Landau-type expansion about an analytically tractable zeroth-order inhomogeneous reference system, and the important presence of chain connectivity contributions is quite evident [295]. In 1996, Freed [296] presented an analytical theory for the competing influences of polymer-surface and polymer-polymer interactions, density and composition variations, and blend asymmetries on the surface profiles of a multi-component polymer blend near an interacting, impenetrable interface. The theory is explicitly applied in the limit of small continuum model polymer-surface interaction parameters, a limit which still enables treating all qualitative behaviors of polymers that individually tend either to aggregate toward or to segregate from the surface [296]. The formulation is based on an analytic combined self-consistent field-density functional theory for inhomogeneous polymer systems, where the compressible polymer system was described with a generic Gaussian chain-random mixing type model, and the bulk phase with a Sanchez-Lacombe-type model [296].

The theory for the interfacial properties of compressible binary polymer systems [294] was generalized to treat phase separated asymmetric binary blends. Lifschitz and Freed [297] used the Sanchez-Lacombe approximation for the homogeneous system free energy and a Cahn-Hilliard-de Gennes expression for the square gradient contribution. These two ingredients enable the same theory to describe both the compositions of the coexisting phases and the interfacial properties [297]. It was found that the coupled equations for the two concentration profiles are numerically unstable [297]. In order to solve this problem a perturbation-variational method for symmetric blends was introduced and extended to treat blends with asymmetry due to differing polymerization indices and/or interaction parameters [297]. The zeroth order approximation, which uses a hyperbolic tangent interpolation between the generally different densities in the coexisting phases, is found to be extremely accurate in predicting the interfacial tensions and widths [297].

Dudowicz et al. [298] investigated a minimal equilibrium polymerization model for the competition between self-assembly on a boundary and in solution that arises when an assembling system is in the presence of an adsorbing interface. Adsorption generally occurs upon cooling, but assembly (equilibrium polymerization) may arise either upon cooling or heating. Both cases are shown to exhibit a coupling between adsorption and self-assembly [298]. When both assembly and adsorption proceed upon cooling, a change in the ratio of the enthalpy of adsorption to the enthalpy of assembly in solution can switch the system between a predominance of self-assembly in solution to assembly on the substrate [298]. If assembly is promoted by heating and adsorption by cooling, as in many self-assembling proteins in aqueous solution, then a self-assembly analog of a closed loop phase boundary is found [298].

This contribution is focused on interfacial properties of binary systems in liquid-liquid equilibrium, which are of considerable practical importance in the description of liquid extraction processes or in all other applications mentioned above. First the interfacial properties will be considered at atmospheric pressure. Keeping the pressure constant, the density gradient theory (DGT) will be blended with an activity coefficient model [278].

The Cahn-Hilliard theory [277] describes the thermodynamic properties of a system where an interface exists between two equilibrium phases. In the liquid-liquid equilibrium of binary systems, consisting of components A and B, at atmospheric pressure p, the temperature T and the compositions of the two bulk phases (I and II) are given by the thermodynamic phase equilibrium conditions. It was assumed that the system is at constant temperature and pressure and to contain a fixed number of moles of each component. If there is a volume change upon mixing, then the total volume V will be a variable. In many practical situations V can be regarded as constant, so that the Helmholtz and Gibbs free energies are equivalent.

In the interface between two phases, only the concentration, expressed in segment fraction, ϕ_B, varies continuously from the bulk concentration of phase I ϕ_B^I to the bulk concentration of phase II ϕ_B^{II}. The phase diagram can be calculated using a free energy model related to the LCT and the Wertheim perturbation theory representing the association forces [188,223].

Following the method suggested in the literature [188,223,278] the interfacial tension σ for a planar interface in a binary system at equilibrium can be computed by:

$$\sigma = 2N \int_{\phi_B^I}^{\phi_B^{II}} \sqrt{\beta \Delta g(\phi_B)}\, d\phi_B \tag{45}$$

where β is the so-called influence parameter and can be fitted to one experimental interfacial tension at one temperature. $\Delta g(\phi_B)$ is the grand thermodynamic potential. The composition profile can be determined by [188,223,278]:

$$z = z_0 + \int_{\phi_{0B}}^{\phi_B} \sqrt{\frac{\beta}{\Delta g(\phi_B)}}\, d\phi_B \tag{46}$$

where z_0 and ϕ_{0B} represent an arbitrarily chosen origin and composition. The z-origin is arbitrarily located at a composition of $\left(\phi_B^I + \phi_B^I\right)/2$. This means $z_0 = 0$ and $\phi_{0B} = \left(\phi_B^I + \phi_B^I\right)/2$. A distance z may be determined for any ϕ_B lying between the bulk concentrations by evaluating the integral numerically.

Knowledge of β and $\Delta g\left(\phi_B\right)$ allows to calculate the interfacial tension with help of Eq. 45 and the composition profile using Eq. 46. In principle, $\Delta g\left(\phi_B\right)$ can be evaluated using any model free energy function, and hence also the LCT or the LCT in combination with the Wertheim theory.

Usually, β is fitted using one interfacial tension at one temperature [278]. Unfortunately, no experimental values of interfacial tension of systems containing hyperbranched polymers are available in the literature. For this reason we discuss not the interfacial tension itself, but the reduced interfacial tension, which is set to $\sigma_{red} = \sigma / \sqrt{\beta}$. The procedure has the advantage, if β is known, for example from one experimental data point; all other interfacial tensions can be calculated. As consequence also the interfacial profiles are given in reduced quantities.

Interfacial Properties - Calculation Examples

Beside the phase equilibrium the DGT allows also to investigate the interfacial properties between the coexisting phases. One important quantity related to the interface is the interfacial tension. Using Eq. 45 in combination with LCT or in combination with LCT and

Wertheim theory permits the calculation of the reduced interfacial tension of the system Boltorn H20 + water. The calculated reduced interfacial tensions as function of the distance from the critical point are plotted in Figure 17.

At the critical point the interfacial tension has to be zero. With increasing distance from the critical point the interfacial tension increases too. Similar to the results obtained for the phase-equilibrium calculation the association forces have an impact on the interfacial tension, but not a dominating influence. The association forces lead to a higher interfacial tension. With decreasing temperature this effect is more pronounced. The same situation could be found for the system propan-1-ol + Boltorn H20 and propan-1-ol + Boltorn U3000 (Figure 18). However; the impact of the association forces is more important in the aqueous polymer solution compared with the solution, where the polymer is dissolved in propan-1-ol. The possible reason for this finding can be the different association strength.

At constant distance from the critical point the interfacial tension of the aqueous polymer solution is much higher than in the system propan-1-ol + Boltorn H20 or propan-1-ol + Boltorn U3000. Using only the LCT the predicted interfacial tension for both systems at a given distance from the critical point is very similar (i.e. $T_C - T = 10\mathrm{K}$, water + Boltorn H20 $\sigma_{red} = 0.15$, propan-1-ol + Boltorn H20 $\sigma_{red} = 0.12$, propan-1-ol + Boltorn U3000 $\sigma_{red} = 0.061$).

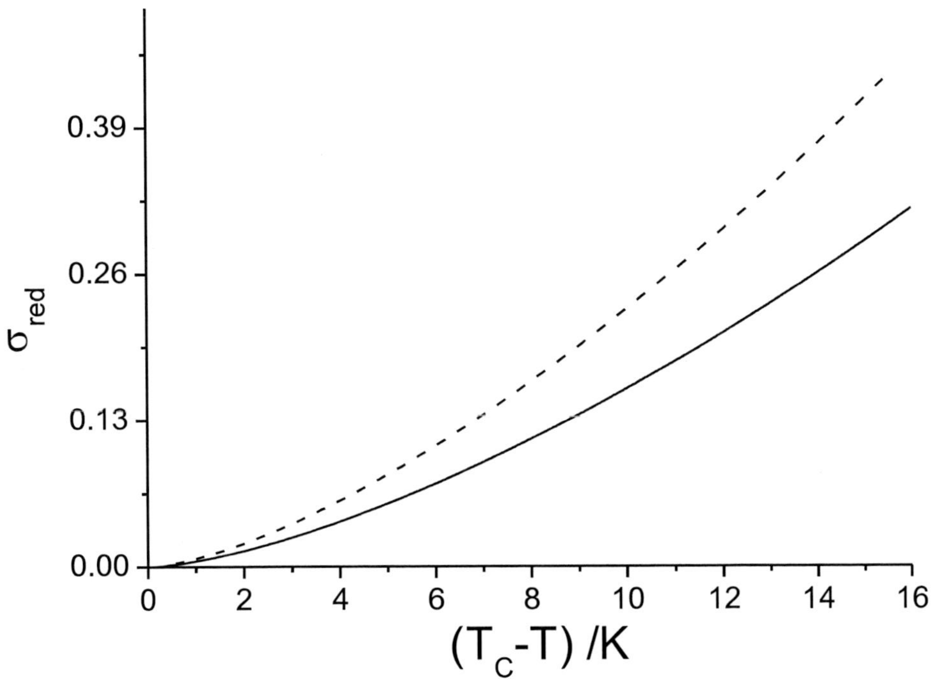

Figure 17. Predicted interfacial tension using LCT (solid line) and LCT + Wertheim theory (broken line) for the system water + Boltorn H20 [188,223].

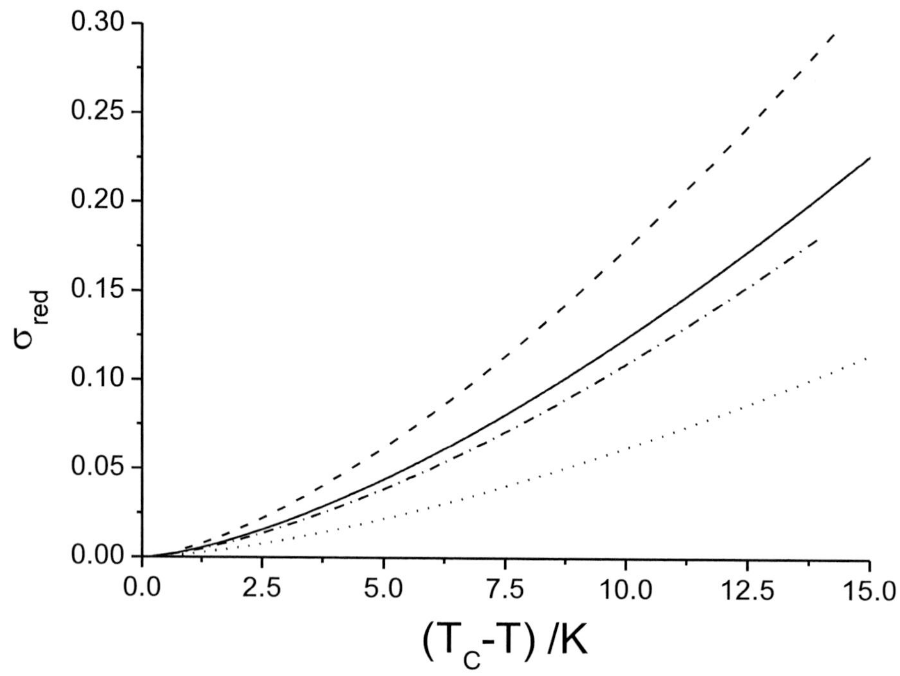

Figure 18. Predicted interfacial tension using LCT (solid line: Boltorn H20 + propan-1-ol [188,223], dashed line: Boltorn U3000 + propan-1-ol) and LCT + Wertheim theory (dotted line: Boltorn H20 + propan-1-ol [188,223], dotted-dashed line: Boltorn U3000 + propan-1-ol).

Taking the association forces into account the solutions differ in the interfacial tension (i.e. $T_C - T = 10\text{K}$, water + Boltorn H20 $\sigma_{red} = 0.22$, propan-1-ol + Boltorn H20 $\sigma_{red} = 0.17$, propan-1-ol + Boltorn U3000 $\sigma_{red} = 0.11$), where the impact of association force is for the aqueous solution more pronounced. The difference of interfacial tension for both polymers (Boltorn H20 and Boltorn U3000) in the same solvent, namely propan-1-ol, can be explained by the different miscibility in terms of length of the tie line (Figure 10 and Figure 11). If experimental data are available, the reduced quantities can be recalculated in interfacial tensions.

Not only the interfacial tension itself, but also the concentration profiles in the interface depend slightly from the intermolecular forces (Figure 19 and Figure 20). The concentration profile of the liquid-liquid interface varies only in the direction normal to the interface, which we take as the z direction. The interfacial profiles are calculated with the help of Eq. 46 in combination with LCT or LCT and Wertheim theory for the system water + Boltorn H20 (Figure 19) and for the system propan-1-ol + Boltorn H20 (Figure 20).

Analyzing the interfacial profiles given in Figure 19 and Figure 20 show again the impact of the association forces. The incorporation of the Wertheim theory leads to a sharper concentration profiles in comparison of the theoretical results, if only the LCT is used. The sharper concentration profiles are strictly connected with a smaller interfacial thickness. No extreme in the concentration profiles could be detected, and hence no relative enrichment in the interface is predicted by the suggested theoretical framework. Sakane et al. [299] measured the cloud point curve and the interfacial tension of phase-separated mixtures of two linear polymers, namely poly(dimethylsiloxane) (PDMS) (1) and poly(1,1,3,3-tetramethyldisiloxanylethylene) (PTMDSE) (2), using the sessile-drop method.

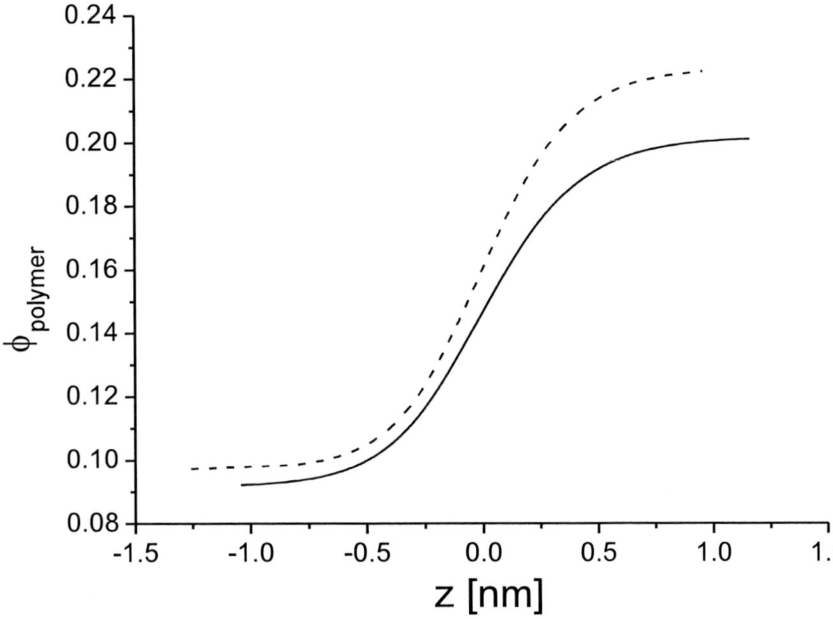

Figure 19. Predicted concentration profile using LCT (solid line) or LCT and Wertheim theory (broken line) for the system water + Boltorn H20 at T=369.2K [188,223].

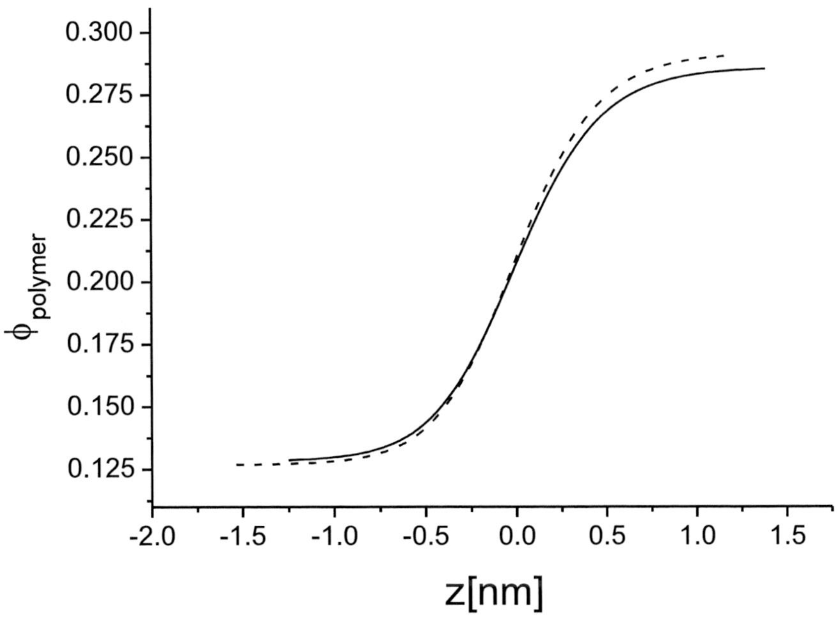

Figure 20. Predicted concentration profile using LCT (solid line) or LCT and Wertheim theory (broken line) for the system propan-1-ol + Boltorn H20 at T=366.9 K[188,223].

Additionally, they adopted also the square-gradient in order to describe the interfacial tension of phase-separated polymer mixtures [299]. In contrast to this contribution Sakane et al. [299], combined the Cahn-Hilliard theory with the classical Flory-Huggins theory. Assuming the monomer of PDMS occupies one lattice site and the other monomer two of them leads to the following characteristic parameters within the LCT:

$$M_1 = 145, \quad M_2 = 218, \quad \varepsilon_{LCT} / k_B = 0.91K \tag{47}$$

where the interaction energy, ε_{LCT}, was fitted to the experimental cloud-point curve taken from the literature [299]. The calculated phase diagram together with the experimental data points are plotted in Figure 21. The comparison shows that the selected theoretical framework leads to a satisfactory agreement with the experimental data, especially if one had in mind that only one adjustable parameter, ε_{LCT}, is used. The phase diagram forms the basic for the calculation of the interfacial tension with the help of Eq. 45. To obtain explicit numerical results for the interfacial tension, we need to fix the value of β. The parameter β was fitted to one experimental interfacial temperature at one temperature. This allows the prediction of the temperature dependency of the interfacial tension.

Figure 22 shows a comparison between the experimental data [299] and the calculated interfacial tension for the binary blend PDMS + PTMDSE. From this result it can be concluded that the combination of the LCT with the density gradient theory can be applied to predict the temperature dependency of the interfacial tension of phase separated polymer blends with a high accuracy. This can be very useful, because the experiments are very challenging due to the high viscosity.

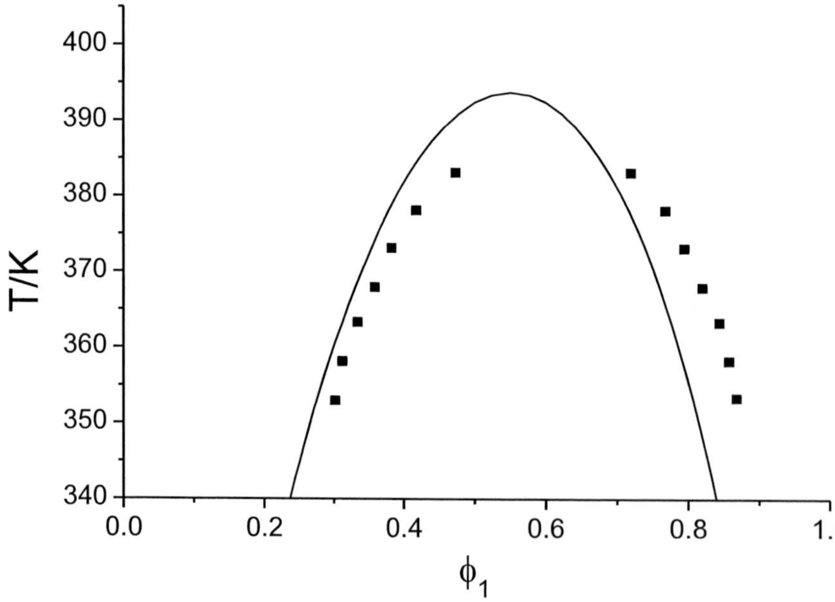

Figure 21. Experimental [299] (squares) and calculated (LCT, line) cloud point curve for the binary blend PDMS (1) + PTMDSE (2).

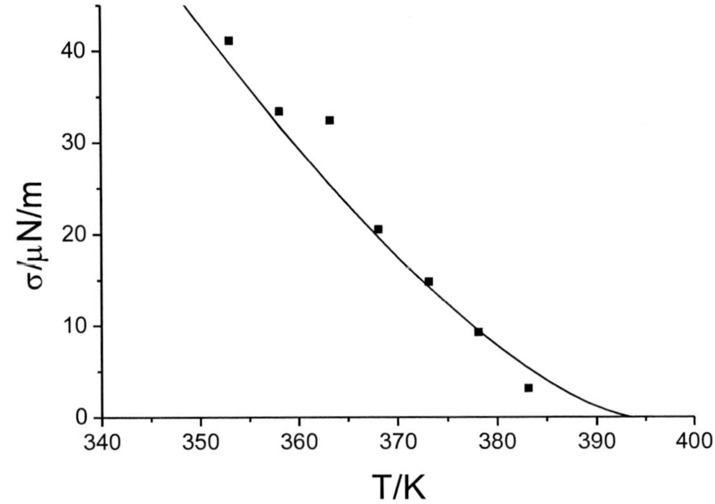

Figure 22. Experimental [299] (squares) and calculated (LCT, line) interfacial tension for the binary blend PDMS (1) + PTMDSE (2).

In Figure 23 two concentration profiles of PDMS within the interface at two different temperatures are plotted versus the coordinate z perpendicular to the interface. This system to feature an UCST behavior (Figure 21) and hence the critical temperature is located at higher temperatures. At the critical temperature the interfacial profile must diverge, because the interfacial thickness must become infinity. Therefore, the interfacial profile at higher temperature ($T = 351K$ in Figure 23) is broader than at lower temperatures ($T = 330K$ in Figure 23). No extremes could by found in the concentration profile meaning that no selective enrichment of one component in the interface occurs.

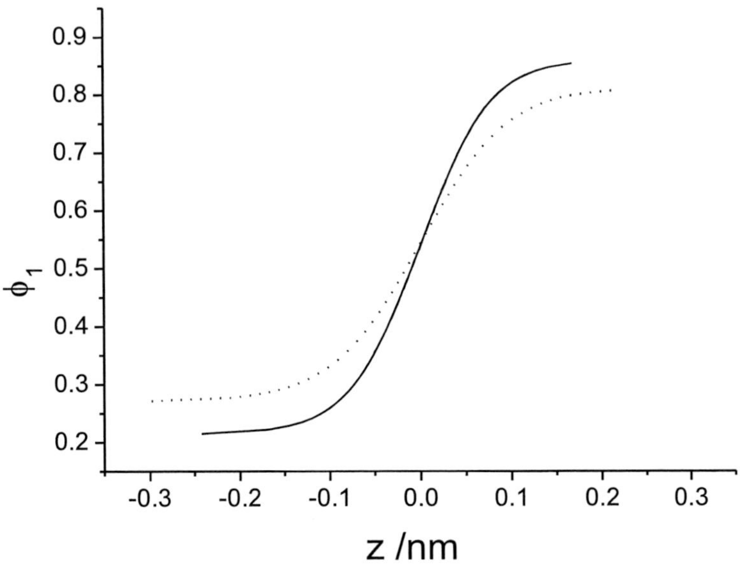

Figure 23. Predicted concentration profile using LCT at two temperatures (solid line: T=330K, dotted line 351K) for the system PDMS (1) + PTMDSE (2).

CONCLUSION

The increasing attention given to hyperbranched polymer for several new applications in different fields ranching from medicine over energy technology until chemical engineering is the driving force of their research. Among others, the development of thermodynamic models is one important point in this direction. The major hurdle in the thermodynamics of hyperbranched polymers is the complexity of these polymers caused by the architecture and the large number of functional groups responsible for specific interaction between the polymer molecules itself and between the polymer and the surrounding solvent. By means of an elaborate lattice cluster theory, allowing describing the polymer architecture, in combination with Wertheim perturbation approach, allowing modeling association forces, both effects could be taken into account in the thermodynamic model. Additionally, the square gradient theory can be utilized for the investigation of inhomogeneous system having an interface. The burden of this theoretical framework is to compute the interfacial properties, like interfacial tension and interfacial profiles, in phase-separated systems.

The endeavor to demonstrate the full potential of this approach we compare the theoretical results in terms of phase equilibria and interfacial tension with experimental data.

Overall, the results indicate how polymer architecture can exert a tremendous influence on the physical and solution properties of a material. A significant benefit of the LCT lies in the algebraic nature of the expressions for thermodynamic properties that apply for all compositions, molecular weights, interaction energies, monomer structure, and temperature. The Gibbs energy of mixing can be calculated using a simple polynomial, which is derived in this contribution. The LCT excess free energy for a binary system in the incompressible limit depends on the interaction energies only through the single dimensionless interaction energy, $\varepsilon_{LCT} = \varepsilon_{11} + \varepsilon_{22} - 2\varepsilon_{12}$.

In the light of this idea, we carried out calculations of phase diagram for polymer solutions containing hyperbranched polymers or polymer blends of linear or hyperbranched polymers.

ACKNOWLEDGMENTS

Authors gratefully acknowledge the financial support of this research by the German Science Foundation (DFG En 291/7-1) and the Max-Buchner Foundation (Grant Nr. 2864), as well Perstorp, Sweden for the polymers.

REFERENCE

[1] R. Koningsveld, W.H. Stockmayer, E. Nies, *Polymer Phase Diagrams*. Oxford University Press, 2001.

[2] K. Kamide, Thermodynamics of Polymer Solutions – Phase Equilibria and Critical Phenomena. in A.D. Jenkins (Ed.) *Polymer Science Library 9,* Elsevier, 1990.

[3] J.M. Prausnitz, Molecular thermodynamics for some applications in biotechnology. *J. Chem. Thermodynamics* 35 (2003) 21-39.

[4] P.A. Albertsson, *Partition of Cell Particles and Macromolecules*, 3rd edition. John Wiley and Sons, New York, 1986.

[5] P.J. Flory, *Principles of Polymer Chemistry*, Cornell University Press, Ithaca, 1953.

[6] J.M.J. Frechet, D.A. Tomalia, *Dendrimers and Other Dendritic Polymers*. Wiley Series in Polymer Science, John Wiley and Sons, Ltd. 2001.

[7] D.A. Tomalia, The dendritic state. *Materials Today* 8 (2005) 34-46.

[8] P.J. Flory, Molecular Size Distribution in Three Dimensional Polymers. VI. Branched Polymers Containing A—R—Bf-1 Type Units. *J. Am. Chem. Soc.* 74 (1952) 2718-2723.

[9] P. A. Gunatilake, G. Odian, D.A. Tomalia, Thermal polymerization of a 2-(carboxyalkyl)-2-oxazoline, *Macromolecules* 21 (1988) 1556-1562.

[10] Y.H. Kim, O.W. Webster, Water-Soluble Hyperbranched Polyphenylene: A Unimolecular Micelle. *J. Am. Chem. Soc.* 112 (1990) 4592-4593.

[11] H.R. Kricheldorf, T. Stukenbrock, New polymer syntheses. XC. A-B-A triblock copolymers with hyperbranched polyester A-blocks, *J. Polym. Sci., Part A: Polym. Chem.* 36 (1998) 31-38.

[12] A. Hult, M. Johansson, E. Malmström, Hyperbranched Polymers, *Adv. Poly. Sci.* 143 (1999) 1-34.

[13] A.M. Fischer, H. Frey, Soluble hyperbranched poly(glycolide) copolymers. *Macromolecules* 43 (2010) 8539-8548.

[14] J. Su, L. Zhang, Q. Wu, Amphiphilic dendritic polymers. *Prog. Chem.* 20 (2008) 1980-1986.

[15] M. Jikei, M.A. Kakimoto, Hyperbranched polymers: A promising new class of materials. *Prog. Polym. Sci.* (Oxford) 26 (2001) 1233-1285.

[16] D. Yan, J. Hou, X. Zhu, J.J. Kosman, H.S. Wu, A new approach to control crystallinity of resulting polymers: Self-condensing ring opening polymerization. *Macrom. Rapid Comm.* 21 (2000) 557-561.

[17] T. Huber, F. Böhme, H. Komber, J. Kronek, J. Luston, D. Voigt, B. Voit, New hyperbranched poly(ether amide)s via nucleophilic ring opening of 2-oxazoline-containing monomers. *Macrom. Chem. Phys.* 200 (1999) 126-133.

[18] M. Trollsås, B. Atthoff, H. Claesson, J.L. Hedrick, Hyperbranched poly(ε-caprolactone)s. *Macromolecules* 31 (1998) 3439-3445.

[19] F. Wurm, H. Schüle, H. Frey, Amphiphilic Linear-Hyperbranched block copolymers with linear poly(ethylene oxide) and hyperbranched poly(carbosilane) block. *Macromolecules* 41 (2008) 9602-9611.

[20] R.V. Srinivasa, A.B. Samui, Effect of molecular architecture and size of mesogen on phase behavior and photoactive properties of photoactive liquid crystalline hyperbranched polyester epoxies containing benzylidene moiety. *J. Polym. Sci., Part A: Polym. Chem.* 46 (2008) 552-563.

[21] V.I. Irzhak, A structural characteristic of hyperbranched polymers. *Polym. Sci. – Ser. B* 51 (2009) 143-148.

[22] D.M.A. Buzza, Power law polydispersity and fractal structure of hyperbranched polymers. *Eur. Phys. J. E 13* (2004) 79-86.

[23] S.D.M.A. Buzza, T.C.B. McLeish, D. Parker, A.J. Keeney, W.J. Feast, Rheology and molecular weight distribution of hyperbranched polymers. *Macromolecules* 35 (2002) 9605-9612.

[24] A.H.E. Müller, D. Yan, M. Wulkow, Molecular parameters of hyperbranched polymers made by self-condensing vinyl polymerization. 1. Molecular weight distribution. *Macromolecules* 30 (1997) 7015-7023.

[25] J.J.J. Frechet, M. Henmi, I. Gitsov, S. Aoshima, M.R. Leduc, R.B. Grubbs, Self-Condensing Vinyl Polymerization: An Approach to Dendritic Materials. *Science* 269 (1995) 1080-1083.

[26] H.T. Chang, J.M.J. Frechet, Proton-Transfer Polymerization: A New Approach to Hyperbranched Polymers. *J. Am. Chem. Soc.* 121 (1999) 2313-2314.

[27] H.C. Kolb, M.G. Finn, K.B. Sharples, Click Chemistry: Diverse Chemical Function from a Few Good Reactions. *Angew. Chem. Int. Ed.* 40 (2001) 2004-2021.

[28] P. Wu, A.K. Feldman, A.K. Nugent, C.J. Hawker, A. Schell, B. Voit, J. Pyun, J.M.J. Frechet, K.B. Sharpless, V.V. Fokin, M.J. Joralemon, R.K. O' Reilly, J.B. Matson, K.L. Wooley, Dendrimers Clicked Together Divergently. *Macromolecules* 38 (2005) 5436-5443.

[29] A. Qin, J.W.Y. Lam, B.Z. Tang, Click polymerization: Progresses, challenges, and opportunities. *Macromolecules* 43 (2010) 8693-8702.

[30] E. Zagar, M. Zigon, Characterization of a Commercial Hyperbranched Aliphatic Polyester Based on 2,2,-Bis(methylol) propionic Acid. *Macromolecules* 35 (2002) 9913-9925.

[31] E. Zagar, M. Zidon, S. Podzimek, Characterization of commercial aliphatic hyperbranched polyesters. *Polymer* 47 (2006) 166-175.

[32] N.K. Jain, U. Gupta, Application of dendrimer-drug complexation in the enhancement of drug solubility and bioavailability. *Expert Opinion on Drug Metabolism and Toxicology* 4 (2008) 1035-1052.

[33] G. Kreutzer, C. Ternat, T.Q. Nguyen, C.J.G. Plummer, J.A.E. Mason, V. Castelletto, I.W. Hamley, F. Sun, S.S. Sheiko, A. Herrmann, L. Ouali, H. Sommer, W. Fieber, M.I. Velazco, H.A. Klok, Water-soluble, unimolecular containers based on amphiphilic multiarm star block copolymers. *Macromolecules* 39 (2006) 4507-4516.

[34] T. Satoh, Unimolecular micelles based on hyperbranched polycarbohydrate cores. *Soft Matter* 5 (2009) 1972-1982.

[35] T. Satoh, Y. Kitajyo, R. Sakai, A. Narumi, T. Takahashi, H. Kaga, N. Kaneko, T. Kakuchi, Size-selective encapsulation property of unimolecular reverse micelle consisting of hyperbranched D-glucan core and L-leucine ethyl ether shell. *Macrom. Symp.* 279 (2009) 145-150.

[36] E. Burakowska, R. Haag, Dendritic polyglycerol core-double-shell architectures: Synthesis and transport properties, *Macromolecules* 42 (2009) 5545-5550.

[37] K.R. Kumar, D.E. Brooks, Comparison of hyperbranched and linear polyglycidol unimolecular reverse micelles as nanoreactors and nanocapsules. *Macrom. Rapid Comm.* 26 (2005) 155-159.

[38] T. Satoh, M. Tamaki, Y. Kitajyo, T. Maeda, H. Ishihara, T. Imai, H. Kaga, T. Kakuchi, Synthesis of unimolecular reversed micelle consisting of a poly(L-lactide) shell and hyperbranched D-mannan core, *J. Polym. Sci., Part A: Polym. Chem.* 44 (2006) 406-413.

[39] Y. Liang, W. Decheng, C. Xiangyu, J. Ming, P. Hongting, Unimolecular micelle derived from hyperbranched polyethylenimine with well-defined hybrid shell of

polyethylene oxide and polystyrene: A versatile nanocapsule. *J. Polym. Sci., Part A: Polym. Chem.* 48 (2010) 681-691.

[40] V. Percec, P. Chu, M. Kawasumi, Toward "willowlike" thermotropic dendrimers, *Macromolecules* 27 (1994) 4441-4453.

[41] V. Percec, P. Chu, G. Ungar, J. Zhou, Rational design of the first nonspherical dendrimer which displays calamitic nematic and smectic thermotropic liquid crystalline phases, *J. Am. Chem. Soc.* 117 (1995) 11441-11454.

[42] Y.H. Kim, Lyotropic liquid crystalline hyperbranched aromatic polyamides. *J. Am. Chem. Soc.* 114 (1992) 4947-4948.

[43] K.L. Wooley, J.M.J. Fréchet, C.J. Hawker, Influence of shape on the reactivity and properties of dendritic, hyperbranched and linear aromatic polyesters. *Polymer* 35 (1994) 4489-4495.

[44] K. Matyjaszweski, J.T. Shigemonto, J.M.J. Frechet, M.C. Leduc, Controlled/"Living" Radical Polymerization with Dendrimers Containing Stable Radicals. *Macromolecules* 29 (1995) 4167-4171.

[45] E. Malmström, A. Hult, Hyperbranched polymers: A review. *J. Macrom. Sci. – Rev. Macrom. Chem. Phys.* 37 (1997) 555-579.

[46] O.A. Matthews, A.N. Shipway, J.F. Stoddart, Dendrimers - Branching out from curiosities into new technologies. *Prog. Polym. Sci.* (Oxford) 23 (1998) 1-56.

[47] M. Fischer, F. Vögtle, Dendrimere: From Design to Application – A Progress Report. *Angew. Chem. Int. Ed.* 38 (1999) 884-905.

[48] B. Voit, Hyperbranched polymers: A chance and a challenge, *Comptes Rendus Chimie* 6 (2003) 821-832.

[49] C. Gao, D. Yan, Hyperbranched polymers: From synthesis to applications, *Prog. Polym. Sci.* (Oxford) 29 (2004) 183-275.

[50] H.J. Huang, S. Ramaswamy, U.W. Tschirner, B.V. Ramarao, A review of separation technologies in current and future biorefineries, *Sep. Purif. Techn.* 62 (2008) 1-21.

[51] M. Irfan, M. Seiler, Encapsulation using hyperbranched polymers: From research and technologies to emerging applications. *Ind. Eng. Chem. Res.* 49 (2010) 1169-1196.

[52] D. Astruc, E. Boisselier, C. Ornelas, Dendrimers designed for functions: From physical, photophysical, and supramolecular properties to applications in sensing, catalysis, molecular electronics, photonics, and nanomedicine, *Chem. Rev.* 110 (2010) 1857-1959.

[53] D. Wilms, S.E. Stiriba, H. Frey, Hyperbranched polyglycerols: From the controlled synthesis of biocompatible polyether polyols to multipurpose applications, *Acc. Chem. Res.* 43 (2010) 129-141.

[54] F. Wurm, H. Frey, Linear-dendritic block copolymers: The state of the art and exciting perspectives. *Prog. Polym. Sci.* (Oxford) 36 (2011) 1-52.

[55] J. Khandare, R. Haag, Pharmaceutically used polymers: Principles, structures, and applications of pharmaceutical delivery systems. *Handbook of Experimental Pharmacology* 197 (2010) 221-250.

[56] R.H. Müller, K. Mäder, S. Dohla, Solid lipid nanoparticles (SLN) for controlled drug delivery - a review of state of the art. *Eur. J. Pharm. Biopharm.* 50 (2000) 161-77.

[57] C.M. Paleos, D. Tsiourvas, Z. Sideratou, L.A. Tziveleka, Drug delivery using multifunctional dendrimers and hyperbranched polymers. *Expert Opinion on Drug Delivery* 7 (2010) 1387-1398.

[58] J. Liu, X. Zhu, D. Yan, Self-assembly of hyperbranched polymers and its biomedical applications. *Adv. Mat.* 22 (2010) 4567-4590.

[59] L.A. Tziveleka, C. Kontoyianni, Z. Sideratou, D. Tsiourvas, C.M. Paleos, Novel functional hyperbranched polyether polyols as prospective drug delivery systems. *Macrom. Biosci.* 6 (2006) 161-169.

[60] P. Kolhe, E. Misra, R.M. Kannan, S. Kannan, M. Lieh-Lai, Drug complexation, in vitro release and cellular entry of dendrimers and hyperbranched polymers. *Intern. J. Pharm.* 259 (2003) 143-160.

[61] P. Kolhe, J. Khandare, O. Pillai, S. Kannan, M. Lieh-Lai, R. Kannan, Hyperbranched polymer-drug conjugates with high drug payload for enhanced cellular delivery. *Pharm. Res.* 21 (2004) 2185-2195.

[62] S. Kannan, P. Kolhe, V. Raykova, M Glibatec, R.M. Kannan, M. Lieh-Lai, D. Bassett, Dynamics of cellular entry and drug delivery by dendritic polymers into human lung epithelial carcinoma cells. *J. Biomat. Sci., Polym. Ed.* 15 (2004) 311-330.

[63] C. Yiyun, X. Tongwen, Dendrimers as Potential Drug Carriers. Part I. Solubilization of Non-Steroidal Anti-Inflammatory Drugs in the Presence of Polyamidoamine Dendrimers. *Eur. J. Med. Chem.* 40 (2005) 1188-1192.

[64] O. Perumal, J. Khandare, P. Kolhe, S. Kannan, M. Lieh-Lai, R.M. Kannan, Effects of branching architecture and linker on the activity of hyperbranched polymer-drug conjugates. *Bioconj. Chem.* 20 (2009) 842-846.

[65] S. Suttiruengwong, J. Rolker, I. Smirnova, W. Arlt, M. Seiler, L. Lüderitz, Y. Pérez De Diego, P.J. Jansens, Hyperbranched polymers as drug carriers: Microencapsulation and release kinetics. *Pharm. Dev. Techn.* 11 (2006) 55-70.

[66] E. Güç, G. Günüdz, U. Gündüz, Fatty acid based hyperbranched polymeric nanoparticles for hydrophobic drug delivery. *Drug Dev. Ind. Pharm.* 36 (2010) 1139-1148.

[67] R.K. Kainthan, C. Mugabe, H.M. Burt, D.E. Brooks, Unimolecular micelles based on hydrophobically derivatized hyperbranched polyglycerols: Ligand binding properties. *Biomacromolecules* 9 (2008) 886-895.

[68] C. Mugabe, B.A. Hadaschik, R.K. Kainthan, D.E. Brooks, A.I. So, M.E. Gleave, H.M. Burt, Paclitaxel incorporated in hydrophobically derivatized hyperbranched polyglycerols for intravesical bladder cancer therapy. *BJU International* 103 (2009) 978-986.

[69] C. Kontoyianni, Z. Sideratou, T. Theodossiou, L.A. Tziveleka, D. Tsiourvas, C.M. Paleos, A novel micellar PEGylated hyperbranched polyester as a prospective drug delivery system for Paclitaxel. *Macrom. Biosci.* 8 (2008) 871-881.

[70] T. Boulikas, I. Tsogas, Microtubule-targeted antitumor drugs: Chemistry, mechanisms and nanoparticle formulations. *Gene Therapy Mol. Biol.* 12 (2008) 313-358.

[71] X. Zhang, X. Zhang, Z. Wu, X. Gao, C. Cheng, Z. Wang, C. Li, A hydrotropic β-cyclodextrin grafted hyperbranched polyglycerol co-polymer for hydrophobic drug delivery. *Acta Biomat.* 7 (2011) 585-592.

[72] L. Zhang, C.H. Hu, S.X. Cheng, R.X. Zhuo, Hyperbranched amphiphilic polymer with folate mediated targeting property. *Coll. Surf. B: Biointerf.* 79 (2010) 427-433.

[73] C. Mugabe, R.T. Liggins, D. Guan, I. Manisali, I. Chafeeva, D.E. Brooks, M. Heller, J.K. Kackson, H.M. Burt, Development and in vitro characterization of paclitaxel and

docetaxel loaded into hydrophobically derivatized hyperbranched polyglycerols. *Intern. J. Pharm.* 2010 in press doi:10.1016/j.ijpharm.2010.010.

[74] S. Reven, J. Grdadolnik, J. Kristl, E. Žagar, Hyperbranched poly(esteramides) as solubility enhancers for poorly water-soluble drug glimepiride. *Intern. J. Pharm.* 396 (2010) 119-126.

[75] M.A. Quadir, M.R. Radowski, F. Kratz, K. Licha, P. Hauff, R. Haag, Dendritic multishell architectures for drug and dye transport, *J. Contr. Release* 132 (2008) 289-294.

[76] T. Wang, M. Li, H. Gao, Y. Wu, Nanoparticle carriers based on copolymers of poly(ε-caprolactone) and hyperbranched polymers for drug delivery. *J. Coll. Interf. Sci.* 353 (2011) 107-115.

[77] K.J. Haxton, H.M. Burt, Hyperbranched polymers for controlled release of cisplatin. *Dalton Transactions* 43 (2008) 5872-5875.

[78] J. Liu, W. Huang, Y. Pang, X. Zhu, Y. Zhou, D. Yan, The in vitro biocompatibility of self-assembled hyperbranched copolyphosphate nanocarriers. *Biomaterials* 31 (2010) 5643-5651.

[79] S. Santra, C. Kaittanis, J.M. Perez, Cytochrome c encapsulating theranostic nanoparticles: A novel bifunctional system for targeted delivery of therapeutic membrane-impermeable proteins to tumors and imaging of cancer therapy. *Mol. Pharm.* 7 (2010) 1209-1222.

[80] T. Wang, Y. Wu, M. Li, Novel hyperbranched poly (amine-ester)-poly (lactide-co-glycolide) polymeric micelles as amphotericin B carriers. *Polym. Bull.* 65 (2010) 425-442.

[81] I.N. Kurniasih, H. Liang, J.P. Rabe, R. Haag, Supramolecular aggregates of water soluble dendritic polyglycerol architectures for the solubilization of hydrophobic compounds. *Macrom. Rapid Comm.* 31 (2010) 1516-1520.

[82] C. Xia, X. Ding, Y. Sun, H. Liu, Y. Li, Hyperbranched-upon-dendritic macromolecules as unimolecular hosts for controlled release. *J. Polym. Sci., Part A: Polym. Chem.* 48 (2010) 4013-4019.

[83] M. Prabaharan, J.J. Grailer, S. Pilla, D.A. Steeber, S. Gong, Amphiphilic multi-arm-block copolymer conjugated with doxorubicin via pH-sensitive hydrazone bond for tumor-targeted drug delivery. *Biomaterials* 30 (2009) 5757-5766.

[84] A. Breitenbach, Y.X. Ti, T. Kissel, Branched biodegradable polyesters for parenteral drug delivery systems. *J. Contr. Release* 64 (2000) 167-178.

[85] J. Rojo, R. Delgado, Glycodendritic structures: Promising new antiviral drugs. *J. Antimicrob. Chemotherapy* 54 (2004) 579-581.

[86] C.M. Paleos, L.A. Tziveleka, Z. Sideratou, D. Tsiourvas, Gene delivery using functional dendritic polymers. *Expert Opinion on Drug Delivery* 6 (2009) 27-38.

[87] C.M. Paleos, D. Tsiourvas, Z. Sideratou, Molecular engineering of dendritic polymers and their application as drug and gene delivery systems. *Mol. Pharm.* 4 (2007) 169-188.

[88] H. Tian, X. Chen, H. Lin, C. Deng, P. Zhang, Y. Wei, X. Jing, Micellization and reversible pH-sensitive phase transfer of the hyperbranched multiarm PEI-PBLG copolymer. *Chemistry - A Eur. J.* 12 (2006) 4305-4312.

[89] V. Russ, H. Elfberg, C. Thoma, J. Kloeckner, M. Ogris, E. Wagner, Novel degradable oligoethylenimine acrylate ester-based pseudodendrimers for in vitro and in vivo gene transfer. *Gene Therapy* 15 (2008) 18-29.

[90] M. Krämer, J.F. Stumbé, G. Grimm, B. Kaufmann, U. Krüger, M. Weber, R. Haag, Dendritic polyamines: Simple access to new materials with defined treelike structures for application in nonviral gene delivery. *ChemBioChem* 5 (2004) 1081-1087.
[91] P. Banerjee, W. Reichardt, R. Weissleder, A. Bogdanov Jr., Novel hyperbranched dendron for gene transfer in vitro and in vivo. *Bioconj. Chem.* 15 (2004) 960-968.
[92] X. Shuai, T. Merdan, F. Unger, M. Wittmar, T. Kissel, Novel Biodegradable Ternary Copolymers hy-PEI-g-PCL-b-PEG: Synthesis, Characterization, and Potential as Efficient Nonviral Gene Delivery Vectors. *Macromolecules* 36 (2003) 5751-5759.
[93] H. Tian, W. Xiong, J. Wei, Y. Wang, X. Chen, X. Jing, Q. Zhu, Gene transfection of hyperbranched PEI grafted by hydrophobic amino acid segment PBLG, *Biomaterials* 28 (2007) 2899-2907.
[94] J.A. Khan, R.K. Kainthan, M. Ganguli, J.N. Kizhakkedathu, Y. Singh, S. Maiti, Water soluble nanoparticles from PEG-based cationic hyperbranched polymer and RNA that protect RNA from enzymatic degradation. *Biomacromolecules* 7 (2006) 1386-1388.
[95] L.A. Tziveleka, A.M.G. Psarra, D. Tsiourvas, C.M. Paleos, Synthesis and evaluation of functional hyperbranched polyether polyols as prospected gene carriers. *Intern. J. Pharm.* 356 (2008) 314-324.
[96] W.J. Kim, A.C. Bonoiu, T. Hayakawa, C. Xia, M.A. Kakimoto, H.E. Pudavar, K.S. Lee, P.N. Prasad, Hyperbranched polysiloxysilane nanoparticles: Surface charge control of nonviral gene delivery vectors and nanoprobes. *Intern. J. Pharm.* 376 (2009) 141-152.
[97] R. Reul, J. Nguyen, T. Kissel, Amine-modified hyperbranched polyesters as non-toxic, biodegradable gene delivery systems. *Biomaterials* 30 (2009) 5815-5824.
[98] J. Chen, C. Wu, D. Oupický, Bioreducible hyperbranched poly(amido amine)s for gene delivery. *Biomacromolecules* 10 (2009) 2921-2927.
[99] X. Wang, Y. He, J. Wu, C. Gao, Y. Xu, Synthesis and evaluation of phenylalanine-modified hyperbranched poly(amido amine)s as promising gene carriers. *Biomacromolecules* 11 (2010) 245-251.
[100] Y.Z. You, Z.Q. Yu, M.M. Cui, C.Y. Hong, Preparation of photoluminescent nanorings with controllable bioreducibility and stimuli-responsiveness. *Angew. Chem. Int. Ed.* 49 (2010) 1099-1102.
[101] R. Reul, J. Nguyen, T. Kissel, Amine-modified hyperbranched polyesters as non-toxic, biodegradable gene delivery systems. *Biomaterials 3*0 (2009) 1-10.
[102] Z. Zhong, Y. Song, J.F.J. Engbersen, M.C. Lok, W.E. Hennink, J. Feijen, A versatile family of degradable non-viral gene carriers based on hyperbranched poly(ester amine)s. *J. Contr. Release* 109 (2005) 317-329.
[103] R.K. Kainthan, D.E. Brooks, In vivo biological evaluation of high molecular weight hyperbranched polyglycerols. *Biomaterials* 28 (2007) 4779-4787.
[104] J. Blacklock, Y.Z. You, Q.H. Zhou, G. Mao, D. Oupický, Gene delivery in vitro and in vivo from bioreducible multilayered polyelectrolyte films of plasmid DNA. *Biomaterials* 30 (2009) 939-950.
[105] O.G. Brown, C. Bergquist, P. Ferm, K.L. Wooley, Unusual, promoted release of guests from amphiphilic cross-linked polymer networks. *J. Am. Chem. Soc.* 127 (2005) 11238-11239.

[106] C. Wiener, M.W. Brechbiel, H. Brothers, R.L. Magin, O.A. Gansow, D.A. Tomalia, P.C. Lauterbur, Dendrimer-based metal chelates: A new class of magnetic resonance imaging contrast agents. *Magn. Reson. Med.* 31 (1994) 1-8.

[107] M.H.M. Oudshoorn, R. Rissmann, J.A. Bouwstra, W.E. Hennink, Synthesis and characterization of hyperbranched polyglycerol hydrogels. *Biomaterials* 27 (2006) 5471-5479.

[108] M.H.M. Oudshoorn, R. Penterman, R. Rissmann, J.A. Bouwstra, D.J. Broer, W.E. Hennink, Preparation and characterization of structured hydrogel microparticles based on cross-linked hyperbranched polyglycerol. *Langmuir* 23 (2007) 11819-11825.

[109] E.G.R. Fernandes, V. Zucolotto, A.A.A. De Queiroz, Electrospinning of hyperbranched poly-L-lysine/polyaniline nanofibers for application in cardiac tissue engineering. *J. Macrom. Sci., Part A: Pure and Applied Chem.* 47 (2010) 1203-1207.

[110] S. Luo, C. Ling, X. Hu, X. Liu, S. Chen, M. Han, J. Xia, Thermoresponsive unimolecular micelles with a hydrophobic dendritic core and a double hydrophilic block copolymer shell. *J. Coll. Interf. Sci.* 353 (2011) 76-82.

[111] S. Pedrón, C. Peinado, P. Bosch, K.S. Anseth, Synthesis and characterization of degradable bioconjugated hydrogels with hyperbranched multifunctional cross-linkers. *Acta Biomat.* 6 (2010) 4189-4198.

[112] Z. Sanping, X. Weilin, Cyclodextrin-containing supramolecular hydrogels. *Prog. Chem.* 22 (2010) 916-926.

[113] S. Pedrón, K. Anseth, J.A. Benton, P. Bosch, C. Peinado, Bioapplications of networks based on photo-cross-linked hyperbranched polymers. *Macrom. Symp.* 291-292 (2010) 307-313.

[114] J. Shepherd, P. Sarker, S. Rimmer, L. Swanson, S. MacNeil, I. Douglas, Hyperbranched poly(NIPAM) polymers modified with antibiotics for the reduction of bacterial burden in infected human tissue engineered skin. *Biomaterials* 32 (2011) 258-267.

[115] S. Xu, Y. Luo, R. Graeser, A. Warnecke, F. Kratz, P. Hauff, K. Licha, R. Haag, Development of pH-responsive core-shell nanocarriers for delivery of therapeutic and diagnostic agents. *Bioorg. Med. Chem. Letters* 19 (2009) 1030-1034.

[116] A. Castonguay, A.K. Kakkar, Dendrimer templated construction of silver nanoparticles. *Adv. Coll. Interf. Sci.* 160 (2010) 76-87.

[117] N. Karak, R. Konwarh, B. Voit, Catalytically active vegetable-oil-based thermoplastic hyperbranched polyurethane/ silver nanocomposites. *Macromol. Mat. Eng.* 295 (2010) 159-169.

[118] S.S. Mahapatra, N. Karak, Silver nanoparticle in hyperbranched polyamine: Synthesis, characterization and antibacterial activity. *Mat. Chem. Phys.* 112 (2008) 1114-1119.

[119] C. Aymonier, U. Schlotterbeck, L. Antonietti, P. Zacharias, R. Thomann, J.C. Tiller, S. Mecking, Hybrids of silver nanoparticles with amphiphilic hyperbranched macromolecules exhibiting antimicrobial properties. *Chem. Comm.* 8 (2002) 3018-3019.

[120] Y. Shen, M. Kuang, Z. Shen, J. Nieberle, H. Duan, H. Frey, Gold nanoparticles coated with a thermosensitive hyperbranched polyelectrolyte: Towards smart temperature and pH nanosensors, *Angew. Chem. Int. Ed.* 47 (2008) 2227-2230.

[121] D. Wan, Q. Fu, J. Huang, Synthesis of amphiphilic hyperbranched polyglycerol polymers and their application as template for size control of gold nanoparticles, *J. Appl. Polym. Sci.* 101 (2006) 509-514.

[122] Y. Zhang, H. Peng, W. Huang, Y. Zhou, D. Yan, Facile preparation and characterization of highly antimicrobial colloid Ag or Au nanoparticles. *J. Coll. Interf. Sci.* 325 (2008) 371-376.

[123] J. Keilitz, M.R. Radowski, J.D. Marty, R. Haag, F. Gauffre, C. Mingotaud, Dendritic polymers with a core - Multishell architecture: A versatile tool for the stabilization of nanoparticles, *Chem. Mat.* 20 (2008) 2423-2425.

[124] J.D. Marty, E. Martinez-Aripe, A.F. Mingotaud, C. Mingotaud. Hyperbranched polyamidoamine as stabilizer for catalytically active nanoparticles in water. *J. Coll. Interf. Sci.* 326 (2008) 51-54.

[125] S.R. Puniredd, C.M. Yin, Y.S. Hooi, P.S. Lee, M.P. Srinivasan, Dendrimer-encapsulated Pt nanoparticles in supercritical medium: Synthesis, characterization, and application to device fabrication. *J. Coll. Interf. Sci.* 332 (2009) 505-510.

[126] X.Y. Liu, F. Cheng, Y. Liu, H.J. Liu, Y. Chen, Preparation and characterization of novel thermoresponsive gold nanoparticles and their responsive catalysis properties. *J. Mat. Chem.* 20 (2010) 360-368.

[127] Q. Tang, F. Cheng, X.L. Lou, H.J. Liu, Y. Chen, Comparative study of thiol-free amphiphilic hyperbranched and linear polymers for the stabilization of large gold nanoparticles in organic solvent. *J. Coll. Interf. Sci.* 337 (2009) 485-491.

[128] X. Liu, K. Matsumura, Y. Tomita, K. Yasui, K. Kojima, K. Chikama, Nonlinear optical responses of nanoparticle-polymer composites incorporating organic (hyperbranched polymer)-metallic nanoparticle complex. *J. Appl. Phys.* 108 (2010) 073102.

[129] H. Li, L. Zhang, J.K. Jo, C.S. Ha, Y.A. Shchipunov, I. Kim, Facile synthesis of titania/hyperbranched polyglycidol nanohybrids with controllable morphologies: from solid spheres, capsules to tubes. *J. Nanopart. Res.* 13 (2011) 1-12.

[130] S. Wei, Y. Zhu, Y. Zhang, J. Xu, Preparation and characterization of hyperbranched aromatic polyamides/Fe3O4 magnetic nanocomposite. *React. Funct. Polym.* 66 (2006) 1272-1277.

[131] X. Gao, X. Zhang, X. Zhang, C. Cheng, Z. Wang, C. Li, Encapsulation of BSA in polylactic acid-hyperbranched polyglycerol conjugate nanoparticles: Preparation, characterization, and release kinetics. *Polym. Bull.* 65 (2010) 787-805.

[132] M. Arkas, D. Tsiourvas, Organic/inorganic hybrid nanospheres based on hyperbranched poly(ethylene imine) encapsulated into silica for the sorption of toxic metal ions and polycyclic aromatic hydrocarbons from water. *J. Hazard. Mat.* 170 (2009) 35-42.

[133] E.G. Oosterom, J.N.H. Reek, P.C.J. Kamer, P.W.N.M. van Leeuwen, Transition metal catalysis using functionalized dendrimers. *Angew. Chem. Int. Ed.* 40 (2001) 1828-1849.

[134] L. Tuchbreiter, S. Mecking, Hydroformylation with dendritic-polymer-stabilized rhodium colloids as catalyst precursors. *Macrom. Chem. Phys.* 208 (2007) 1688-1693.

[135] G. Zhang, X. Fan, J. Kong, Y. Liu, Y. Xie, Hyperbranched organosilicon polymers. *Prog. Chem.* 20 (2008) 326-338.

[136] K. Tahara, H. Shimakoshi, A. Tanaka, Y. Hisaeda, Synthesis, characterization and catalytic function of a B 12-hyperbranched polymer. *Dalton Transactions* 39 (2010) 3035-3042.

[137] M. Gladitz, S. Reinemann, H.J. Radusch, Preparation of silver nanoparticle dispersions via a dendritic-Polymer template approach and their use for antibacterial surface treatment, *Macromol. Mat. Eng.* 294 (2009) 178-189.

[138] J. Fang, H. Kita, K.I. Okamoto, Gas permeation properties of hyperbranched polyimide membranes. *J. Membr. Sci.* 182 (2001) 245-256.

[139] J. Rolker, M. Seiler, L. Mokrushina, W. Arlt, Potential of Branched Polymers in the Filed of Gas Absorption: Experimental Gas Solubilities and Modeling. *Ind. Eng. Chem. Res.* 46 (2007) 6572-6583.

[140] H. Gao, D. Wang, W. Jiang, S. Guan, Z. Jiang, Gas permeability of fluorinated hyperbranched polyimide. *J. Appl. Polym. Sci.* 109 (2008) 2341-2346.

[141] Y. Yin, L. Yang, M. Yoshino, J. Fang, K. Tanaka, H. Kita, K.I. Okamoto, Synthesis and gas permeation properties of star-like poly(ethylene oxide)s using hyperbranched polyimide as central core. *Polymer J.* 36 (2004) 294-302.

[142] T. Suzuki, Y. Yamada, Physical and gas transport properties of novel hyperbranched polyimide - Silica hybrid membranes. *Polym. Bull.* 53 (2005) 139-146.

[143] T. Suzuki, Y. Yamada, K. Itahashi, 6FDA-TAPOB hyperbranched polyimide-silica hybrids for gas separation membranes. *J. Appl. Polym. Sci.* 109 (2008) 813-819.

[144] T. Suzuki, Y. Yamada, Y. Tsujita, Gas transport properties of 6FDA-TAPOB hyperbranched polyimide membrane. *Polymer* 45 (2004) 7167-7171.

[145] J. Fang, H. Kita, K.I. Okamoto, Gas Permeation Properties of hyperbranched polyimide membranes. *J. Membr. Sci.* 182 (2001) 245-56.

[146] D.M. Sterescu, D.F. Stamatialis, M. Wessling, Boltorn-modified polyimide gas separation membranes. *J. Membr. Sci.* 310 (2008) 512-521.

[147] D.M. Sterescu, D.F. Stamatialis, E. Mendes, J. Kruse, K. Rätzke, F. Faupel, M. Wessling, Boltom-modified poly(2,6-dimethyl-1,4-phenylene oxide) gas separation membranes, *Macromolecules* 40 (2007) 5400-5410.

[148] L. Boogh, B. Pettersson, J.A.E. Manson, Dendritic hyperbranched polymers as tougheners for epoxy resins, *Polymer* 40 (1999) 2249-2261.

[149] P. Ghosh, W.M. Lackowski, R.M. Crooks, Two new approaches for patterning polymer films using templates prepared by microcontact printing. *Macromolecules* 34 (2001) 1230-1236.

[150] Y. Hong, S.J. Coombs, J.J. Cooper-White, M.E. MacKay, C.J. Hawker, E. Malmström, N. Rehnberg, Film blowing of linear low-density polyethylene blended with a novel hyperbranched polymer processing aid. *Polymer* 41 (2000) 7705-7713.

[151] Y. Hong, J.J. Cooper-White, M.E. Mackay, C.J. Hawker, E. Malmström, N. Rehnberg, A novel processing aid for polymer extrusion: Rheology and processing of polyethylene and hyperbranched polymer blends. *J. Rheol.* 43 (1999) 781-793.

[152] S.M. Burkinshaw, P.E. Fröhling, M. Mignanelli, The effect of hyperbranched polymers on the dyeing of polypropylene fibres. *Dyes and Pigments* 53 (2002) 229-235.

[153] P.E. Fröhling, Dendrimers and dyes - A review. *Dyes and Pigments* 48 (2001) 187-195.

[154] R. Allabashi, M. Arkas, G. Hörmann, D. Tsiourvas, Removal of some organic pollutants in water employing ceramic membranes impregnated with cross-linked silylated dendritic and cyclodextrin polymers. *Water Res.* 41 (2007) 476-486.

[155] M. Arkas, D. Tsiourvas, C.M. Paleos, Functional dendritic polymers for the development of hybrid materials for water purification. *Macromol. Mat. Eng.* 295 (2010) 883-898.

[156] T. Itoh, M. Ikeda, N. Hirata, Y. Moriya, M. Kubo, O. Yamamoto, Ionic conductivity of the hyperbranched polymer-lithium metal salt systems. *J. Power Sources* 81-82 (1999) 824-829.

[157] E. Barriau, H. Frey, A. Kiry, M. Stamm, F. Gröhn, Negatively charged hyperbranched polyether-based polyelectrolytes. *Coll. Polym. Sci.* 284 (2006) 1293-1301.
[158] T. Itoh, N. Hirata, Z. Wen, M. Kubo, O. Yamamoto, Polymer electrolytes based on hyperbranched polymers. *J. Power Sources* 97-98 (2001) 637-640.
[159] Z. Wang, M. Ikeda, N. Hirata, M. Kubo, T. Itoh, O. Yamamoto, Thermal, electrochemical, and spectroscopic characterizations of hyperbranched polymer electrolyte. *J. Electrochem. Soc.* 146 (1999) 2209-2215.
[160] T. Itoh, S. Horii, S. Hashimoto, T. Uno, M. Kubo, Polymer electrolytes plasticized with hyperbranched polymer for lithium polymer batteries. *Ionics* 10 (2004) 450-457.
[161] L. Rakesh, A. Müller, P. Chhetri, Development of a hyperbranched fuel cell membrane material for improved proton conductivity. *Fluid Dynam. Mat. Proc.* 6 (2010) 179-202.
[162] T. Itoh, K. Hirai, M. Tamura, T. Uno, M. Kubo, Y. Aihara, Synthesis and characteristics of hyperbranched polymer with phosphonic acid groups for high-temperature fuel cells. *J. Solid State Electrochem.* 14 (2010) 2179-2189.
[163] T. Itoh, T. Sakakibara, Y. Takagi, M. Tamura, T. No, M. Kubo, Y. Aihara, Proton-conducting electrolyte membranes based on hyperbranched polymer with a sulfonic acid group for high-temperature fuel cells. *Electrochim. Acta* 55 (2010) 1419-1424.
[164] D.M. Tigelaar, M.A.B. Meador, W.R. Bennett, Composite electrolytes for lithium batteries: Ionic liquids in APTES cross-linked polymers. *Macromolecules* 40 (2007) 4159-4164.
[165] F.M. Wang, H.C. Wu, C.S. Cheng, C.L. Huang, C.R. Yang, High ionic transfer of a hyperbranched-network gel copolymer electrolyte for potential electric vehicle (EV) application. *Electrochim. Acta* 54 (2009) 3788-3793.
[166] F. Wu, T. Feng, C. Wu, Y. Bai, L. Ye, J. Chen, Thermally stable hyperbranched polyether-based polymer electrolyte for lithium-ion batteries. *J. Phys. D: Appl. Phys.* 43 (2010) 035501.
[167] I. Gur, N.A. Fromer, C.P. Chen, A.G. Kanaras, A.P. Alivisatos, Hybrid solar cells with prescribed nanoscale morphologies based on hyperbranched semiconductor nanocrystals. *Nano Letters* 7 (2007) 409-414.
[168] J. Qiao, C. Yang, Q. He, F. Bai, Y. Li, Hyperbranched conjugated polymers for photovoltaic applications. *J. Appl. Polym. Sci.* 92 (2004) 1459-1466.
[169] Y. Tang, P. Shen, T. Ding, H. Huang, B. Zhao, S. Tan, Hyperbranched conjugated polymers with donor-π-acceptor architecture as organic sensitizers for dye-sensitized solar cells. *Eur. Polym. J.* 46 (2010) 2033-2041.
[170] Q. Cao, P. Liu, Hyperbranched polyurethane as novel solid–solid phase change material for thermal energy storage. *Eur. Polym. J.* 42 (2006) 2931-2939.
[171] Q. Cao, L. Liao, H. Xu, Study on the influence of thermal characteristics of hyperbranched polyurethane phase change materials for energy storage. *J. Appl. Polym. Sci.* 115 (2010) 2228-2235.
[172] L. Liao, Q. Cao, H. Liao, Investigation of a hyperbranched polyurethane as a solid-state phase change material. *J. Mat. Sci.* 45 (2010) 2436-2441.
[173] Y.H. Kim, Hyperbranched Polymers 10 Years After. *J. Polym. Sci. Part A: Polym. Chem.* 36 (1998) 1685-1698.
[174] L.A. Kleintjens, H.M. Schoffeleers, R. Koningsveld, Entmischung in hochmolekularen Vielstoffsystemen, XVII. Eine Wechselwirkungsfunktion für Lösungen verzweigter Polymerer. Ber. Bunsenges. *Phys. Chem.* 81 (1977) 980-985.

[175] T.W. de Loos, W. Poot, R.N. Lichtenthaler, The Influence of Branching on High-Pressure Vapor-Liquid Equilibria in Systems of Ethylene and Polyethylene. *J. Supercrit. Fluids* 5 (1992) 282-286.

[176] A. Striolo, J.M. Prausnitz, Osmotic second virial coefficient for linear and star poly(ethylene oxide). *Polymer* 42 (2001) 4773-4775.

[177] J. W. Kennedy, M. Gordon and R. Koningsveld, Generalization of the Flory-Huggins treatment of polymer solutions. *J. Polym. Sci., Part C: Polym. Symp.* 39 (1972) 43–69.

[178] L.A. Kleintjens, R. Koningsveld, Liquid-Liquid Phase Separation in Multicomponent Polymer Systems. 18. Effect of Short-Chain Branching. *Macromolecules* 13 (1980) 303-310.

[179] S. Chen, M. Banaszak, M. Radosz, Phase Behavior of Poly(ethylene -1 butene) in Subcritical and Supercritical Propane: Ethyl Branches Reduce Segment Energy and Enhance Miscibility. *Macromolecules* 28 (1995) 1812-1817.

[180] F.J. Blas, L.F. Vega, Thermodynamic properties and phase equilibria of branched chain fluids using first- and second-order Wertheim's thermodynamic perturbation theory. *J. Chem. Phys.* 115 (2001) 3906-3915.

[181] M.K. Kozlowska, J.F. Bas, C.S. Schacht, J. Gross, T.W. de Loos, Phase Behavior of Hyperbranched Polymer Systems: Experiments and Application of the Perturbed-Chain Polar SAFT Equation of State. *J. Phys. Chem. B* 113 (2009) 1022-1029.

[182] U. Domańska, K. Paduszyński, Z. Ołek-Tryznowska, (Liquid + liquid) equilibria of binary systems containing hyperbranched polymer Boltorn® H2004 - Experimental study and modelling in terms of lattice-cluster theory. *J. Chem. Thermodynamics* 43 (2011) 167-171.

[183] U. Domańska, K. Paduszyński, Z. ZoŁek-Tryznowska, Liquid-liquid phase equilibria of binary systems containing hyperbranched polymer B-U3000: Experimental study and modeling in terms of lattice cluster theory. *J. Chem. Eng. Data* 55 (2010) 3842-3846.

[184] M. Seiler, J. Rolker, W. Arlt, Phase Behavior and Thermodynamic Phenomena of Hyperbranched Polymer Solutions. *Macromolecules* 36 (2003) 2085-2092.

[185] V. Munoz Portela, E.J.M. Straver, T.W. de Loos, High-Pressure Phase Behavior of the System Propane - Boltorn H320. *J. Chem. Eng. Data* 54 (2009) 2593-2598.

[186] M. Seiler, W. Arlt, H. Kautz, H. Frey, Experimental data and theoretical considerations on vapor-liquid and liquid-liquid equilibria of hyperbranched polyglycerol and PVA solutions. *Fluid Phase Equilibria* 201 (2002) 359-379.

[187] J.G. Jang, Y.C. Bae, Phase behaviors of hyperbranched polymer solutions. *Polymer 40* (1999) 6761-6768.

[188] T. Zeiner, P. Schrader, S. Enders, D. Browarzik, Phase- and Interfacial Behavior of Hyperbranched Polymer Solutions. *Fluid Phase Equilibria* 2010 doi:10.1016/j.fluid.2010.07.021

[189] U. Domańska, Z. Zołek-Tryznowska, M.M. Tshibangu, D. Ramjugernath, T.M. Letcher, Separation of an alcohol and a tetrahydrofuran, methyl tert-butyl ether, or ethyl tert-butyl ether by solvent extraction with a hyperbranched polymer at T = 298.15 K. J. *Chem. Eng. Data* 55 (2010) 2879-2885.

[190] U. Domanska, Z. Zolek-Tryznowska, A. Pobudkowskas, Separation of Hexane/Ethanol Mixtures. LLE of Ternary Systems (Ionic Liquid or Hyperbranched Polymer + Ethanol + Hexane) at T=298.15. *J. Chem. Eng. Data* 54 (2009) 972-976.

[191] C.S. Schacht, S. Bahramali, D. Wilms, H. Frey, J. Gross, T.W. de Loos, Phase Behavior of the system hyperbranched polyglycerol + methanol + carbon dioxide. *Fluid Phase Equilibria* 299 (2010) 252-258.

[192] J.G. Lieu, M. Liu, J.M.J. Frechet, J.M. Prausnitz, Vapor-Liquid Equilibria for Dendritic-Polymer Solutions. *J. Chem. Eng. Data* 44 (1999) 613-620.

[193] C. Mio, S. Kiritsov, Y. Thio, R. Brafman, J. Prausnitz, C. Hawker, E.E. Malmstrom, Vapor-Liquid Equilibria for Solutions of Dendritic Polymers. *J. Chem. Eng. Data* 43 (1998) 541-550.

[194] U. Domanska, Z. Zolek-Tryznowska, Solubility of hyperbranched polymer, Boltorn W-3000, in alcohols, ethers and hydrocarbons. *J. Chem. Thermodynamics* 42 (2010) 1304-1309.

[195] S.J. Ko, Y.C. Bae, S.H. Kong, S.O. Ryu, Phase Behaviors of Hyperbranched Solid Polymer Electrolyte/Salt Systems: The Structure Effect. *J. Polym. Sci., Part B: Polym. Phys.* 42 (2004) 2254-2262.

[196] A. Polese, C. Mio, A. Bertucco, Infinite-dilution activity coefficients of polar and nonpolar solvents in solutions of hyperbranched polymers. *J. Chem. Eng. Data* 44 (1999) 839-845.

[197] G.S. Dritsas, K. Karatasos, C. Panayiotou, Investigation of Thermodynamic Properties of Hyperbranched Poly(ester amide) by Inverse Gas Chromatography. *J. Polym. Sci., Part B: Polym. Phys.* 46 (2008) 2166-2172.

[198] G.S. Dritsas, K. Karatasos, C. Panayiotou, Investigation of thermodynamic properties of hyperbranched aliphatic polyestern by inverse gas chromatography. *J. Chromatography* A 1216 (2009) 8979-8985.

[199] U. Domańska, Z. Zołek-Tryznowska, Thermodynamic properties of hyperbranched polymer, Boltorn U3000, using inverse gas chromatography. *J. Phys. Chem.* B 113 (2009) 15312-15321.

[200] U. Domanska, Z. Zolek-Tryznowska, Measurements of mass-fraction activity coefficient at infinite dilution of aliphatic and aromatic hydrocarbon, thiophene, alcohols, water, ethers, and ketones in hyperbranched polymer, Boltorn H2004, using inverse gas chromatography. *J. Chem. Thermodynamics* 42 (2010) 363-370.

[201] U. Domańska, Z. Zołek-Tryznowska, Mass-fraction activity coefficients at infinite dilution measurements for organic solutes and water in the hyperbranched polymer Boltorn W3000 using inverse gas chromatography, *Journal J. Chem. Eng. Data* 55 (2010) 1258-1265.

[202] M. Xu, X. Yan, R. Cheng, X. Yu, Investigation into the solution properties of hyperbranched polymer. *Polymer International* 50 (2001) 1338-1345.

[203] U. Domańska, Z. Zołek-Tryznowska, Measurement of the density and viscosity of binary mixtures of (hyper-branched polymer B-H2004 + 1-butanol, or 1-hexanol, or octanol, or methyl tert-butyl ether). *J. Chem. Thermodynamics* 42 (2010) 651-658.

[204] U. Domańska, Z. Zołek-Tryznowska, Temperature and composition dependence of the density and viscosity of binary mixtures of (hyperbranched polymer, B-U3000 + 1-alcohol, or ether), *J. Chem. Thermodynamics* 41 (2009) 821-828.

[205] I.B. Rietveld, D. Bedeaux, J.A.M. Smit, Osmotic compressibility of poly(propylene imine) dendrimers in deuterated methanol. *J. Coll. Interf. Sci.* 232 (2000) 317-325.

[206] H. Cheng, S. Xie, Y. Zhou, W. Huang, D. Yan, J. Yang, B. Ji, Effect of degree of branching on the thermoresponsive phase transition behaviors of hyperbranched

multiarm copolymers: Comparison of systems with LCST transition based on coil-to-globule transition or hydrophilic-hydrophobic Balance. *J. Phys. Chem.* B 114 (2010) 6291-6299.

[207] X. Liu, F. Cheng, H. Liu, Y. Chen, Unusual salt effect on the lower critical solution temperature of hyperbranched thermoresponsive polymers. *Soft Matter* 4 (2008) 1991-1994.

[208] T.D. Martter, M.D. Foster, T. Yoo, S. Xu, G. Lizzaraga, R.P. Quirk, P.D. Butler, Nonuniversal Behavior of the Thermodynamic Interaction Parameter in Blends of Star and Linear Polybutadiene. *Macromolecules* 35 (2002) 9763-9772.

[209] D.J. Massa, K.A. Shriner, S.R. Turner, B.I. Voit, Novel Blends of Hyperbranched Polyesters and Linear Polymers. *Macromolecules* 28 (1995) 3214-3220.

[210] A.M. Nemirovski, M.G. Bawendi, K.F. Freed, Lattice models of polymer solutions: monomers occupying several lattice sites. *J. Chem. Phys.* 87 (1987) 7272-7284.

[211] K.F. Freed, M.G. Bawendi, Lattice theories of polymeric fluids. *J. Phys. Chem.* 93 (1989) 2194-2203.

[212] J. Dudowicz, K.F. Freed, Effect of Monomer Structure and Compressibility on the Properties of Multicomponent Polymer Blends and Solutions: 1. Lattice Cluster Theory of Compressible Systems. *Macromolecules* 24 (1991) 5076-5095.

[213] J. Dudowicz, M.S. Freed, K.F. Freed, Effect of Monomer Structure and Compressibility on the Properties of Multicomponent Polymer Blends and Solutions: 2. Application to Binary Blends. *Macromolecules* 24 (1991) 5096-5111.

[214] J. Dudowicz, K.F. Freed, Effect of Monomer Structure and Compressibility on the Properties of Multicomponent Polymer Blends and Solutions: 3. Application to PS(D)/PVME Blends. *Macromolecules* 24 (1991) 5112-5123.

[215] K.F. Freed, J. Dudowicz, Influence of monomer molecular structure on the miscibility of polymer blends. *Adv. Poly. Sci.* 183 (2005) 63-126.

[216] K.W. Foreman, K.F. Freed, Generalizations of Huggins-Guggenheim-Miller-type theories to describe the architecture of branched lattice chains. *J. Chem. Phys.* 102 (1995) 4663-4672.

[217] K.W. Foreman, K.F. Freed, I.M. Ngola, Nonrandom mixing in polymer blends: Implications for phase behavior, *J. Chem. Phys.* 107 (1997) 4688-4704.

[218] J. Dudowicz, K.F. Freed, W.G. Madden, Role of molecular structure on the thermodynamic properties of melts, blends, and concentrated polymer solutions: comparison of Monte Carlo simulations with the cluster theory for the lattice model. *Macromolecules* 23 (1990) 4803-4819.

[219] D. Buta, K.F. Freed, I. Szleifer, Thermodynamic properties of lattice polymers: Monte Carlo simulations and mean-field theories. *J. Chem. Phys.* 112 (2000) 6040-6048.

[220] D. Buta, K.F. Freed, I. Szleifer, Monte Carlo test of the lattice cluster theory: Thermodynamic properties of binary polymer blends. *J. Chem. Phys.* 114 (2001) 1424-1431.

[221] D. Buta, K.F. Freed, Mixtures of lattice polymers with structured monomers. *J. Chem. Phys.* 120 (2004) 6288-6298.

[222] W.G. Madden, A.I. Pesci, K.F. Freed, Phase equilibria of lattice polymer and solvent: Tests of theories against simulations. *Macromolecules* 23 (1990) 1181-1191.

[223] T. Zeiner, *Phase- and Interfacial Behaviour of Hyperbranched Polymer Solutions*. Ph-D thesis, Technical University Berlin, 2010.

[224] T. Zeiner, D. Browarzik, S. Enders, Calculation of the liquid-liquid equilibrium of aqueous solutions of hyperbranched polymers. *Fluid Phase Equilibria* 286 (2009) 127-133.

[225] T. Zeiner, S. Enders, Flüssig-Flüssig Gleichgewichte von hyperverzweigten Polymeren in Lösungsmittelmischungen. CIT 82 (2010) 1378.

[226] T. Zeiner, S. Enders, Phase-Behavior of Hyperbranched Polymer Solutions in Mixed Solvents. *Chem. Eng. Sci.* submitted September 2010.

[227] T. Zeiner, C. Browarzik, D. Browarzik, S. Enders, Calculation of the liquid-liquid equilibrium of solutions of hyperbranched polymers with the lattice-cluster theory combined with an association model. *J. Chem. Thermodynamics* submitted 2011.

[228] J. Dudowicz, M. Lifschitz, K.F. Freed, J.F. Douglas, How far is far from critical point in polymer blends? Lattice cluster theory computations for structured monomer, compressible systems. *J. Chem. Phys.* 99 (1993) 4804-4820.

[229] J. Dudowicz, K.F. Freed, Pressure dependence of polymer fluids: Application of the lattice cluster theory. *Macromolecules* 28 (1995) 6625-6641.

[230] W. Li, K.F. Freed, Lattice cluster theory for phase behavior of rectangular mesogens. II. Nearest-neighbor interactions, phase diagrams, and competitive nematic orderings. *J. Chem. Phys.* 103 (1995) 5693-5711.

[231] J. Dudowicz, K.F. Freed, J.F. Douglas, Lattice model of equilibrium polymerization. IV. Influence of activation, chemical initiation, chain scission and fusion, and chain stiffness on polymerization and phase separation. *J. Chem. Phys.* 119 (2003) 12645-12666.

[232] K.W. Foreman, K.F. Freed, Influence of stiffness, monomer structure, and energetic asymmetries on polymer blend miscibilities: Applications to polyolefins. *Macromolecules* 30 (1997) 7279-7295.

[233] K.W. Foreman, K.F. Freed, Microscopic parameters influencing the phase separation in compressible binary blends of linear semiflexible polymers. *J. Chem. Phys.* 106 (1997) 7422-7437.

[234] K.F. Freed, J. Dudowicz, Influence of short chain branching on the miscibility of binary polymer blends: Application to polyolefin mixtures. *Macromolecules* 29 (1996) 625-636.

[235] J. Dudowicz, K.F. Freed, J.F. Douglas, Entropy theory of polymer glass formation revisited. I. General formulation. *J. Chem. Phys.* 124 (2006) 064901.

[236] E.B. Stukalin, J.F. Douglas, K.F. Freed, Application of the entropy theory of glass formation to poly(α -olefins). *J. Chem. Phys.* 131 (2009) 114905.

[237] E.B. Stukalin, J.F. Douglas, K.F. Freed, Plasticization and antiplasticization of polymer melts diluted by low molar mass species. *J. Chem. Phys.* 132 (2010) 084504.

[238] J. Dudowicz, K.F. Freed, J.F. Douglas, Generalized entropy theory of polymer glass formation. *Adv. Chem. Phys.* 137 (2008) 125-222.

[239] J. Dudowicz, K.F. Freed, J.F. Douglas, Direct computation of characteristic temperatures and relaxation times for glass-forming polymer liquids. *J. Chem. Phys.* 123 (2005) 1-4.

[240] J. Dudowicz, K.F. Freed, Molecular influences on miscibility patterns in random copolymer/homopolymer binary blends. *Macromolecules* 31 (1998) 5094-5104.

[241] J. Dudowicz, K.F. Freed, Lattice cluster theory for copolymer blends: General theory in the incompressible system, long chain limit. *Polish Journal of Chemistry* 75 (2001) 527-545.

[242] J. Dudowicz, K.F. Freed, Lattice cluster theory for pedestrian. 2. Random copolymer systems. *Macromolecules* 33 (2000) 3467-3477.

[243] K.F. Freed, J. Dudowicz, Lattice cluster theory for pedestrians: Models for random copolymer blends. *Macrom. Symp.* 149 (2000) 11-16.

[244] J. Dudowicz, K.F. Freed, Explanation for the unusual phase behavior of polystyrene-b-poly(n-alkyl methacrylate) diblock copolymers: specific interactions. *Macromolecules* 33 (2000) 5292-5299.

[245] J. Dudowicz, K.F. Freed, Energetically driven asymmetries in random copolymer miscibilities and their pressure dependence, *Macromolecules* 30 (1997) 5506-5519.

[246] J. Dudowicz, K.F. Freed, Molecular origin of the free energy dependence on the monomer sequence in random copolymer systems. *Macromolecules* 29 (1996) 7826-7827.

[247] K.F. Freed. Extension of lattice cluster theory to strongly interacting, self-assembling polymeric systems. *J. Chem. Phys*. 130 (2009) 061103.

[248] J. Dudowicz, J.F. Douglas, K.F. Freed, Equilibrium polymerization models of re-entrant self-assembly. *J. Chem. Phys*. 130 (2009) 164905.

[249] M.S. Wertheim, Fluids with highly directional attractive forces. I. Statistical thermodynamics. *J. Stat. Phys*. 35 (1984) 19-34.

[250] M.S. Wertheim, Fluids with highly directional attractive forces. II. Thermodynamic perturbation theory and integral equations. *J. Stat. Phys*. 35 (1984) 35-47.

[251] M.S. Wertheim, Fluids with highly directional attractive forces. III. Multiple attraction sites. *J. Stat. Phys*. 42 (1986) 459-476.

[252] M. S. Wertheim, Fluids with highly directional attractive forces. IV. Equilibrium polymerization. *J. Stat. Phys*. 42 (1986) 477-492.

[253] T. Morita, K. Hiroike, A New Approach to the Theory of Classical Fluids. III. General Treatment of Classical Systems. *Progr. Theor. Phys*. 25 (1961) 537-578.

[254] G. Stell, *The Equilibrium Theory of Classical Fluids*; Benjamin; New York (1964).

[255] E. Nies, T. Li, H. Berghmans, R.K. Heenan, S.M. King, Upper Critical Solution Temperature Phase Behavior, Composition Fluctuations, and Complex Formation in Poly(Vinyl Methyl Ether)/DO Solutions: Small-Angle Neutron-Scattering Experiments and Wertheim Lattice Thermodynamics Perturbation Theory Predictions. *J. Phys. Chem.* B 110 (2006) 5321-5329.

[256] E. Nies, A. Ramzi, H. Berghmans, T. Li, R.K. Heenan, S. M. King, Composition Fluctuations, Phase Behavior, and Complex Formation in Poly(vinyl methyl ether)/DO Investigated by Small-Angle Neutron Scattering. *Macromolecules* 38 (2005) 915-924.

[257] J.G. Jang, Y. C. Bae, Phase behavior of hyperbranched polymer solutions with specific interactions. *J. Chem. Phys*. 114 (2001) 5034-5042.

[258] P.F. Arce, M. Aznar, Modeling of thermodynamic behavior of PVT properties and cloud point temperatures of polymer blends and polymer blend + carbon dioxide systems using non-cubic equations of state. *Fluid Phase Equilibria* 286 (2009) 17-27.

[259] C.S. Gudipati, J.A. Finlay, J.A. Callow, M.E. Callow, K.L. Wooley, The antifouling and fouling-release perfomance of hyperbranched fluoropolymer (HBFP)-poly(ethylene

glycol) (PEG) composite coatings evaluated by adsorption of biomacromolecules and the green fouling alga ulva. *Langmuir* 21 (2005) 3044-3053.
[260] M.A. Elrehim, B. Voit, B. Bruchmann, K.J. Eichhorn, K. Grundke, C. Bellmann, Structural and end-group effects on bulk and surface properties of hyperbranched poly(urea urethane)s. *J. Polym. Sci., Part A: Polym. Chem.* 43 (2005) 3376-3393.
[261] S. Peleshanko, R. Gunawidjaja, S. Petrash, V.V. Tsukruk, Synthesis and interfacial behavior of amphiphilic hyperbranched polymers: Poly(ethylene oxide)-polystyrene hyperbranches. *Macromolecules* 39 (2006) 4756-4766.
[262] M.E. Mackay, G. Carmezini, B.B. Sauer, W. Kampert, On the surface properties of hyperbranched polymers. *Langmuir* 17 (2001) 1708-1712.
[263] E. Žagar, M. Žigon, Aliphatic hyperbranched polyesters based on 2,2-bis(methylol) propionic acid - Determination of structure, solution and bulk properties. *Prog. Polym. Sci.* (Oxford) 36 (2011) 53-88.
[264] A. Sidorenko, X.W. Zhai, S. Peleshanko, A. Greco, V.V. Shevchenko, V.V. Tsukruk, Hyperbranched polyesters on solid surfaces. *Langmuir* 17 (2001) 5924-5931.
[265] Y. Mikhaylova, V. Dutschk, M. Müller, K. Grundke, K.J. Eichhorn, B. Voit, Study of the solid-liquid interface of hydroxyl-terminated hyperbranched aromatic polyesters (HBP-OH) in aqueous media. II. Adsorption of model proteins *Coll. Surf. A: Physicochem. Eng. Aspects* 297 (2007) 19-29.
[266] Y. Mikhaylova, V. Dutschk, C. Bellmann, K. Grundke, K.J. Eichhorn, B. Voit, Study of the solid-liquid interface of hydroxyl-terminated hyperbranched aromatic polyesters (HBP-OH) in aqueous media. I. Characterisation of the properties and swelling behaviour of HBP-OH thin films. *Coll. Surf. A: Physicochem. Eng. Aspects* 279 (2006) 20-27.
[267] Y. Mikhaylova, E. Pigorsch, K. Grundke, K.J. Eichhorn, B. Voit, Surface properties and swelling behaviour of hyperbranched polyester films in aqueous media. *Macrom. Symp.* 210 (2004) 271-280.
[268] G. Jannerfeldt, L. Boogh, J.A.E. Manson, Tailored interfacial properties for immiscible polymers by hyperbranched polymers. *Polymer* 41 (2000) 7627-7630.
[269] A.M. Díez-Pascual, A. Compostizo, A. Crespo-Colín, R.G. Rubio, R. Miller, Adsorption of water-soluble polymers with surfactant character. Adsorption kinetics and equilibrium properties, *J. Coll. Interf. Sci.* 307 (2007) 398-404.
[270] U. Domańska, Z. Zolek-Tryznowska, Effect of temperature and composition on the surface tension and thermodynamic properties of binary mixtures of Boltorn U3000 with alcohols and ether. *J. Sol. Chem.* 39 (2010) 864-876.
[271] A. Ziemer, M. Azizi, D. Pleul, F. Simon, S. Michel, M. Kreitschmann, P. Kierkus, B. Voit, K. Grundke. Influence of hyperbranched polyesters on the surface tension of polyols. *Langmuir* 20 (2004) 8096-8102.
[272] S. Peleshanko, V.V. Tsukruk, The architectures and surface behavior of highly branched molecules. *Prog. Polym. Sci.* (Oxford) 33 (2008) 523-580.
[273] I.C. Sanchez, *Physics of Polymer Surface and Interface, Butterworth-Heinemann*: Boston, London, Oxford (1992).
[274] Z. Qian, V.S. Minnikanti, L.A. Archer, Surface segregation of highly branched polymer additives in linear hosts. *J. Polym. Sci., Part B: Polym. Phys.* 46 (2008) 1788-1801.
[275] J. D. van der Waals, Over de Continuiteit van den Gasen Vloeistoftoestand. University of Leiden, 1873 (thesis).

[276] W.S. Lord Rayleigh, On the Theory of Surface Forces. *Phil. Mag.* 30 (1890) 285-298.

[277] J.W. Cahn, J.E. Hilliard, Free Energy of a Nonuniform System. I. Interfacial Free Energy. *J. Chem. Phys.* 28 (1958) 258-267.

[278] S. Enders, K. Quitzsch, Calculation of Interfacial Properties of Demixed Fluids Using Density Gradient Theory. *Langmuir* 14 (1998) 4606-4614.

[279] H. Kahl, S. Enders, Calculation of Surface Properties of Pure Fluids Using Density Gradient Theory and SAFT-EOS. *Fluid Phase Equilibria* 172 (2000) 27-42.

[280] D. Dittmar, R. Eggers, H. Kahl, S. Enders, Measurement and Modelling of the Interfacial Tensions of Triglyceride Mixtures in Contact with Dense Gases. *Chem. Eng. Sci.* 57 (2002) 355-363.

[281] S. Enders, H. Kahl, Interfacial properties of binary mixtures. *Phys. Chem. Chem. Phys.* 4 (2002) 931-936.

[282] S. Enders, H. Kahl, J. Winkelmann, Interfacial Properties of Polystyrene in Contact with Carbon Dioxide. *Fluid Phase Equilibria* 228-229 (2005) 511-522.

[283] S. Enders, H. Kahl, Interfacial Properties of Water + Alcohol Mixtures. *Fluid Phase Equilibria* 263 (2008) 160-167.

[284] O.G. Nino-Amezquita, S. Enders, Ph. T. Jaeger, R. Eggers, Measurement and Prediction of Interfacial Tension of Binary Mixtures. *Ind. Eng. Chem. Res.* 49 (2010) 592-601.

[285] G. Nino-Amezquita, S. Enders, Prediction of Interfacial Tension of Binary Mixtures. *Computer-Aided Chem. Eng.* 28 (2010) 85-90.

[286] O.G. Niño Amézquita, S. Enders, P.T. Jaeger, R. Eggers, Interfacial properties of mixtures containing supercritical gases, *J. Supercrit. Fluids* 55 (2010) 724-34.

[287] Z.G. Wang, A.M. Nemirovsky, K.F. Freed, Interaction of a polymer chain with an asymmetric liquid-liquid interface, *J. Chem. Phys.* 85 (1986) 3068-3077.

[288] W.E. McMullen, K.F. Freed, A density functional theory of polymer phase transitions and interfaces, *J. Chem. Phys.* 92 (1990) 1413-1426.

[289] W.E. McMullen, K.F. Freed, A density functional theory of polymer phase transitions and interfaces. II. Block copolymers. *J. Chem. Phys.* 93 (1990) 9130-9144.

[290] M. Lifschitz, K.F. Freed, H. Tang, Hyperbolic tangent variational approximation for interfacial profiles of binary polymer blends. *J. Chem. Phys.* 103 (1995) 3767-3781.

[291] H. Tang, K.F. Freed, Interfacial studies of incompressible binary blends. *J. Chem. Phys.* 94 (1991) 6307-6322.

[292] X.C. Zeng, D.W. Oxtoby, H. Tang, K.F. Freed, On square-gradient theories of polymer blend interfaces. *J. Chem. Phys.* 96 (1992) 4816-4817.

[293] K.R. Myers, A.M. Nemirovsky, K.F. Freed, Surface tension of dilute polymer solutions. I. A renormalization group approach. *J. Chem. Phys.* 97 (1992) 2790-2803.

[294] M. Lifschitz, K.F. Freed, Interfacial behavior of compressible polymer blends. *J. Chem. Phys.* 98 (1993) 8994-9013.

[295] K.F. Freed, Interrelation between density functional and self-consistent-field formulations for inhomogeneous polymer systems. *J. Chem. Phys.* 103 (1995) 3230-3239.

[296] K.F. Freed, Analytic theory of surface segregation in compressible polymer blends. *J. Chem. Phys.* 105 (1996) 10572-10582.

[297] M. Lifschitz, K.F. Freed, Interfacial behavior of phase separated asymmetric compressible binary polymer blends. *J. Chem. Phys.* 105 (1996) 1633-1645.

[298] J. Dudowicz, J.F. Douglas, K.F. Freed, Competition between self-assembly and surface adsorption, *J. Chem. Phys*. 130 (2009) 084903.

[299] Y. Sakane, K. Inomata, H. Morita, T. Kawakatsu, M. Doi, T. Nose, Effects of low-molecular-weight additives on interfacial tension of polymer blends: experiments for poly(dimethylsiloxane)/poly(tetramethyldisiloxanylethylene) 1 oligo(dimethylsiloxane), and comparison with mean-field calculations, *Polymer* 42 (2001) 3883-3891.

In: Advances in Polymer Research
Editors: Losif Sidorov, Fedor Nikitin, pp. 259-285
ISBN: 978-1-61942-223-0

PHASE SEPARATION, PHASE DISSOLUTION AND CRYSTALLIZATION IN POLY (ε-CAPROLACTONE)/ POLY (STYRENE-CO-ACRYLONITRILE) BLENDS

Petr Svoboda*
Centre of Polymer Systems, Faculty of Technology,
Tomas Bata University in Zlin, Nam. T.G. Masaryka 5555,
760 01 Zlin, Czech Republic.

ABSTRACT

A blend of poly(ε-caprolactone) (PCL) and poly(styrene-co-acrylonitrile) (SAN) containing 27.5 wt% of acrylonitrile having the critical composition (80/20 PCL/SAN) was studied. This PCL/SAN blend having a lower critical solution temperature (LCST) phase boundary at 122°C offered an excellent opportunity to investigate, firstly the kinetics of phase separation above LCST (125-180°C), and secondly the kinetics of phase dissolution below LCST (50-115°C). The blend underwent a temperature-jump above LCST where spinodal decomposition (SD) proceeded, yielding a regularly phase-separated structure (SD structure). Then, it was quenched to the temperatures below LCST when the phase dissolution proceeded. Optical microscopy was used to observe the spinodal decomposition qualitatively while light scattering was used to characterize the phase separation and phase dissolution quantitatively. It was found that during phase dissolution the peak maximum moved towards a smaller angle (wavelength of concentration fluctuations increases) while the peak intensity decreased. This behavior was explained by a model. Also it was found that the fastest phase dissolution kinetics at 80°C, which was characterized by an apparent diffusion coefficient, was about 10 times slower than the kinetics of phase separation at 180°C. Crystallization after various levels of spinodal decomposition was observed by optical microscopy. Order parameter of the lamellae inside the spherulites was evaluated with the help of H_v light scattering. Transmission electron microscopy revealed interesting lamellar structure after spinodal decomposition.

Keywords: Spinodal Decomposition; Phase Dissolution; Crystallization; Polymer Blends; PCL; SAN.

---

* Tel. 420–576 031 335, fax. 420–577 210 172. E-mail address: svoboda@ft.utb.cz

INTRODUCTION

Poly(ε-caprolactone) (PCL) is a linear semicrystalline aliphatic polyester that became available commercially following efforts at Union Carbide to identify synthetic polymers that could be degraded by microorganisms. The high solubility of PCL, its low melting point (59-64°C), and its exceptional ability to form blends has stimulated research on its application as a biomaterial. PCL degrades at a slower pace than PLA (polylactic acid) and can therefore be used in drug delivery devices that remain active for over 1 year [1]. Industrially produced copolymer of styrene and acrylonitrile (SAN) usually contains 15 to 35 wt% of acrylonitrile and is prepared by radical polymerization. The higher the acrylonitrile content, the greater the heat and chemical resistance, impact strength, toughness, scratch resistance and barrier properties. SAN is frequently used in households as well as in industry. Quite many scientists have studied PCL/SAN blends, several examples are listed here [2-17]. PCL and SAN are miscible only when SAN contains 8 ~ 28 wt% of acrylonitrile. Near the boarders of this "miscibility window", the lower critical solution temperature (LCST) behavior was observed. In this study SAN with 27.5 wt% of acrylonitrile (AN) was used. Being driven by thermal noises, local concentration in a polymer blend fluctuates continually. When temperature is changed, the distribution of wavelength of concentration fluctuations changes until it reaches a new equilibrium. If a temperature change exceeds a certain limit, concentration fluctuations grow enormously and the blend undergoes phase separation through spinodal decomposition (SD) mechanism. Dynamics of fluctuation growth during SD has been extensively investigated in the last decades because of its theoretical as well as practical importance, e.g. by Hashimoto et al. [18-22]. When the temperature of a blend undergoing SD is changed to a value in the single-phase region, concentration fluctuations that had grown at one time in the spinodal region, gradually attenuate. Dynamics of such attenuation of concentration fluctuations is expected to follow the same mechanism as the growth of fluctuations during SD [23]. However, only a few experiments have been made to investigate the dynamics of attenuation of concentration fluctuations [23-29], which makes a marked contrast with the number of experimental studies performed to clarify the dynamics of fluctuation growth during the phase separation. Therefore, our research focused on this not so thoroughly understood field.

In this study PCL/SAN-27.5 blend exhibiting LCST behavior with critical composition (80/20) was investigated. Firstly the blend underwent phase separation at various temperatures above LCST giving us an opportunity to study the spinodal decomposition by optical microscopy and time-resolved light scattering (TRLS). Then the phase-separated blend was quenched to various temperatures below LCST where we could study the process of phase dissolution. Dynamics of attenuation of concentration fluctuations was experimentally investigated by light scattering. Also crystallization after spinodal decomposition was studied by light scattering and by optical microscopy.

EXPERIMENTAL

The PCL was a commercial polymer of Union Carbide Co. (PCL-767; M_w =40,400 g mol^{-1}, M_w/M_n=2.61). SAN containing 27.5 wt% of acrylonitrile (SAN-27.5) was synthesized at 60°C using ethylbenzene as a solvent and azoisobutyronitrile (AIBN) at a concentration of 0.02 mol/l as an initiator; M_w=169,000 g mol^{-1}, M_w/M_n=2.09.

The solution cast method was used to prepare PCL/SAN blends with various compositions. PCL and SAN-27.5 were dissolved at 5 wt% of total polymer in 1,2-dichloroethane. The solution was cast onto a cover glass and the solvent was allowed to evaporate for 24 h at a room temperature. The dried film was annealed at 80°C (below LCST that is at around 122°C) for 1 min to achieve a single-phase state, and then it was subjected to a temperature jump to 125–180°C (>LCST) to allow isothermal SD for the desired time (3–240 min). The decomposed sample was subjected to a temperature drop to temperatures below LCST (50–115°C) where it was isothermally annealed to observe the phase dissolution.

For the optical microscopy observations, the specimen was initially annealed on first hot stage at desired temperature. The melted specimen was then placed onto a LINKAM hot stage of the microscope. Structural development during the isothermal annealing was observed under both the optical and the polarizing microscope (Olympus BH-2) equipped with a video recording system and exposure control unit (Olympus PM-20).

For the TEM analysis, the specimen was microtomed to an ultrathin section about 70 nm thick using a Reichert-Jung ultracryomicrotome with a diamond knife at -80°C and then the section was stained with RuO_4 vapor at room temperature for 2 h. The structure was observed by electron microscope, JEM 100CX (100 kV) or HITACHI 700.

Real time analyses of the structure developments above and below LCST were carried out by light scattering method. A polarized He-Ne gas laser of 632.8 nm wavelength was applied vertically to the film specimen. The scattered light passed through an analyzer and then onto a highly sensitive CCD camera with 567 x 382 pixels in a sensor of dimension 13.3 x 8.8 mm (Princeton Instruments, Inc.). This allowed the time-resolved measurement of a one-dimensional angular distribution of scattered light with 576 points in a time scale of 0.14 s. The input data from the CCD camera was digitized by a ST-13X controller. The digitized data was stored in a personal computer equipped with Dyna-100 software for further analysis.

Results and Discussion

Figure 1 shows the phase diagram of PCL/SAN-27.5 blend. All specimens of PCL/SAN-27.5 blends were optically clear, and no structure was observed by the microscope at 80°C (above the melting temperature of PCL). When the specimens were heated above LCST, they became opaque, and a two-phase structure was observed.

As described in the experimental section, the blend was first annealed below LCST in one-phase region (point A). Then the sample was transferred onto another hot stage to various temperatures set above LCST (point B) and annealed isothermally there. Phase separation proceeded to various levels depending on temperature and duration of annealing at the point B. Finally, the phase-separated sample was quenched to various phase dissolution temperatures (50-115°C) below LCST (point C).

In Figure 2, the structural development in 80/20 PCL/SAN-27.5 after the temperature jump from 80 to 130°C is shown. Initially no structure was seen. After 5 min, a two-phase structure with a unique periodicity and phase connectivity was detected. The contrast of the structure became higher with the time of annealing (30 min). Then the structure became coarser and larger with time (60, 90 and 120 min).

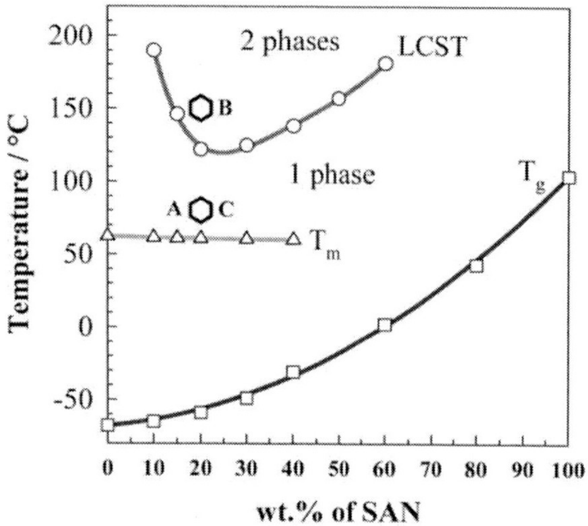

Figure 1. Phase diagram of PCL/SAN-27.5 blend.

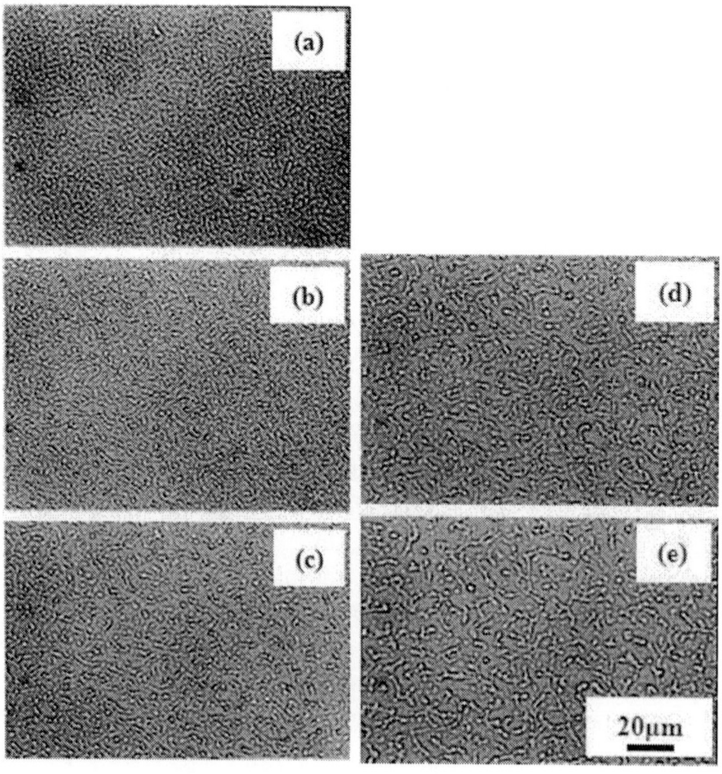

Figure 2. Optical micrographs of PCL/SAN-27.5 (80/20) blend showing phase separation at constant temperature 130°C developing in time: (a) 5 min, (b) 30 min, (c) 60 min, (d) 90 min, (e) 120 min.

In this way one is able to divide the SD process into three stages. In the early stage the position of the maximum q_m stays almost constant. Then in the intermediate and late stages a scaling analysis is being used. Binder and Stauffer [33] proposed scaling rules, assuming that clusters aggregate by a diffusion process and coalesce into larger clusters and described the self-similar growth of the structure by the time evolution of $q_m(t,T)$ and $I_m(t,T)$ as

$$q_m(t,T) \sim t^{-\alpha} \tag{3}$$

and

$$I_m(t,T) \sim t^{\beta} \tag{4}$$

Various models and various α and β exponents were also reported, e.g. by Langer, Baron, Miller [34] or Siggia [35]. In the intermediate stage a relation $\beta > 3\alpha$ is observed. The late stage is reached when $\beta = 3\alpha$ [31,36-40]. In the late stage the two phases attain equilibrium compositions, and self-similar growth of the structure can be observed. Summary of the analysis presented in Figure 4:

1) In the *early stage,* there is no increase in wavelength Λ_m, (q_m stays constant) concentration fluctuations $\Delta\phi(t)$ (ϕ is volume concentration) are increasing and the slope of the line (α) is 0. The duration for our case was found 0 – 25 min. This is in agreement with observation by optical microscopy (see Figs. 2a and 2b) when there is only very small increase in the wavelength Λ_m between 5 and 30 min.
2) In the *intermediate stage,* both the amplitude of the concentration fluctuations $\Delta\phi(t)$ and wavelength $\Lambda_m(t)$ of the fluctuations grow with time, and the position of maximum q_m in Figure 3 is decreasing to lower angles, the slope of the line α is gradually increasing. Because β is almost a constant, β/α is gradually decreasing until it reaches value 3. The duration was found to be 25 – 70 min.
3) In the *late stage,* where the value $\Delta\phi$ reaches an equilibrium value $\Delta\phi_e$ determined by the coexistence curve and the phase separation temperature (in Figure 1), only Λ_m grows with time. The duration of the late stage was 70 – 220 min and the ratio $\beta/\alpha = 3$.

The variation of Λ_m in time for PCL/SAN-27.5 (80/20) blends at two different temperatures 130 and 150°C is plotted in Figure 5. The values of Λ_m can be used for designing the desired size of domains. And indeed we used this plot to prepare two different samples with apparently the same wavelength of the concentration fluctuations Λ_m being about 4.3 μm, the first one prepared by spinodal decomposition at 130°C for 220 min and the other at 150°C for 38 min. These two supposedly same samples were then subjected to phase dissolution (PD) at 90°C.

The results are shown in Figures 6 and 7. The light scattering machine was set to measure precisely the intensity at smaller angles (compared to Figure 3).

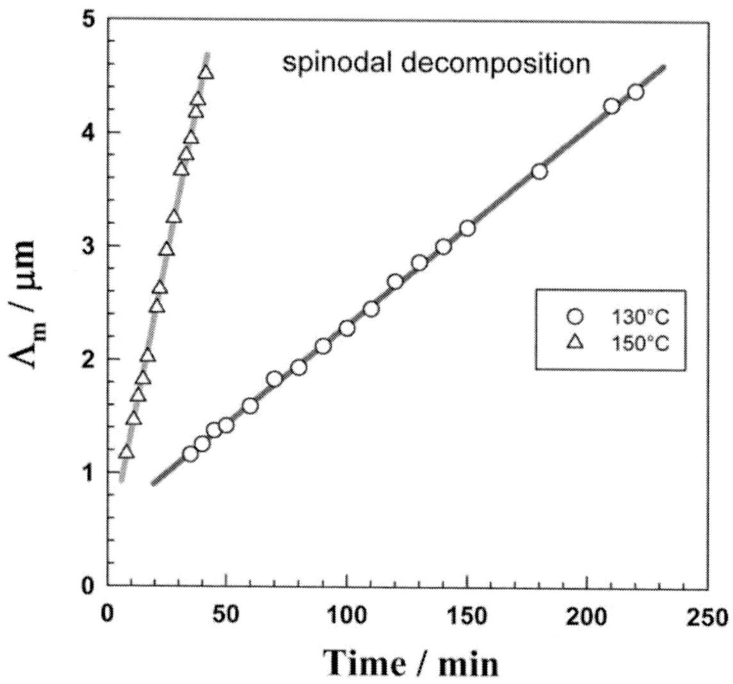

Figure 5. Wavelength of concentration fluctuations Λ_m (from V_v scattering) vs. time for PCL/SAN-27.5 (80/20) blend. Samples were heated to 80°C for 1 min (one-phase region) and then transferred to another hot stage at 130°C or 150°C (two-phase region).

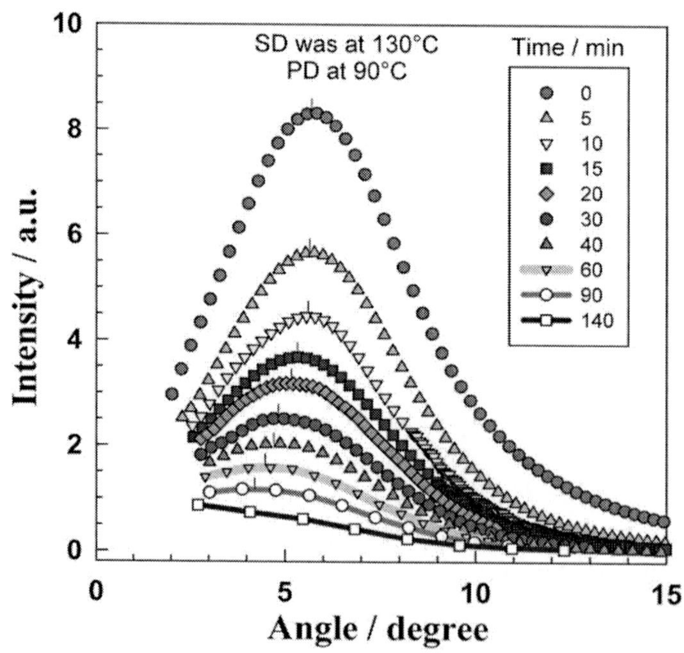

Figure 6. Change of light scattering profile of PCL/SAN-27.5 (80/20) blend during isothermal phase dissolution at 90°C after phase separation at 130°C for 220 min.

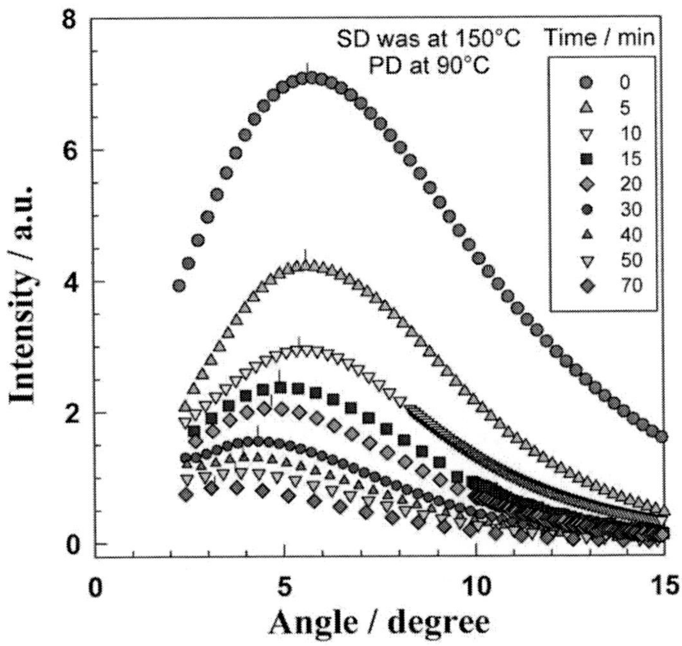

Figure 7. Change of light scattering profile of PCL/SAN-27.5 (80/20) blend during isothermal phase dissolution at 90°C after phase separation at 150°C for 38 min.

At first please note that the initial peak at 0 min is a little broader in the case of the sample phase separated at 150°C compared to that separated at 130°C. This indicates quite narrow particle size distribution of the phase-separated structure created at 130°C and a coexistence of larger and smaller particles created at 150°C.

In both cases, the light scattering peak is decreasing and the position of the maximum is shifting towards the lower angle. There are several possible interpretations of the shift of the peak maximum in Figures 6 and 7, one of which is the faster decay of the high-q Fourier components in the concentration pattern. Only the slower components with long wavelengths (small q) survive at longer times. This results in an effective peak shift towards small q (small angles) during the dissolution process.

This is schematically shown in Figure 8, particles 1 and 2. While the large particle no. 1 survives up to Figure 8c particle no. 2 is in Figure 8c already dissolved. Light scattering at late stage of phase dissolution detects only the remaining particle no. 1. Another possible interpretation of the peak shift towards smaller angle is coalescence of close particles no. 3 and 4. The actual growth of the particles during phase dissolution is possible too, as it was observed directly by TEM by Okada [22,23] and calculated by using a model by Cheng [26]. This is schematically shown by particle no. 5. In our particular PCL/SAN (80/20) system the particles are formed by SAN-rich phase. SAN has rather high T_g (about 103°C) while for PCL T_g=−66°C and T_m≅60°C. Also molecular weight of SAN (M_w=169,000 g mol^{-1}) is higher compared to PCL (M_w =40,400 g mol^{-1}). Above mentioned facts might contribute to much higher mobility of PCL compared to SAN at phase dissolution temperature (90°C). Thus it is possible that PCL molecules diffuse into SAN particle faster than SAN molecules are leaving. This would result in swelling of SAN particles.

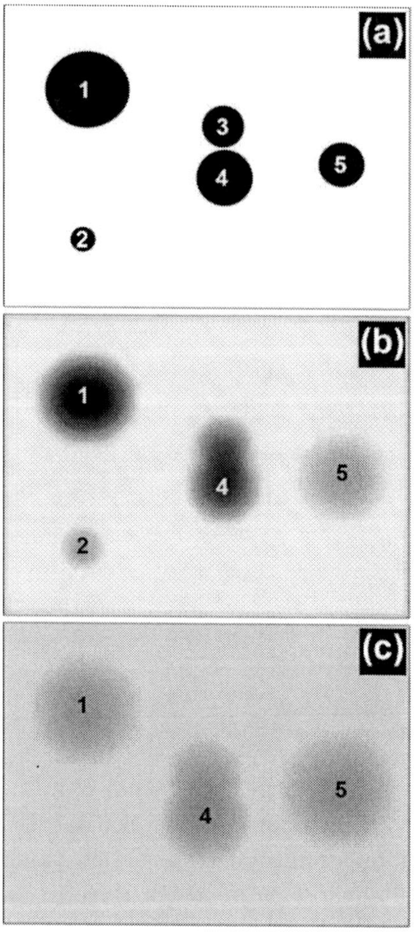

Figure 8. Model of phase dissolution development in time derived from light scattering data shown in Figures 6 and 7.

Concerning *phase separation*: the existence of scattering peak that is moving to lower values of scattering angle (already shown Figure 3) indicates that the demixing proceeds by the spinodal decomposition (SD) mechanism. To confirm this point, we analyzed the early stage on the basis of the linearized Cahn-Hilliard theory [32, 41]. Cahn in 1965 [32] laid a basis for mathematical treatment of spinodal decomposition. He started with expression for free energy of inhomogeneous solution whose composition everywhere differs only slightly from the average composition, and with small composition gradients.

$$F = \int [f(c) + \kappa(\nabla c)^2] dV \tag{5}$$

Here *f(c)* is the free-energy density of homogeneous material of composition *c*, $\kappa(\nabla c)^2$ is the additional free-energy density if the material is in a gradient in composition. In the development of the theory it became convenient to consider the Fourier components of the composition rather than the composition. Because of the orthogonality of the Fourier components, ΔF is the sum of contributions from each Fourier component separately. The kinetics of the initial stages of phase separation can be obtained by solving the diffusion equation. We define a mobility M which is (minus) the ratio of diffusional flux to the gradient in chemical potential

$$J_B = -J_A = M\nabla(\mu_A - \mu_B) \tag{6}$$

Simple thermodynamic considerations show that M must be positive if diffusion which results spontaneously from the chemical potential gradient is to result in a decrease in free energy. We obtain μ_A-μ_B from the variational derivative of F in Eq. (5),

$$\mu_A - \mu_B = \left(\frac{\delta F}{\delta c_A}\right) = \left(\frac{\partial f}{\partial c_A}\right) - 2\kappa\nabla^2 c_A + \textit{ terms higher in order and degree} \tag{7}$$

We may discard the higher order terms if we are only interested in the initial stages. By substituting Eq. (7) into Eq. (6) and taking the divergence, and keeping only first degree terms, we obtain the following diffusion equation:

$$\frac{\partial c}{\partial t} = M\left(\frac{\partial^2 f}{\partial c^2}\right)\nabla^2 c - 2M\kappa\nabla^4 c \tag{8}$$

Since the coefficient of ∇^2c may be identified with a diffusion coefficient we see that at the spinodal the diffusion coefficient changes sign. The solution to Eq. (8) is

$$c - c_0 = e^{[R(\beta)t]}\cos(\beta\,\mathbf{r}) \tag{9}$$

where β is wavenumber and R(β) is given by

$$R(\beta) = -M\left(\frac{\partial^2 f}{\partial c^2}\right)\beta^2 - 2M\kappa\beta^4 \tag{10}$$

Cahn has developed his model for small molecules in metallurgy. His equations were slightly modified for polymer blends. These days we are using g instead of f and q instead of β. Light scattering became popular in quantitative evaluation of spinodal decomposition in polymer blends. Here, in the initial stage of phase separation (or phase dissolution), the intensity of scattered light (I) is expected to increase (or decrease) exponentially with time t.

$$I_{Vv}(q,t) \propto e^{2R(q)t} \tag{11}$$

where $R(q)$ is the growth rate (or decay rate) of concentration fluctuation at the scattering vector q. Today $R(q)$ is given by (essentially the same as Eq. 10 in 1965)

$$R(q) = -Mq^2\left(\frac{\partial^2 g}{\partial c^2} + 2\kappa q^2\right) \tag{12}$$

where M is the mobility, g is the free-energy density of the blend with composition c, and κ is the gradient energy coefficient.

Cahn started with describing the Gibbs free energy of fluctuating system by a sum of the local free energy of mixing and the square of concentration gradient term. The gradient term remains in Eq.12 as the second term.

The first term of Eq.12 is the second derivative of g which is the thermodynamic driving force for fluctuation growth (i.e., the degree of instability of homogeneously mixed state).

The second term of Eq.12 suggests that a very sharp gradient with large q (short wavelength of fluctuation) suppresses the fluctuation growth. In contrast, the fluctuation with small q (long wavelength) grows slowly, because the molecules should diffuse long distance for the fluctuation growth. Then, there should exist a peak of rate constant *R(q)* at a particular q.

As it was mentioned earlier and described by Eq. 11 in the early stage of spinodal decomposition the intensity should increase exponentially with time. This was indeed observed and is shown in Figure 9. We used natural logarithm of intensity and therefore the plot yields straight lines in the initial stage. Then one can estimate the growth (or decay) rate *R(q)* of concentration fluctuations from the initial slope of a $\ln(I_{Vv})$ vs. t plot (in Figure 9 for phase separation and in Figure 10 for phase dissolution) for different angles (or q values).

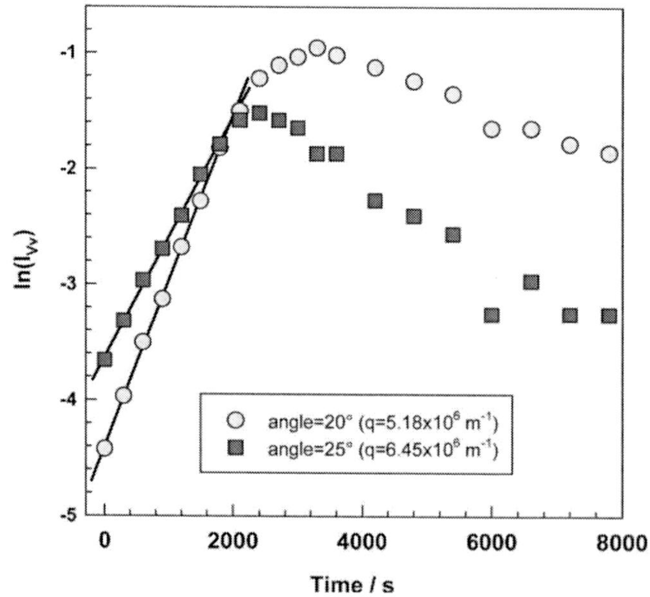

Figure 9. Change of $\ln(I_{Vv})$ in time during spinodal decomposition at 130°C (from light scattering). Evaluation of slopes for various angles.

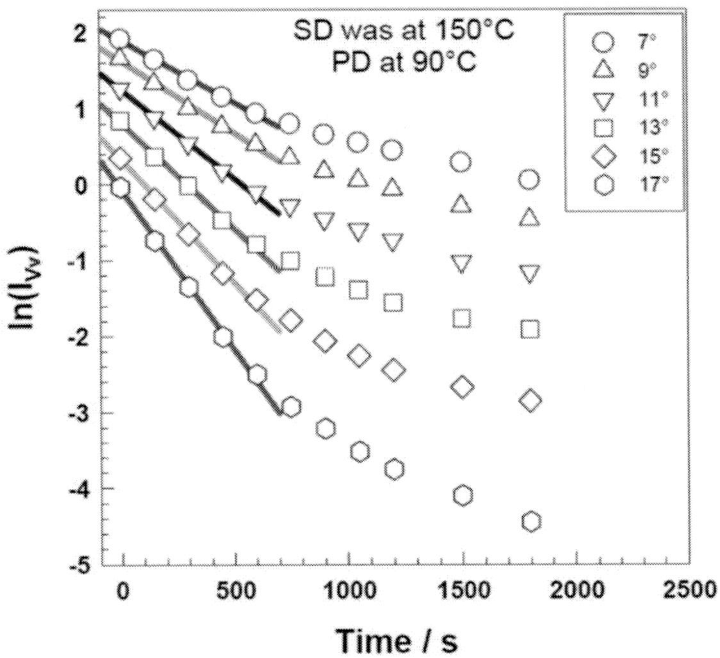

Figure 10. Change of ln(I_{Vv}) in time during phase dissolution at 90°C (from light scattering). Evaluation of slopes for various angles.

After such detailed analysis, it is possible to calculate the $R(q)/q^2$ vs. q^2 plot as it is shown in Figure 11 for SD and Figure 12 for PD for various temperatures. The apparent diffusion coefficient D_{app} is defined by

$$D_{app} = -M\left(\frac{\partial^2 g}{\partial c^2}\right) \tag{13}$$

and can be obtained from $R(q)/q^2$ vs. q^2 plot as the intercept of the $R(q)/q^2$ axis in Figures 11 and 12.

The apparent diffusion coefficient D_{app} obtained in this way was positive for phase dissolution and negative for phase separation. It is interesting to compare the absolute values of D_{app} for SD and PD in one plot, see Figure 13.

The D_{app} for phase dissolution was found to have values between 3 and 11 x 10^{-17} m^2s^{-1} in the temperature range 50-115°C. Closer to LCST the values are decreasing and in range 50-65°C they are low as well. The diffusion should be slower at lower temperature due to a lower mobility. In our case, also the presence of a virtual upper critical solution temperature (UCST) around 40°C, found earlier for similar systems [42], should cause the decrease in D_{app}. Comparable values of D_{app} for phase separation (just with opposite sign – negative values) were found in the temperature range 125-150°C. By a further increase of the temperature the spinodal decomposition proceeded much faster, an example being value of D_{app} at 180°C which is more than 10 times higher than the fastest phase dissolution. From the exponential-like curve one could assume that this difference would be probably even much higher (such as 100x or even 1000x) if we increased the temperature to e.g. 200 or 220°C.

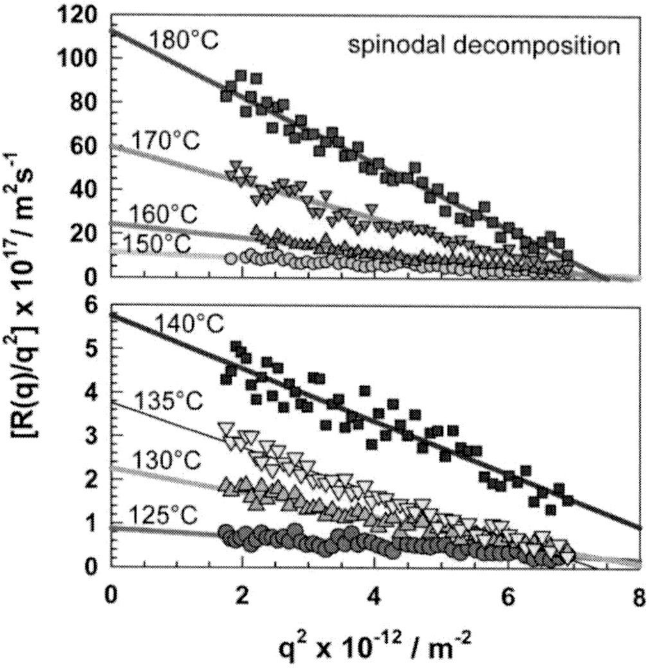

Figure 11. Plot of $R(q)/q^2$ vs. q^2 of PCL/SAN-27.5 (80/20) blend: spinodal decomposition at 125-180°C. *R(q)* values were obtained by analysis of slopes shown by Figure 10.

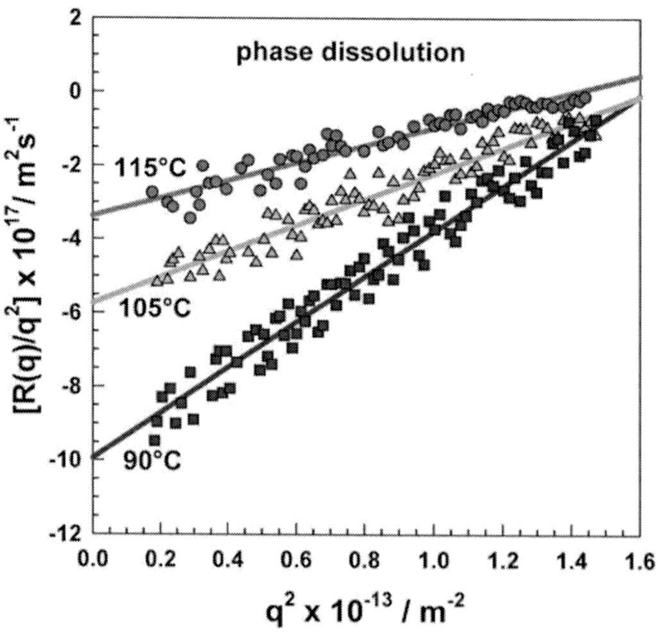

Figure 12. Plot of $R(q)/q^2$ vs. q^2 of PCL/SAN-27.5 (80/20) blend: phase dissolution at 90, 105 and 115°C after previous spinodal decomposition at 150°C for 15 min. *R(q)* values were obtained by analysis of slopes shown in Figures 11 and 12.

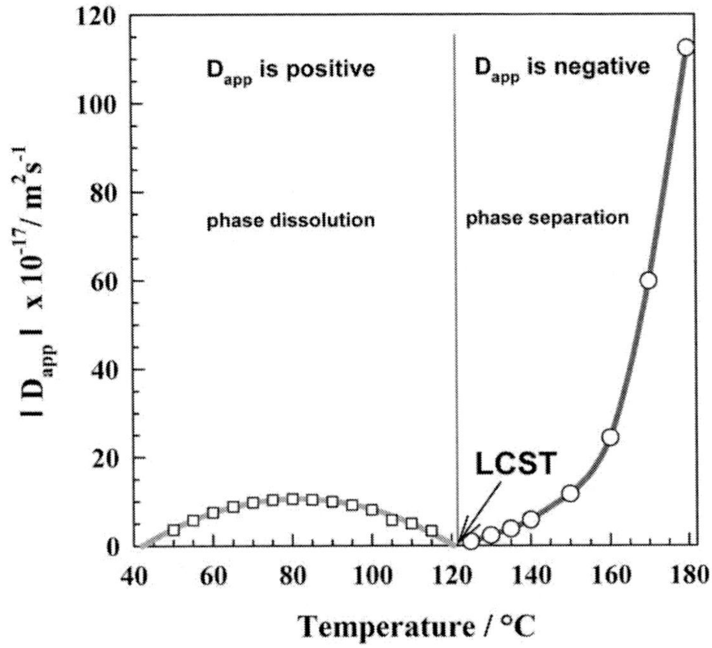

Figure 13. Apparent diffusion coefficient for phase dissolution (squares) and phase separation (circles) as a function of annealing temperature. Lower critical solution temperature (LCST) is shown by arrow.

Crystallization Kinetics by Optical Microscopy

The specimens were forced to undergo demixing into different levels, that was quantitatively evaluated using the V_v light scattering technique described above; then they were crystallized by quenching to a temperature below T_m. The starting point of the crystallization kinetics analysis was crystallization of PCL/SAN-27.5 (80/20) from the homogeneous state (one phase) attained by annealing the sample at 100°C for 1 min. After this annealing, the sample was transferred onto another hot stage at 40°C, where the time development of spherulites was observed by optical microscopy. An example of growth of spherulites in time is given in Figure 14. The radius of the spherulites vs. time of crystallization for this case is plotted in Figure 16. As the size and shape of spherulites slightly differ, about three spherulites were usually averaged. As can be seen, a linear dependence was always obtained, so the slope of the line is equivalent to the spherulite growth rate G.

In addition the samples were forced to undergo SD to different levels, represented by different times of SD at 150° and then quenched to crystallization temperature of 40°C. An example of growth of spherulites from these phase-separated structures is given in Figure 15. The first observation leads to the conclusion that after SD the number of small spherulites is higher, implying more nucleation centers. The size of the spherulites is clearly larger than the size of the domains. This means that crystallization proceeds through the PCL-rich matrix somehow ignoring the obstacles represented in this case by SAN-rich domains. Quantitative and more detailed conclusions are difficult to obtain using just optical microscopy.

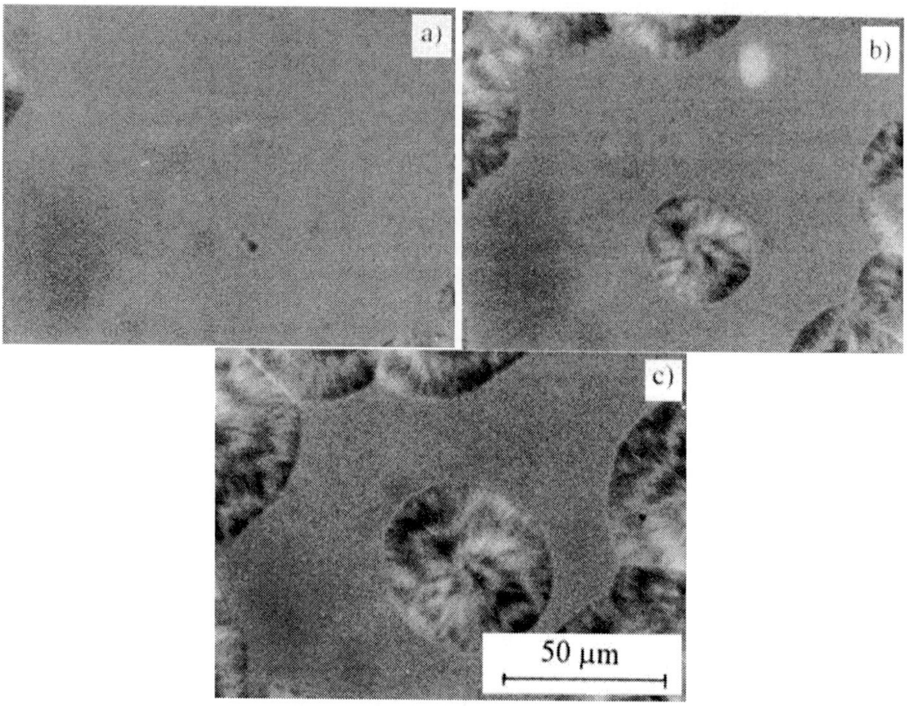

Figure 14. Optical micrographs of PCL/SAN-27.5 (80/20) blend that was heated in the one-phase region (100° for 1 min) and then quenched to the crystallization temperature (40°C). Spherulite growth times (a) 14 min, (b) 26 min, (c) 34 min.

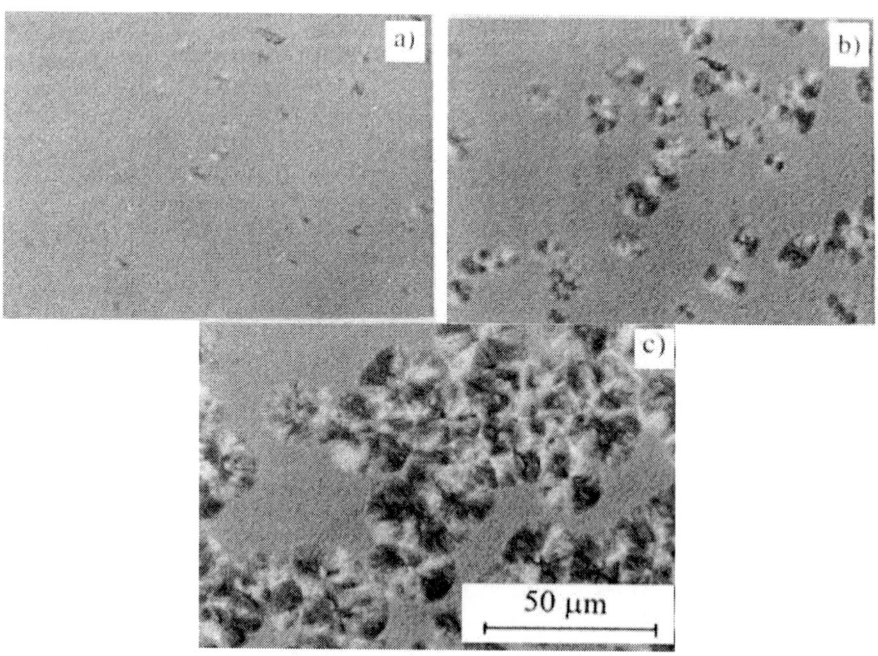

Figure 15. Optical micrographs of PCL/SAN-27.5 (80/20) blend that was heated in the one phase region (100°C for 1 min), then phase-separated 40 min at 150°C and finally crystallized at 40°C. Time of development: (a) 2 min, (b) 4 min, (c) 8 min.

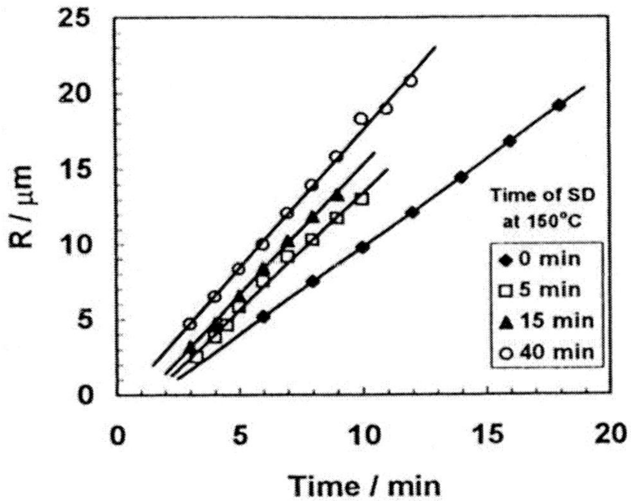

Figure 16. Variation of spherulite radius R with crystallization time for PCL/SAN-27.5 (80/20) at 40°C by optical microscopy. Comparison is made for sample crystallized from the one-phase region (0 min) with samples that have undergone different times of spinodal decomposition at 150°C.

Therefore, light scattering and TEM techniques were used for this purpose, as will be described later. The variations in the radius of the spherulites are plotted in Figure 16. Again the growth is linear with time. Apparently, the growth rate is accelerated by SD. The further SD proceeds, the faster is the crystallization. Even though crystallization does not proceed in SAN-rich domains, the crystallization is not stopped at the SAN-rich borders because of interconnectivity of the PCL-rich phase. It proceeds in the PCL-rich phase.

As shown in Figure 17, crystallization kinetics is accelerated in the early and middle stage of SD.

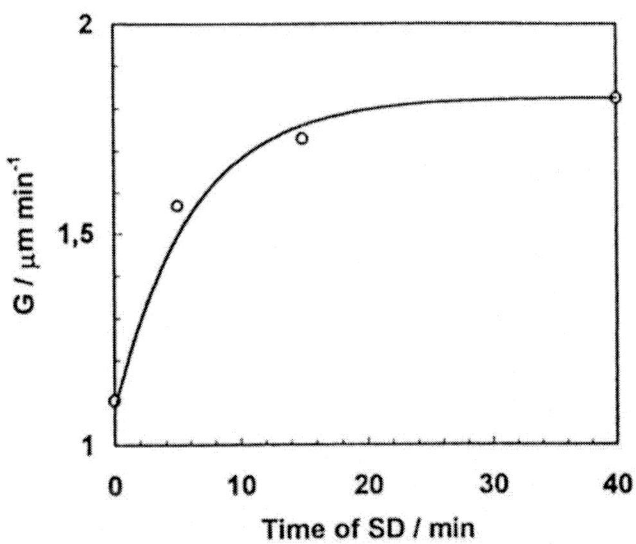

Figure 17. Growth rate of spherulites, G, at 40°C vs. time of spinodal decomposition at 150°C.

The increase of the G may be ascribed to the increase in the concentration of PCL in PCL-rich region with the demixing time of SD of the evolution of concentration fluctuations. On the other hand, G is almost constant with demixing time in the late stage of SD. This is due to no further evolution of the concentration fluctuations in the late stage of SD (only Λ_m increases with demixing time).

An interesting phenomenon is also the ordering of the PCL lamellae inside the spherulites. A deeper insight into this problem is discussed in the next two sections.

CRYSTALLIZATION BY H_v LIGHT SCATTERING

The H_v light scattering is a powerful technique for investigation of crystallization in polymer blends. Compared to optical microscopy, where always has to choose from limited number of spherulites, H_v light scattering gives information about the whole system.

For this study a crystallization temperature of 40°C was chosen. The observed scattering patterns were of the four-leaf-clover type. The clover patterns became smaller with time of crystallization. Figure 18 shows the one dimensional H_v scattering profiles at an azimuthal angle of 45° for a sample crystallized after 20 min of SD. The average radius of the spherulites R_{Hv} can be calculated from the position of the maximum in scattering intensity, θ_m, by using this equation:

$$4.09 = 4\pi(R_{Hv}/\lambda)sin(\theta_m/2) \quad (14)$$

In order to discuss the time variation of the internal spherulitic structure, it is convenient to employ the scattering profiles at an azimuthal angle 45° using a reduced scattering angle w.

$$w = \left(\frac{2\pi}{\lambda}\right) R_{Hv} sin\theta \quad (15)$$

Using w, the scattering profiles are corrected to the reduced profiles with the peak maximum at w=4 as shown in Figure 19. The sharpness of the peaks is different for the samples with different time of SD. Increased radial disorder leads to increased relative I_{Hv} scattered intensity at angles larger than that of the maximum. To compare quantitatively the order inside the spherulite, it is convenient to define a so-called order parameter (Pr).

$$Pr = \frac{I(4)}{I(x)} \quad (16)$$

where I(4) represents intensity at w=4 and I(x) is the intensity at w=x. In the literature two values of x were used: 8 and 15. The situation is schematically shown in Figure 19. How the order parameter is changing during crystallization is shown in Figure 20. At the beginning of crystallization the ordering inside the spherulite is higher and then it decreases and remains at an almost constant level. This behavior was observed for the sample crystallized from the homogeneous phase and also for samples crystallized after different times of SD.

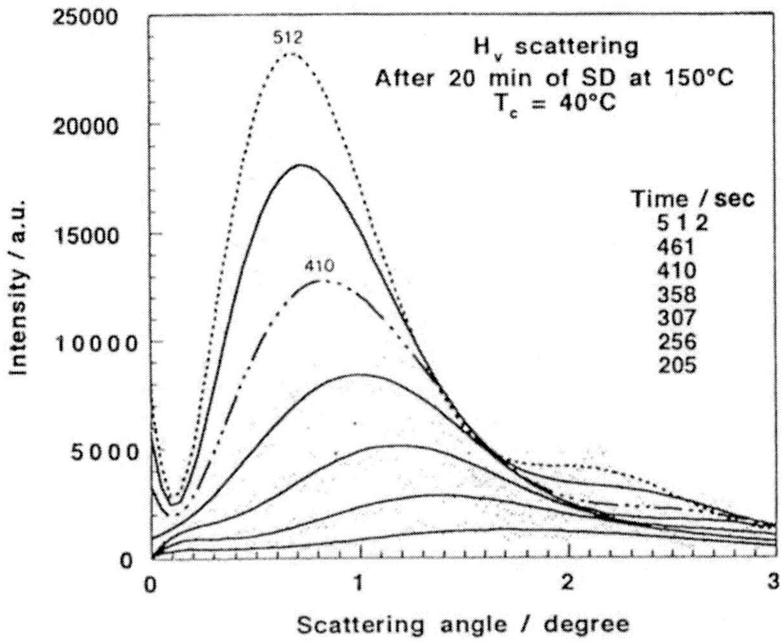

Figure 18. Change of H_v scattering profiles with time after quenching the PCL/SAN-27.5 (80/20) blend to 40°C after 20 min of spinodal decomposition at 150°C, azimuthal angle μ=45°.

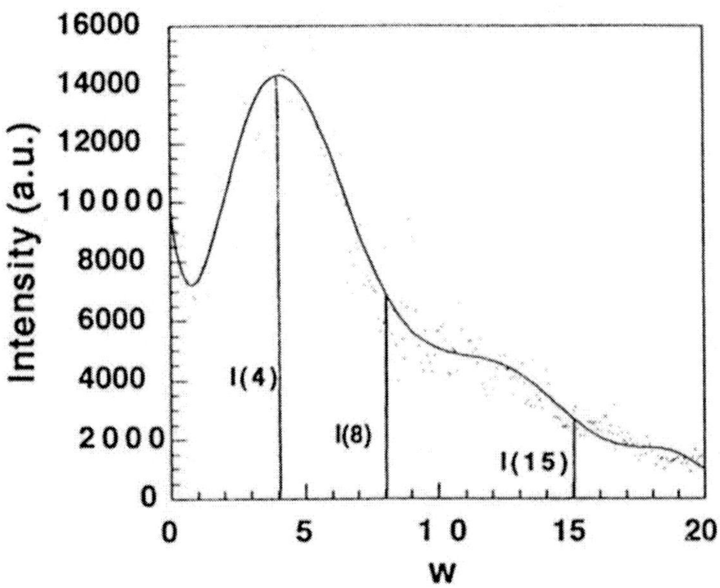

Figure 19. Schematic picture of determination of order parameter I(4)/I(8) and I(4)/I(15).

This phenomenon can be explained in this way. At the beginning of the crystallization, lamellae of PCL grow from the center of the spherulite in the radial direction, and the order parameter is high. After a short time branching starts in the regions where the concentration of PCL is lower. This branching decreases the order parameter. Then the radial growth and the branching reach an equilibrium, and the order parameter becomes constant. This constant value was used to compare samples crystallized from different stages of SD, as shown in Figure 21.

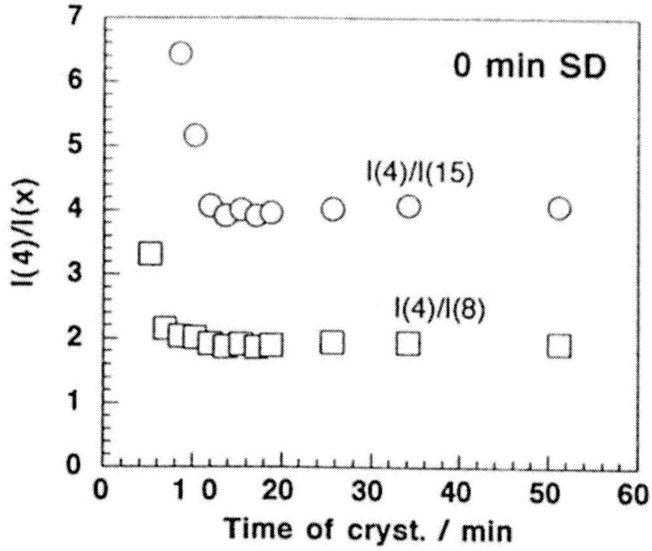

Figure 20. Order parameter vs. time of crystallization at 40°C for (80/20) blend quenched from homogeneous state (100°C for 1 min).

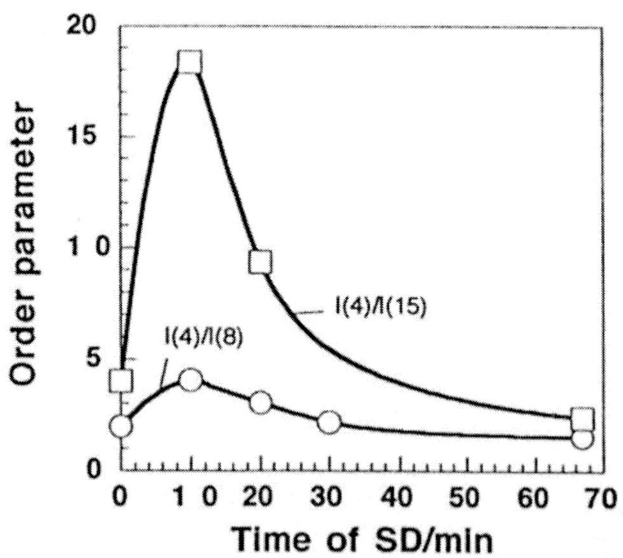

Figure 21. Order parameter vs. time of spinodal decomposition at 150°C.

In this graph, two order parameters [I(4)/I(8) and I(4)/I(15)] are plotted as a function of time of SD at 150°C. All samples were crystallized at constant temperature 40°C. Regardless of which order parameter is considered, an interesting phenomenon appeared. At the beginning the order parameter increases (about 10 min) and then slowly decreases with increasing time of SD. By the V_v light scattering analysis at 150°C one can find a change from the middle to late stages of SD at about 10 min. Then from Figure 21, it is apparent that ordering of the spherulites increases for the samples crystallized after SD in the early and middle stages. Coming to the late stage, the ordering parameter decreases continually with increasing time of SD. In order to understand this phenomenon fully, it was necessary to make a TEM analysis.

TEM Analysis

For the transmission electron microscopy (TEM) analysis, the following samples were chosen. All the samples have the same blend composition PCL/SAN-27.5 (80/20) and were crystallized at 40°C for 14 h after different prior history. The first sample (Figure 22) was crystallized from the homogeneous phase that was attained by annealing the sample at 80°C for 1 min. The lamellae are short and oriented in many directions, differing from the radial growth of spherulites. It looks like the SAN is partly segregated from the growing front and partly remains in an amorphous mixture with PCL between the PCL lamellae. As RuO_4 stains benzene rings, that is, SAN, the broad dark region might be the areas where the concentration of SAN is increased due to segregation during crystallization. Many times the lamellae end in these areas and start to grow in new directions. This could explain a low order parameter.

Three additional samples with different degrees of phase separation (Figure 23 - 5 min, Fig 24 - 15 min, Figure 25 - 40 min) were also examined. A clear two phase system can be seen in all three cases. Figure 23 shows a micrograph for 5 min of SD (middle stage). The lamellae are highly oriented in one direction. This corresponds with a high order parameter. An interesting phenomenon in this case is that lamellae grow through the SAN-rich areas (dark) without change of direction.

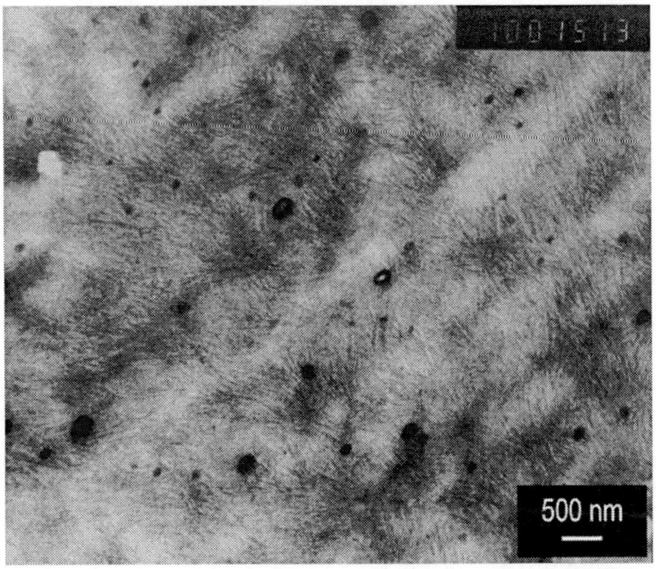

Figure 22. TEM micrograph of PCL/SAN-27.5 (80/20) sample crystallized at 40°C for 14 h after this history: 80°C for 1 min.

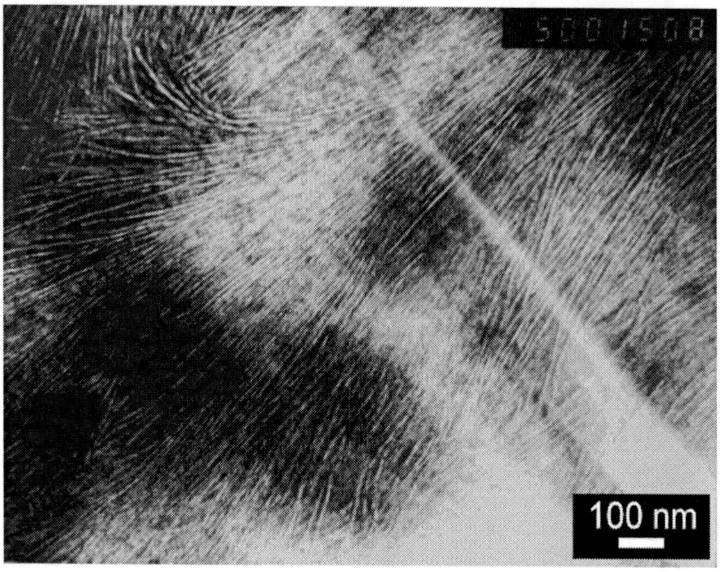

Figure 23. TEM micrograph of PCL/SAN-27.5 (80/20) sample crystallized at 40°C for 14 h after this history: 100°C for 1 min, 150°C for 5 min.

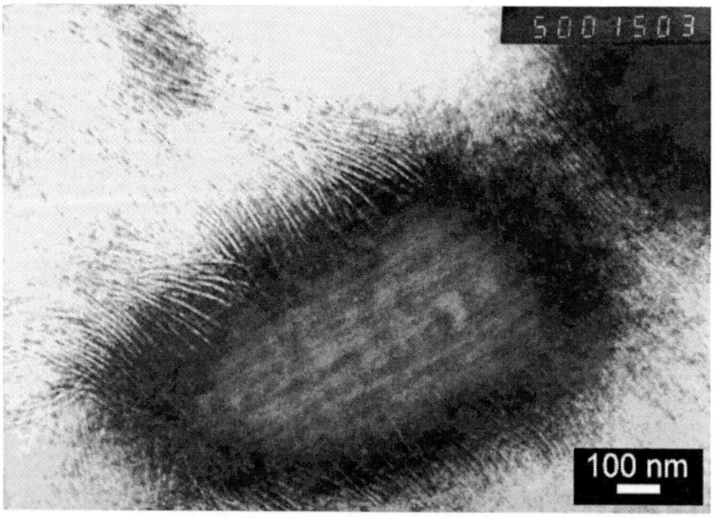

Figure 24. TEM micrograph of PCL/SAN-27.5 (80/20) sample crystallized at 40°C for 14 h after this history: 100°C for 1 min, 150°C for 15 min.

The number of lamellae in these areas is, as expected, smaller because of the smaller concentration of PCL in these SAN-rich areas. The next micrograph (Figure 24) shows the system phase separated for 15 min before crystallization (late stage). The concentration inside the SAN-rich domains has already reached the equilibrium value. The difference is also in the size of SAN-rich domains that are in this case larger.

As the concentration of PCL inside the SAN-rich domains crossed the critical level that would allow crystallization, the lamellae are not able to penetrate these domains. This is probably caused by the mobility effect.

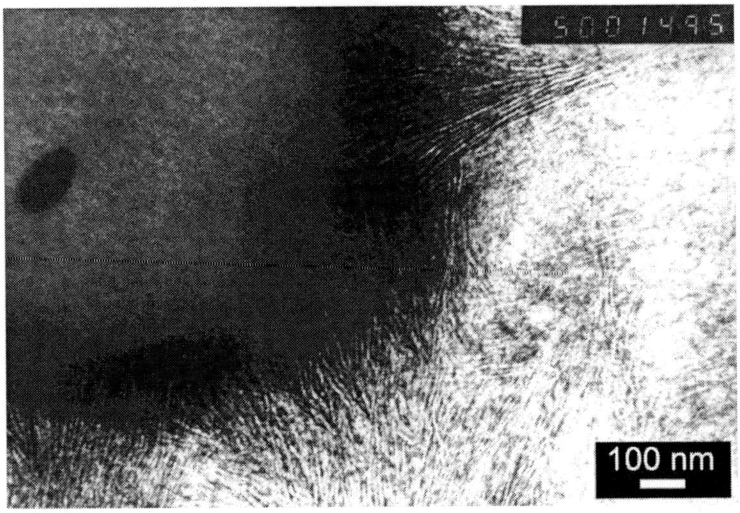

Figure 25. TEM micrograph of PCL/SAN-27.5 (80/20) sample crystallized at 40°C for 14h after this history: 100°C for 1 min, 150°C for 40 min.

The T_g of these domains is probably higher than the crystallization temperature and the mixture of PCL and SAN is frozen inside these domains. The alignment of lamellae in PCL-rich matrix is rather high, which corresponds with a high order parameter. The growth of spherulite continues through the bridges between SAN-rich domains. The last sample (Figure 25) had undergone 40 min of SD (again late stage). The size of the domains had gradually increased. The difference compared with Figure 24 is that the alignment of lamellae is much worse; lamellae grow in all directions. This corresponds with a lower order parameter. In both cases (Figures 24 and 25), the surface of the SAN-rich domain is darker, suggesting that in this interface the concentration of SAN is higher than inside the domain. This could be explained by crystallization of almost all the PCL from this area while SAN remains. Inside the domains the concentration of amorphous PCL is higher than at the surface of domains.

In SAN-rich phase, the crystallization growth rate is much slower (or the crystallization can be prevented because of the mobility effect) than in the PCL-rich phase. This is due to increasing T_g of the blend when the content of SAN is increasing (T_g of pure PCL is about -60° and T_g of SAN is about 105°C). However, the PCL-rich phase is highly interconnected, so lamellae are growing in this phase much faster, somehow ignoring the slow SAN-rich domains. At this blend composition (PCL/SAN 80/20), the growth of large spherulites is not prevented by spinodal decomposition. By the formation of the PCL-rich phase during spinodal decomposition, the concentration of SAN in this phase is gradually decreasing down to an equilibrium value; thus the T_g of this phase is decreasing, which finally results in a higher crystallization rate.

CONCLUSION

PCL/SAN-27.5 blend gave us an excellent opportunity to investigate the phase dissolution kinetics and to add some experimental data (that are still quite rare) to this subject. Firstly, we focused on the phase separation above LCST (being at 122°C), and studied it qualitatively by optical microscopy (at 130°C), and then quantitatively by light scattering at various temperatures (125-180°C). The spinodal decomposition mechanism of phase separation was confirmed by the detailed data analysis based on Cahn-Hilliard theory. Secondly, the phase separated structure was quenched to various temperatures below LCST (50-115°C) to study the phase dissolution. The phase dissolution was affected by the temperature of previously performed spinodal decomposition. Based on light scattering data, a model of phase dissolution mechanism was created. There are several possible interpretations of the shift of the peak maximum during phase dissolution, one of which is the faster decay of the high-q Fourier components in the concentration pattern. Only the slower components with long wavelengths (small q) survive at longer times. This results in an effective peak shift towards small q (small angles) during the dissolution process. Again, for the PD, we were able to use linearized theory to obtain an apparent diffusion coefficient D_{app}. Finally, the D_{app} values were compared for SD and PD with the result showing that the fastest PD was about 10 times slower than the fastest SD in measured temperature range.

By the use of optical microscopy, the crystallization kinetics was observed for samples crystallized after decomposition into different stages of SD. A major acceleration of crystallization kinetics was observed for the samples that had undergone SD in the early and middle stages. This can be explained by a gradual change of concentration in the PCL-rich domains. However, in the late stage, the concentration reaches an equilibrium value and the crystallization kinetics reach almost constant level.

The H_v light scattering technique enabled a quantitative evaluation of the ordering of PCL lamellae inside the spherulites which, in our case, were always bigger than the size of the phase-separated domains. The result of this analysis showed that the radial ordering of lamellae is increasing for samples crystallized after SD to the early and middle stages and then gradually decrease for samples separated to the late stage of SD.

TEM analysis explained rather surprising results of the previous analysis. In the sample crystallized from the homogeneous phase, the ordering of lamellae is rather poor. During crystallization SAN is probably partly segregated. When the amount of segregated SAN exceeds a certain level, lamellae start to grow in new directions. For the samples crystallized after separation to the middle stage of SD, the ordering of lamellae is much higher. Lamellae grow through the SAN-rich areas without changing direction. For the samples crystallized after SD to the late stage, the lamellae of PCL cannot penetrate the SAN-rich domains. Crystallization stops at the borders of the SAN rich domains. As the PCL-rich phase is interconnected, crystallization can proceed through bridges between SAN-rich domains. Then the growth of lamellae continues by branching, which, of course, decreases the radial ordering of lamellae inside the spherulite. The more decomposition proceeds, the less ordered are the lamellae.

Review

This work was reviewed by Professor Jörg Kressler from University Halle Wittenberg, Fachbereich Chemie, Institut für Physikalische Chemie, D-06099 Halle (Saale), Germany.

ACKNOWLEDGEMENT

This work has been supported by the Ministry of Education of the Czech Republic as a part of the project No. VZ MSM 7088352102 and also with support of Operational Programme Research and Development for Innovations co-funded by the European Regional Development Fund (ERDF) and national budget of Czech Republic within the framework of the Centre of Polymer Systems project (reg.number: CZ.1.05/2.1.00/03.0111).

REFERENCES

[1] Ratner BD, Hoffman AS, Schoen FJ, Lemons EJ. *Biomaterials Science*. New York: Elsevier Academic Press, 2004.

[2] Madbouly SA, Abdou NY, Mansour AA. Isothermal crystallization kinetics of poly(ε-caprolactone) with tetramethyl polycarbonate and poly(styrene-co-acrylonitrile) blends using broadband dielectric spectroscopy. *Macromolecular Chemistry and Physics* 2006;207(11):978-986.

[3] Kiersnowski A, Piglowski J. Polymer-layered silicate nanocomposites based on poly(ε-caprolactone). *European Polymer Journal* 2004;40(6):1199-1207.

[4] Men Y, Strobl G. Critical strains in poly(ε-caprolactone) and blends with poly(vinyl methyl ether) and poly(styrene-co-acrylonitrile). *Macromolecules* 2003;36(6):1889-1898.

[5] Aoki Y, Arendt O Melt rheology of poly(ε-caprolactone)/ poly(styrene-co-acrylonitrile) blends. *Journal of Applied Polymer Science* 2001;82(8):2037-2041.

[6] Kim SW, Jo WH, Lee MS, Ko MB, Jho JY. Preparation of clay-dispersed poly(styrene-co-acrylonitrile) nanocomposites using poly(ε-caprolactone) as a compatibilizer. *Polymer* 2001;42(24):9837-9842.

[7] Hong Z, Shaw MT, Weiss RA. Structure evolution and phase behavior of polymer blends under the influence of shear. *Polymer* 2000;41(15):5895-5902.

[8] Keroack D, Zhao Y, Prud'homme RE. Molecular orientation in crystalline miscible blends. *Polymer 1*999;40(1):243-251.

[9] Wang Z, Wang X, Yu D, Jiang B. The formation of ring-banded spherulites of poly(ε-caprolactone) in its miscible mixtures with poly(styrene-co-acrylonitrile). *Polymer* 1997;38(23):5897-5901.

[10] Oudhuis AACM, Thiewes HJ, van Hutten PF, ten Brinke G. A Comparison between the morphology of semicrystalline polymer blends of poly(ε-caprolactone)/poly(vinyl methyl ether) and poly(ε-caprolactone)/(styrene-acrylonitrile). *Polymer* 1994;35(18):3936-3942.

[11] Li W, Prud'homme RE. Orientation and miscibility of poly(ε-caprolactone)/poly(styrene-co-acrylonitrile) mixtures. *Polymer* 1994;35(15):3260-3267.

[12] Gan PP, Paul DR, Padwa AR. Phase behaviour of blends of homopolymers with α-methylstyrene/acrylonitrile copolymers. *Polymer* 1994;35(7):1487-1502.

[13] Janathanan V, Kressler J, Karasz FE, Macknight WJ. Intramolecular interactions in Poly(styrene-co-acrylonitrile)/Poly(ε-caprolactone) blends. *Journal of Polymer Science, Part B: Polymer Physics* 1993;31(8):1013-1017.

[14] Kressler J, Kammer HW, Silvestre C, Di Pace E, Cimmino S, Martuscelli E. Crystallization kinetics in blends of poly-ε-caprolactone and poly(styrene-ran-acrylonitrile). *Polymer networks & blends* 1991;1(4):225-233.

[15] Shah VS, Keitz JD, Paul DR, Barlow JW. Miscible ternary blends containing polycarbonate, SAN, and aliphatic polyesters. *Journal of Applied Polymer Science* 1986;32(3):3863-3879.

[16] Chiu SC, Smith TG. Compatibility of poly(epsilon-caprolactone) (PCL) and poly(styrene-*co*-acrylonitrile) (SAN) blends – II. The influence of the AN content in SAN copolymer upon blend copatibility. *Journal of Applied Polymer Science* 1984;29(5):1797-1814.

[17] Schulze K, Kressler J, Kammer HW. Phase behaviour of poly(ε-caprolactone)/ poly(styrene-ran-acrylonitrile) blends exhibiting both liquid-liquid unmixing and crystallization. *Polymer* 1993;34(17):3704-3709.

[18] Hashimoto T, Tsukahara Y, Kawai H. Dynamic Small-Angle X-ray Scattering Studies on Diffusion of Macromolecules in Bulk. 2. Principle and Preliminary Experimental Results. *Macromolecules* 1981;14(3):708-715.

[19] Hashimoto T, Kumaki J, Kawai H. Time-Resolved Light Scattering Studies on Kinetics of Phase Separation and Phase Dissolution of Polymer Blends. 1. Kinetics of Phase Separation of a Binary Mixture of Polystyrene and Poly(vinyl methyl ether). *Macromolecules* 1983;16(4):641-648.

[20] Inaba N, Sato K, Suzuki S, Hashimoto T. Morphology Control of Binary Polymer Mixtures by Spinodal Decomposition and Crystallization. 1. Principle of Method and Preliminary Results on PP/EPR. *Macromolecules* 1986;19(6):1690-1695.

[21] Nakai A, Shiwaku T, Hasegawa H, Hashimoto T. Spinodal Decomposition of Polymer Mixtures with a Thermotropic Liquid Crystalline Polymer as One Component. *Macromolecules* 1986;19(12):3008-3010.

[22] Inaba N, Yamada T, Suzuki S, Hashimoto T. Morphology Control of Binary Polymer Mixtures by Spinodal Decomposition and Crystallization. 2. Further Studies on Polypropylene and Ethylene-Propylene Random Copolymert. *Macromolecules* 1988;21(2):407-414.

[23] Okada M, Tao J, Nose T. Attenuation of concentration fluctuations after a quench to a temperature in the single-phase region. *Polymer* 2002;43(2):329-335.

[24] Okada M, Tao J, Nose T. Dissolution of phase-separated domains after a jump to a temperature in the single-phase region. *Polymer* 2002;43(26):7429-7432.

[25] Akcasu AZ, Bahar I, Erman B, Feng Y, Han CC. Theoretical and experimental study of dissolution of inhomogeneities formed during spinodal decomposition in polymer mixtures. *J. Chem. Phys.* 1992;97(8):5782-5793.

[26] Cheng MH, Nauman EB. Non-linear diffusion with concentration-driven flows in miscible systems. *Polymer* 2003;44(21):6707-6712.

[27] Okamoto M, Shiomi K, Inoue T. LCST-type phase behaviour and structure development during melt processing in a polycarbonate/poly(styrene-co-acrylonitrile) blend. *Polymer* 1995;36(1):87-91.

[28] Sato T, Han CC. Dynamics of concentration fluctuation in a polymer blend on both sides of the phase boundary. *J. Chem. Phys.* 1988;88(3):2057-2065.

[29] Takagi Y, Ougizawa T, Inoue T. Phase dissolution in polymer blends. Kinetics of dissolution and related problems in rubber technology. *Polymer* 1987;28(1):103-108.

[30] Svoboda P, Svobodova D, Chiba T, Inoue T. Competition of phase dissolution and crystallization in poly(ε-caprolactone)/poly(styrene-co-acrylonitrile) blend. *European Polymer Journal* 2008;44(2):329-341.

[31] Ribbe AE, Hashimoto T. Real Space Analysis of the Structural Evolution of a Polymer Blend via Spinodal Decomposition. *Macromolecules* 1997:30(14):3999-4009.

[32] Cahn JW. Phase Separation by Spinodal Decomposition in Isotropic Systems. *J. Chem. Phys.* 1965;42(1):93-99.

[33] Binder K, Stauffer D. Theory for the Slowing Down of the Relaxation and Spinodal Decomposition of Binary Mixtures. *Physical Review Letters* 1974:33(17):1006-1009.

[34] Langer J, Baron M, Miller HD. New computational method in theory of spinodal decomposition. *Physical Review A* 1975:11(4):1417-1429.

[35] Siggia ED. Late stages of spinodal decomposition in binary mixtures. *Physical Review* A 1979:20(2):595-605.

[36] Fujita H, Hashimoto T, Takenaka M. Time change in the maximum of scattering intensity during spinodal decomposition. *Macromolecules* 1989:22 (12):4663-4664.

[37] Hashimoto T, Izumitani T. Effect of a block copolymer on the kinetics of spinodal decomposition of polymer blends. 1. Nonuniversality in scaled characteristic quantities versus reduced time. *Macromolecules* 1993:26(14):3631-3638.

[38] Izumitani T, Hashimoto T. Effect of a Block Copolymer on the Kinetics of Spinodal Decomposition of Polymer Blends. 2. Scaled Structure Factor. *Macromolecules* 1994:27(7):1744-1750.

[39] Takeno H, Iwata M, Takenaka M, Hashimoto T. Combined Light Scattering and Laser Scanning Confocal Microscopy Studies of a Polymer Mixture Involving a Percolation-to-Cluster Transition. *Macromolecules* 2000:33(26):9657-9665.

[40] Jinnai H, Hasegawa H, Hashimoto T, Han CC. Time-Resolved Small-Angle Neutron Scattering in Intermediate and Late-Stage Spinodal Decomposition of DPB/HPI Blends. *Macromolecules* 1991:24(1):282-289.

[41] Cahn JW, Hilliard JE. Free Energy of a Nonuniform System. I. Interfacial Free Energy. *Journal of Chemical Physics* 1958:28(2):258-267.

[42] Svoboda P, Kressler J, Chiba T, Inoue T, Kammer HW. Light-scattering and TEM analyses of virtual upper critical solution temperature behaviour in PCL/SAN blends. *Macromolecules* 1994;27(5):1154-1159.

INDEX

A

D

E

F

G

J

K

L

M

N

O

P

Q

R

S

T

U

V

W

X

Y